ANNUAL EDITIONS

GEOGRAPHY
Sixteenth Edition

01/02

EDITOR

Gerald R. Pitzl
Macalester College

Gerald R. Pitzl, professor of geography at Macalester College, received his bachelor's degree in secondary social science education from the University of Minnesota in 1964 and his M.A. (1971) and Ph.D. (1974) in geography from the same institution. He teaches a wide array of geography courses and is the author of a number of articles on geography, the developing world, and the use of the Harvard case method.

McGraw-Hill/Dushkin
530 Old Whitfield Street, Guilford, Connecticut 06437

Visit us on the Internet
http://www.dushkin.com

World Map

This map has been developed to give you a graphic picture of where the countries of the world are located, the relationship they have with their region and neighbors, and their positions relative to the superpowers and power blocs. We have focused on certain areas to more clearly illustrate these crowded regions.

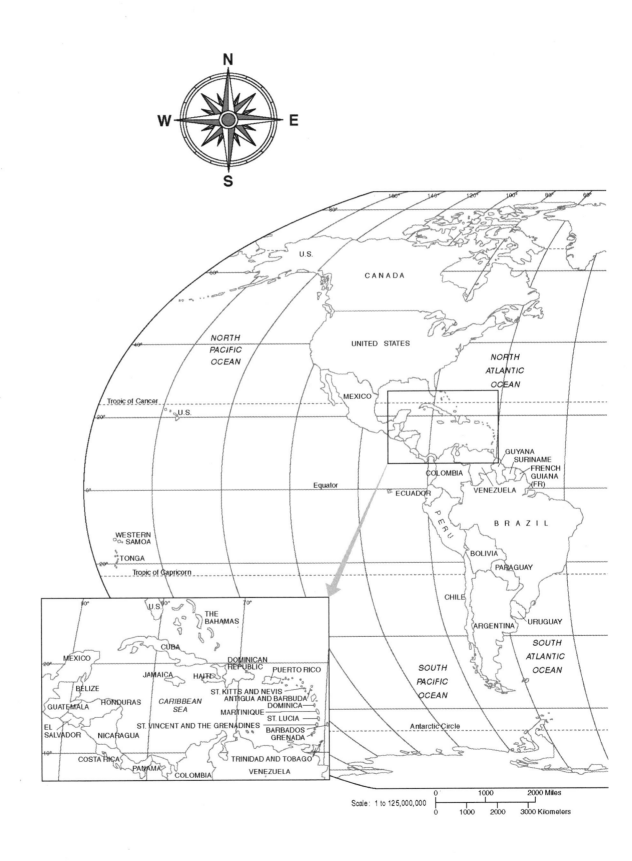

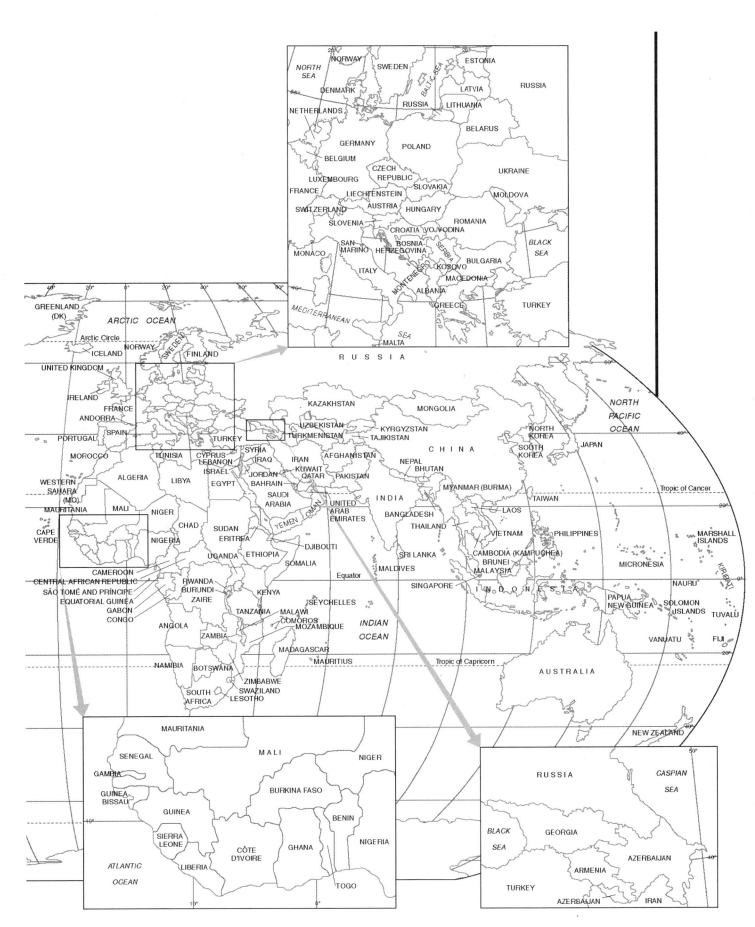

Credits

1. Geography in a Changing World
Unit photo—© 2001 by PhotoDisc, Inc.
2. Human-Environment Relationships
Unit photo—United Nations photo.
3. The Region
Unit photo—United Nations photo.
4. Spatial Interaction and Mapping
Unit photo—Courtesy of Digital Stock.
5. Population, Resources, and Socioeconomic Development
Unit photo—Courtesy USDA.

Copyright

Cataloging in Publication Data
Main entry under title: Annual Editions: Geography. 2001/2002.
 1. Geography—Periodicals. 2. Anthropo-geography—Periodicals. 3. Natural resources—Periodicals. I. Pitzl, Gerald R., *comp.* II. Title: Geography.
ISBN 0–07–243564–X   87–641715   ISSN 1091–9937   910'.5

© 2001 by McGraw-Hill/Dushkin, Guilford, CT 06437, A Division of The McGraw-Hill Companies.

Copyright law prohibits the reproduction, storage, or transmission in any form by any means of any portion of this publication without the express written permission of McGraw-Hill/Dushkin, and of the copyright holder (if different) of the part of the publication to be reproduced. The Guidelines for Classroom Copying endorsed by Congress explicitly state that unauthorized copying may not be used to create, to replace, or to substitute for anthologies, compilations, or collective works.

Annual Editions® is a Registered Trademark of McGraw-Hill/Dushkin, A Division of The McGraw-Hill Companies.

Sixteenth Edition

Cover: Photo of Yellowstone National Park by Sweet By and By/Cindy Brown

Printed in the United States of America   1234567890BAHBAH54321   Printed on Recycled Paper

Editors/Advisory Board

Members of the Advisory Board are instrumental in the final selection of articles for each edition of ANNUAL EDITIONS. Their review of articles for content, level, currentness, and appropriateness provides critical direction to the editor and staff. We think that you will find their careful consideration well reflected in this volume.

EDITOR

Gerald R. Pitzl
Macalester College

ADVISORY BOARD

Paul S. Anderson
Illinois State University

Robert S. Bednarz
Texas A&M University

Sarah Witham Bednarz
Texas A&M University

Roger Crawford
San Francisco State University

James Fryman
University of Northern Iowa

Allison C. Gilmore
Mercer University

J. Michael Hammett
Azusa Pacific University

Miriam Helen Hill
Auburn University–Main

Lance F. Howard
Clemson University

Artimus Keiffer
Valparaiso University

Vandana Kohli
California State University Bakersfield

David J. Larson
California State University Hayward

Tom L. Martinson
Auburn University–Main

John T. Metzger
Michigan State University

Peter O. Muller
University of Miami

Robert E. Nunley
University of Kansas

Ray Oldakowski
Jacksonville University

Thomas R. Paradise
University of Arkansas

James D. Proctor
University of California Santa Barbara

Donald M. Spano
University of Southern Colorado

Eileen Starr
Valley City State University

Wayne G. Strickland
Roanoke College

John A. Vargas
Pennsylvania State University–DuBois

Daniel Weiner
West Virginia University

Randy W. Widdis
University of Regina

Staff

EDITORIAL STAFF

Ian A. Nielsen, Publisher
Roberta Monaco, Senior Developmental Editor
Dorothy Fink, Associate Developmental Editor
Addie Raucci, Senior Administrative Editor
Robin Zarnetske, Permissions Editor
Joseph Offredi, Permissions Assistant
Diane Barker, Proofreader
Lisa Holmes-Doebrick, Senior Program Coordinator

TECHNOLOGY STAFF

Richard Tietjen, Senior Publishing Technologist
Jonathan Stowe, Director of Technology
Janice Ward, Software Support Analyst
Ciro Parente, Editorial Assistant

PRODUCTION STAFF

Brenda S. Filley, Director of Production
Charles Vitelli, Designer
Laura Levine, Graphics
Mike Campbell, Graphics
Tom Goddard, Graphics
Eldis Lima, Graphics
Nancy Norton, Graphics
Juliana Arbo, Typesetting Supervisor
Marie Lazauskas, Typesetter
Karen Roberts, Typesetter
Jocelyn Proto, Typesetter
Larry Killian, Copier Coordinator

To the Reader

In publishing ANNUAL EDITIONS we recognize the enormous role played by the magazines, newspapers, and journals of the public press in providing current, first-rate educational information in a broad spectrum of interest areas. Many of these articles are appropriate for students, researchers, and professionals seeking accurate, current material to help bridge the gap between principles and theories and the real world. These articles, however, become more useful for study when those of lasting value are carefully collected, organized, indexed, and reproduced in a low-cost format, which provides easy and permanent access when the material is needed. That is the role played by ANNUAL EDITIONS.

The articles in this sixteenth edition of *Annual Editions: Geography* represent the wide range of topics associated with the discipline of geography. The major themes of spatial relationships, regional development, the population explosion, and socioeconomic inequalities exemplify the diversity of research areas within geography.

The book is organized into five units, each containing articles relating to geographical themes. Selections address the conceptual nature of geography and the global and regional problems in the world today. This latter theme reflects the geographer's concern with finding solutions to these serious issues. Regional problems, such as food shortages in the Sahel and the greenhouse effect, concern not only geographers but also researchers from other disciplines.

The association of geography with other fields is important, because expertise from related research will be necessary in finding solutions to some difficult problems. Input from the focus of geography is vital in our common search for solutions. This discipline has always been integrative. That is, geography uses evidence from many sources to answer the basic questions, "Where is it?" "Why is it there?" and "What is its relevance?" The first group of articles emphasizes the interconnectedness not only of places and regions in the world but of efforts toward solutions to problems as well. No single discipline can have all of the answers to the problems facing us today; the complexity of the issues is simply too great.

The writings in unit 1 discuss particular aspects of geography as a discipline and provide examples of the topics presented in the remaining four sections. Units 2, 3, and 4 represent major themes in geography. Unit 5 addresses important problems faced by geographers and others.

Annual Editions: Geography 01/02 will be useful to both teachers and students in their study of geography. The anthology is designed to provide detail and case study material to supplement the standard textbook treatment of geography. The goals of this anthology are to introduce students to the richness and diversity of topics relating to places and regions on Earth's surface, to pay heed to the serious problems facing humankind, and to stimulate the search for more information on topics of interest.

I would like to express my gratitude to Barbara Wells-Howe for her continued help in preparing this material for publication. Her typing, organization of materials, and many helpful suggestions are greatly appreciated. Without her diligence and professional efforts, this undertaking could not have been completed. Special thanks are also extended to Ian Nielsen for his continued encouragement during the preparation of this new edition, and to the rest of the editorial staff for coordinating the production of the reader. A word of thanks must go as well to all those who recommended articles for inclusion in this volume and who commented on its overall organization. Sarah Witham Bednarz, Miriam Helen Hill, Artimus Keiffer, Peter O. Muller, Tom Paradise, Wayne J. Strickland, and Randy W. Widdis were especially helpful in that regard.

In order to improve the next edition of *Annual Editions: Geography*, we need your help. Please share your opinions by filling out and returning to us the postage-paid *article rating form* on the last page of this book. We will give serious consideration to all your comments.

Gerald R. Pitzl
Editor

Contents

World Map	ii
To the Reader	vi
Topic Guide	4
⦿ Selected World Wide Web Sites	6
Overview	8

1. **Rediscovering the Importance of Geography,** Alexander B. Murphy, *The Chronicle of Higher Education,* October 30, 1998. — 10

    Geography's renaissance in U.S. education is the key theme of this piece. The author insists that geography be recognized not as an exercise in place names, but because it addresses physical and human processes and sheds light on the nature and meaning of changing ***spatial arrangements*** and ***landscapes.***

2. **The Four Traditions of Geography,** William D. Pattison, *Journal of Geography,* September/October 1990. — 12

    This key article, originally published in 1964, was reprinted, with the author's later comments, in the 75-year retrospective of the *Journal of Geography.* It is a classic in the ***history of geography.*** William Pattison discusses the four main themes that have been the focus of work in the discipline of geography for centuries—the spatial concept, area studies, land-human relationships, and earth science.

3. **The American Geographies,** Barry Lopez, *Orion,* Autumn 1989. — 16

    The American ***landscape*** is nearly incomprehensible in depth and complexity, according to Barry Lopez. To truly understand American ***geography,*** one must seek out local experts who have an intimate knowledge of ***place***—people who have a feel for their locale that no outsider could possibly develop.

4. **Human Domination of Earth's Ecosystems,** Peter M. Vitousek, Harold A. Mooney, Jane Lubchenco, and Jerry M. Melillo, *Science,* July 25, 1997. — 21

    "Human alteration of the Earth," the authors contend, "is substantial and growing." Earth's ***ecosystems*** are being affected by human intervention through ***land transformation*** and biotic and chemical changes. Recommendations are made to reduce the rate of change, to better understand Earth's ecosystems, and to assume more responsibility for managing the planet.

5. **Seas and Soils Emerge as Keys to Climate,** William K. Stevens, *New York Times,* May 16, 2000. — 27

    Following the 1997 ***Kyoto Protocol,*** the fight to reduce ***global warming*** has continued in earnest. William Stevens details the controversy surrounding the role of ***carbon sequestration*** approaches.

6. **Plying a Fabled Waterway: Melting May Open the Northwest Passage,** Tim Appenzeller, *U.S. News & World Report,* August 28, 2000. — 30

    Global warming is softening the Arctic ice pack. Tim Appenzeller speculates that the ***climatic change*** could result in the

UNIT 1

Geography in a Changing World

Eight articles discuss the discipline of geography and the extremely varied and wide-ranging themes that define geography today.

The concepts in bold italics are developed in the article. For further expansion please refer to the Topic Guide and the Index.

opening of the **Northwest Passage,** a **transportation corridor** sought after for centuries.

7. **Lead in the Inner Cities,** Howard W. Mielke, *American Scientist,* January/February 1999.   32

   Howard Mielke, program director of a multimedia study of **metals in urban settings,** conclusively points to the dangers to children of accumulated lead in soils. His article highlights a classic example of a negative impact on the **ecosystem** and a plan to begin solving the problem.

8. **Uneasy Collaborators: To China, Taiwan Investors Are Both Welcome and Suspect,** *Business Week,* August 14, 2000.   46

   Economic activities between China and Taiwan are increasing at a high rate. High technology products are the favored products as the "two Chinas" combine to modify the **geopolitical** climate in the East Asian **region.**

UNIT 2

Human-Environment Relationships

Eleven articles examine the relationship between humans and the land on which we live. Topics include global warming, water management, urban sprawl, pollution, and the effects of human society on the global environment.

Overview   48

9. **Global Warming: The Contrarian View,** William K. Stevens, *New York Times,* February 29, 2000.   50

   William Stevens's article focuses on a debate about the causes of **global warming.** The contrarian view suggests that the effect of **greenhouse gas** accumulations in the atmosphere are overestimated.

10. **The Great Climate Flip-Flop,** William H. Calvin, *The Atlantic Monthly,* January 1998.   54

    A group of scientists are suggesting that **global warming** may well lead to a period of abrupt global cooling. Such a cooling could take place if the North Atlantic Current changes its course, something that has occurred in the past. William Calvin goes into detail about the possible consequences of cooling and ways to prevent this shift.

11. **Breaking the Global-Warming Gridlock,** Daniel Sarewitz and Roger Pielke Jr., *The Atlantic Monthly,* July 2000.   63

    The authors contend that the debate on **global warming** is misdirected. The main focus of attention needs to address proper responses to severe **weather** disruptions and practical approaches to these immediate problems.

12. **Beyond the Valley of the Dammed,** Bruce Barcott, *Utne Reader,* May/June 1999.   70

    A number of **dams** in the United States are coming down. Bruce Barcott reports on recent **water** management thinking, resulting in the removal of some of the 75,000 dams. What is known now is that sediments deposited in the reservoirs become polluted with chemicals and toxic waste, which is damaging to the **environment.**

13. **A River Runs Through It,** Andrew Murr, *Newsweek,* July 12, 1999.   75

    Dams are scheduled to be removed from California to Maine. In the case of Maine's Edwards Dam, Andrew Murr writes that the 0.1 percent of Maine's power it produced no longer justifies the **environmental** damage it did by preventing fish from reaching their spawning grounds.

14. **There Goes the Neighborhood,** Fen Montaigne, *Audubon*, March/April 2000.   76
   Urban sprawl in the Atlantic metropolitan area has doubled in the last 30 years. Large chunks of previously rural **landscape** have changed to built-up areas, reports Fen Montaigne.

15. **The Race to Save Open Space,** Weston Kosova, *Audubon*, March/April 2000.   81
   A number of initiatives have been launched nationwide to increase **green spaces** on the edges of fast-growing urban areas to stem **urban sprawl.** Weston Kosova indicates that this open-space movement has a strong momentum.

16. **Past and Present Land Use and Land Cover in the USA,** William B. Meyer, *Consequences*, Spring 1995.   83
   Cities, towns, and villages now cover slightly more than 4 out of every 100 acres of the land in the continental United States, and every day the percentage grows. William Meyer looks at both the positive and negative **consequences of changes in land use.**

17. **Operation Desert Sprawl,** William Fulton and Paul Shigley, *Governing*, August 1999.   92
   Population growth in Las Vegas has been staggering (Clark County has grown from 790,000 in 1900 to over 1.2 million in 1998). Las Vegas is one of the fastest growing **urban** places in the United States. Its **economic** base of tourism has not been damaged, but strains are showing in **transportation** and **water delivery systems.**

18. **Turning the Tide: A Coastal Refuge Holds Its Ground as Sprawl Spills Into Southern Maine,** Richard M. Stapleton, *Land and People*, Spring 2000.   95
   Although **wetlands** in coastal Maine are protected as part of the Rachel Carson National Wildlife Refuge, the adjacent uplands are vulnerable to **development** and **urban sprawl.** Richard Stapleton outlines approaches taken to preserve the area's wildlife.

19. **A Greener, or Browner, Mexico?** *The Economist*, August 7, 1999.   98
   The North American Free Trade Association (NAFTA) was claimed to be the world's first **environmentally** friendly trade treaty. Critics suggest that NAFTA has resulted in a more **polluted** Mexico. Employment in Mexico's maquiladoras has doubled since 1991, but so have the levels of toxic waste.

Overview   100

20. **The Rise of the Region State,** Kenichi Ohmae, *Foreign Affairs*, Spring 1993.   102
   The **global** market has dictated a new, more significant set of **region states** that reflect the real flow of economic activity. The old idea of the **nation-state** is considered an anachronism, an unnatural and dysfunctional unit for organizing either human activity or emerging economic systems.

21. **Continental Divide,** Torsten Wohlert, *World Press Review*, January 2000.   106
   Torsten Wohlert reports that the **geopolitical** picture in Europe may change greatly as the European Union (EU) considers applications for membership from Central European countries. One of the controversies is the EU's insistence on intensive **agricultural** practices, a process not beneficial to the Central European **region.**

UNIT 3

The Region

Ten selections review the importance of the region as a concept in geography and as an organizing framework for research. A number of world regional trends, as well as the patterns of area relationships, are examined.

The concepts in bold italics are developed in the article. For further expansion please refer to the Topic Guide and the Index.

22. **Russia's Fragile Union,** Matthew Evangelista, *The Bulletin of the Atomic Scientists,* May/June 1999. 108
Russia has experienced ***economic*** decline and ***regional*** crises in recent years. These problems have led to deterioration in the basic social and economic infrastructure of the country and a realignment of its ***geopolitical*** situation.

23. **Beyond the Kremlin's Walls,** *The Economist,* May 20, 2000. 114
Russia's president has launched a plan to enhance ***economic*** growth and development in many of the country's ***regions.*** These moves have ***geopolitical*** implications as attempts are made to strengthen Russia's central government.

24. **The Delicate Balkan Balance,** *The Economist,* August 19, 2000. 116
Elections in the autumn of 2000 could reshape the Balkan ***region.*** The ***geopolitical*** situation in Bosnia and Kosovo could be significantly affected. Recently, 200,000 people, mostly minorities, have returned to their homes. Acceptance of ethnic ***diversity*** may be occuring.

25. **Sea Change in the Arctic,** Richard Monastersky, *Science News,* February 13, 1999. 119
Scientific expeditions to the Arctic ice pack found that ice thicknesses have decreased significantly in the ***region.*** A melting ice cap could have a dramatic impact on ***climate*** worldwide. The melt is due partly to ***El Niño*** and partly to ***greenhouse gases.***

26. **Greenville: From Back Country to Forefront,** Eugene A. Kennedy, *Focus,* Spring 1998. 123
Eugene Kennedy traces the rise of Greenville, South Carolina, from a back country ***region*** in its early days, to "Textile Center of the World" in the 1920s, to a supremely successful, diversified ***economic*** stronghold in the 1990s. Greenville's location within a thriving area midway between Charlotte and Atlanta gives it tremendous growth potential.

27. **Flood Hazards and Planning in the Arid West,** John Cobourn, *Land and Water,* May/June 1999. 129
This article identifies two main types of ***flooding*** in the ***arid*** West. A number of proposals are made for addressing ***drought, water*** quality, ***urban*** growth, and a ***sustainable*** ecosystem in the ***region.***

28. **The Rio Grande: Beloved River Faces Rough Waters Ahead,** Steve Larese, *New Mexico,* June 2000. 132
As Steve Larese reports, the Rio Grande is in danger as the American Southwest ***region*** comes into a protracted period of ***drought.*** Natural occurrences and the ***damming*** of the river have constricted the flow of ***water*** in Albuquerque and areas farther downstream.

29. **Does It Matter Where You Are?** *The Economist,* July 30, 1994. 139
As the computer affects society, it appears that it is possible to run a business from any location. The truth, however, as this article points out, is that people and businesses are most effective when they have a center or base of operations; where one is located does, indeed, matter greatly.

Overview	140

30. Transportation and Urban Growth: The Shaping of the American Metropolis, Peter O. Muller, *Focus,* Summer 1986. — 142

Peter Muller reviews the importance of **transportation** in the growth of American **urban places.** The highly compact urban form of the mid-nineteenth century was transformed in successive eras by electric streetcars, highways, and expressways. The city of the future may rely more on **communication** than on the actual movement of people.

31. GIS Technology Reigns Supreme in Ellis Island Case, Richard G. Castagna, Lawrence L. Thornton, and John M. Tyrawaski, *Professional Surveyor,* July/August 1999. — 150

A 160-year-old territorial dispute between New Jersey and New York over Ellis Island was settled through the use of **GIS.** The resulting **map** eliminated a long-standing ambiguity surrounding this boundary disagreement.

32. Teaching About Karst Using U.S. Geological Survey Resources, Joseph J. Kerski, *Focus,* Fall 1998. — 154

Joseph Kerski's article highlights the advantages in using USGS **maps** and other resources rather than **landscape** analysis. Through use of USGS materials, students can trace the route and effect of **pollutants** within the **environment.**

33. Gaining Perspective, Molly O'Meara Sheehan, *World Watch,* March/April 2000. — 161

This in-depth article discusses important uses for satellite imagery to monitor **weather** patterns, **flooding,** and problems in **agriculture.** Molly Sheehan describes how **mapping** has been greatly enhanced through satellite technology.

34. Counties With Cash, John Fetto, *American Demographics,* April 2000. — 170

This county-based **choropleth** map illustrates the great range of per-capita income in the United States. A table of counties with the highest dollar values include **metropolitan** and rural counties.

35. Do We Still Need Skyscrapers? William J. Mitchell, *Scientific American,* December 1997. — 172

The need for skyscrapers, the signature structure in America's great **urban** places, is called into question. With rapid **communication** through e-mail, the Internet, phone, and fax, is it necessary to bring people face-to-face? William Mitchell answers with a resounding "yes" and speculates that even taller buildings are coming.

36. Bio Invasion, Janet Ginsburg, *Business Week,* September 11, 2000. — 175

Janet Ginsburg reviews recent increases in **global economic** activity and tourism, which have led to a higher incidence of imported diseases in humans and animals. Improved **transportation systems,** illegal aliens, millions of imported farm animals, and 20 million wild animals brought to the United States just in the year 2000 results in a literal "bioinvasion" of **diseases.**

UNIT 4

Spatial Interaction and Mapping

Seven articles discuss the key theme in geographical analysis: place-to-place spatial interaction. Human diffusion, transportation systems, and cartography are some of the themes examined.

The concepts in bold italics are developed in the article. For further expansion please refer to the Topic Guide and the Index.

UNIT 5

Population, Resources, and Socioeconomic Development

Ten articles examine the effects of population growth on natural resources and the resulting socioeconomic level of development.

Overview   **180**

37. Before the Next Doubling, Jennifer D. Mitchell, *World Watch,* January/February 1998.   **182**

The article warns of the possibility of another doubling of world **population** in the next century. Food shortages, inadequate supplies of fresh water, and greatly strained **economic** systems will result. Family planning efforts need to be increased and a change in the desire for large families should be addressed.

38. A Turning-Point for AIDS? *The Economist,* July 15, 2000.   **189**

The spread of AIDS, particularly devastating in the sub-Sahara **region,** is addressed. A picture of its effects on Botswana is shown in superimposed **population** pyramids for 2020, one showing the country's population without the AIDS epidemic and one with AIDS.

39. The End of Cheap Oil, Colin J. Campbell and Jean H. Laherrere, *Scientific American,* March 1998.   **193**

Petroleum, a prominent source of **energy,** will become a scarce **resource** in coming years. Some experts, however, suggest that undiscovered reserves will eventually be found, that advanced technologies will make extraction of oil more efficient, and that oil shales and "heavy oil" can someday be exploited. A decline in oil production would have serious **economic** and **geopolitical** repercussions.

40. Gray Dawn: The Global Aging Crisis, Peter G. Peterson, *Foreign Affairs,* January/February 1999.   **198**

Countries in the **developed** world will experience a decided increase in the number of elderly in the **population.** This **demographic transition** will have dramatic **economic** impact. By 2020, one in four people in the developed world will be 65 years old or older.

41. A Rare and Precious Resource, Houria Tazi Sedeq, *The UNESCO Courier,* February 1999.   **203**

Fresh **water** is fast becoming a scarce **resource.** Water, necessary for every human being, is extensively used in **agriculture** and industry. The availability of fresh water is a **regional** problem. As **population** increases, more regions are facing water shortages.

42. Rethinking Flood Prediction: Does the Traditional Approach Need to Change? William D. Gosnold Jr., Julie A. LeFever, Paul E. Todhunter, and Leon F. Osborne Jr., *Geotimes,* May 2000.   **207**

The authors strongly suggest that the traditional approaches to **flood** prediction are outdated. Changes occurring in **land use** and **watershed** management also contribute to the need for new research efforts, which will lead to better plannning and preparation.

43. Risking the Future of Our Cities, Thomas Hackett, *Ford Foundation Report,* Fall 1999.   **212**

Expanded **transportation systems** have allowed **urban** areas to grow far beyond their original boundaries, leaving a poorer inner city. Now **suburban** areas are experiencing similar downturns as the spread outward continues. Thomas Hackett reviews several proposals for addressing this uncoordinated growth.

The concepts in bold italics are developed in the article. For further expansion please refer to the Topic Guide and the Index.

44. **The End of Urban Man? Care to Bet?** *The Economist,* 215
December 31, 1999.
Urban places are chosen by people because they serve their needs. Any speculations about the demise of *cities* is put to rest in this editorial. With all its problems, the city is here to stay.

45. **China's Changing Land: Population, Food Demand and Land Use in China,** Gerhard K. Heilig, *Options,* 218
Summer 1999.
This article focuses on trends in *land transformation* in the world's most populated country. Paramount in importance are reforestation, expansion of surface *transportation* to enhance *accessibility,* expansion and refinement of the agricultural system, and a goal to reduce the number of people in *poverty.* A dramatic decline in *population* is predicted for the twenty-first century.

46. **Helping the World's Poorest,** Jeffrey Sachs, *The* 221
Economist, August 14, 1999.
Jeffrey Sachs identifies 42 Highly Indebted Poor Countries (HIPCs) in the *developing world* and contrasts their woeful position relative to the 30 richest countries in the *developed world.* Socioeconomic conditions continue to deteriorate in the HIPCs because of *population* increase and adverse *climate* change. Technology in the rich countries must be applied to problem areas in the poor countries, or all will suffer.

Index 225
Test Your Knowledge Form 228
Article Rating Form 229

The concepts in bold italics are developed in the article. For further expansion please refer to the Topic Guide and the Index.

Topic Guide

This topic guide suggests how the selections in this book relate to the subjects covered in your course.

The Web icon (☻) under the topic articles easily identifies the relevant Web sites, which are numbered and annotated on the next two pages. By linking the articles and the Web sites by topic, this ANNUAL EDITIONS reader becomes a powerful learning and research tool.

TOPIC AREA	TREATED IN	TOPIC AREA	TREATED IN
Accessibility	45. China's Changing Land ☻ 19		7. Lead in the Inner Cities ☻ 2, 3, 5, 6, 7, 8, 9
Agriculture	21. Continental Divide 33. Gaining Perspective 41. Rare and Precious Resource ☻ 5, 15, 25, 28	**El Niño**	25. Sea Change in the Arctic ☻ 2, 3, 5, 8, 9
		Energy	39. End of Cheap Oil ☻ 1, 2, 3, 5, 6
Arid Lands	27. Flood Hazards and Planning ☻ 10, 11, 12	**Environment**	12. Beyond the Valley of the Dammed 13. River Runs Through It 19. Greener, or Browner, Mexico? 32. Teaching About Karst ☻ 2, 3, 5, 9, 11, 15, 16, 17, 29
Carbon Sequestration	5. Seas and Soils ☻ 5		
Climatic Change	5. Seas and Soils 6. Plying a Fabled Waterway 9. Global Warming 10. Great Climate Flip-Flop 11. Breaking the Global Warming Gridlock 25. Sea Change in the Arctic 46. Helping the World's Poorest ☻ 5, 6, 7, 8	**Flooding**	27. Flood Hazards and Planning 33. Gaining Perspective 42. Rethinking Flood Prediction ☻ 2, 5
		Geographic Information Systems (GIS)	31. GIS Technology Reigns Supreme ☻ 1, 3, 20, 21, 24
		Geography	1. Rediscovering the Importance of Geography 3. American Geographies 29. Does It Matter Where You Are? ☻ 1, 2, 19
Communication	30. Transportation and Urban Growth 35. Do We Still Need Skyscrapers? ☻ 18, 19		
Dams	12. Beyond the Valley of the Dammed 13. River Runs Through It 28. Rio Grande ☻ 5	**Geopolitical**	8. Uneasy Collaborators 21. Continental Divide 22. Russia's Fragile Union 23. Beyond the Kremlin's Walls 24. Delicate Balkan Balance 39. End of Cheap Oil ☻ 15, 19
Demographic Transition	40. Gray Dawn ☻ 25, 28		
Development	18. Turning the Tide 40. Gray Dawn 46. Helping the World's Poorest ☻ 10, 11, 12, 25	**Global Issues**	5. Seas and Soils 6. Plying a Fabled Waterway 9. Global Warming 10. Great Climate Flip-Flop 11. Breaking the Global Warming Gridlock 20. Rise of the Region State 29. Does It Matter Where You Are? 36. Bio Invasion ☻ 11, 15, 19, 28
Diversity	24. Delicate Balkan Balance ☻ 10, 19, 26		
Drought	27. Flood Hazards and Planning 28. Rio Grande ☻ 2, 5, 12, 19		
Economic Issues	8. Uneasy Collaborators 17. Operation Desert Sprawl 22. Russia's Fragile Union 23. Beyond the Kremlin's Walls 26. Greenville 29. Does It Matter Where You Are? 36. Bio Invasion 37. Before the Next Doubling 39. End of Cheap Oil 40. Gray Dawn ☻ 1, 5, 10, 11, 16, 19, 28	**Green Spaces**	15. Race to Save Open Space ☻ 16, 18, 19
		Greenhouse Effect	9. Global Warming 25. Sea Change in the Arctic ☻ 2, 3, 5
		History of Geography	2. Four Traditions of Geography ☻ 2, 3
		Kyoto Protocol	5. Seas and Soils ☻ 6
Ecosystem	4. Human Domination of Earth's Ecosystems		

TOPIC AREA	TREATED IN	TOPIC AREA	TREATED IN
Land Transformation	4. Human Domination of Earth's Ecosystems 42. Rethinking Flood Prediction 45. China's Changing Land ● 9	**Region State**	20. Rise of the Region State ● *15, 19*
		Resources	39. End of Cheap Oil 41. Rare and Precious Resource ● *2, 25, 28*
Landscape	1. Rediscovering the Importance of Geography 3. American Geographies 14. There Goes the Neighborhood 16. Past and Present Land Use 32. Teaching About Karst ● *2, 3, 8*	**Spatial Arrangements**	1. Rediscovering the Importance of Geography ● *1, 2, 8, 29*
		Suburban	43. Risking the Future of Our Cities ● *13, 23*
Location	29. Does It Matter Where You Are? ● *2, 9, 19*	**Sustainable Development**	27. Flood Hazards and Planning ● *15, 19*
Maps	31. GIS Technology Reigns Supreme 32. Teaching About Karst 33. Gaining Perspective 34. Counties With Cash ● *20, 21, 22, 23, 24*	**Transportation Systems**	6. Plying a Fabled Waterway 17. Operation Desert Sprawl 30. Transportation and Urban Growth 36. Bio Invasion 43. Risking the Future of Our Cities 45. China's Changing Land ● *18, 19*
Nation-State	20. Rise of the Region State ● *19*		
Northwest Passage	6. Plying a Fabled Waterway ● *2, 5*	**Urban**	7. Lead in the Inner Cities 14. There Goes the Neighborhood 15. Race to Save Open Space 17. Operation Desert Sprawl 18. Turning the Tide 27. Flood Hazards and Planning 30. Transportation and Urban Growth 34. Counties With Cash 35. Do We Still Need Skyscrapers? 43. Risking the Future of Our Cities 44. End of Urban Man? 45. China's Changing Land ● *14, 16, 18, 19*
Place	3. American Geographies ● *2, 3, 9*		
Pollution	19. Greener, or Browner, Mexico? 32. Teaching About Karst ● *11, 14, 15, 16, 18*		
Population	17. Operation Desert Sprawl 37. Before the Next Doubling 38. Turning-Point for AIDS? 40. Gray Dawn 41. Rare and Precious Resource 45. China's Changing Land 46. Helping the World's Poorest ● *20, 21, 22, 23, 24, 25, 29*	**Water**	12. Beyond the Valley of the Dammed 17. Operation Desert Sprawl 27. Flood Hazards and Planning 28. Rio Grande 41. Rare and Precious Resource 42. Rethinking Flood Prediction ● *5, 6*
		Weather	11. Breaking the Global Warming Gridlock 25. Sea Change in the Arctic 33. Gaining Perspective ● *2, 3, 5, 8, 9, 10, 11*
Poverty	45. China's Changing Land ● *23, 25, 26*		
Region	8. Uneasy Collaborators 21. Continental Divide 22. Russia's Fragile Union 23. Beyond the Kremlin's Walls 24. Delicate Balkan Balance 25. Sea Change in the Arctic 26. Greenville 27. Flood Hazards and Planning 28. Rio Grande 29. Does It Matter Where You Are? 38. Turning-Point for AIDS? 41. Rare and Precious Resource ● *9, 13, 19, 23, 27, 29*	**Wetlands**	18. Turning the Tide ● *5, 15*

AE: Geography

The following World Wide Web sites have been carefully researched and selected to support the articles found in this reader. The sites are cross-referenced by number and the Web icon (◉) in the topic guide. In addition, it is possible to link directly to these Web sites through our DUSHKIN ONLINE support site at http://www.dushkin.com/online/.

The following sites were available at the time of publication. Visit our Web site—we update DUSHKIN ONLINE regularly to reflect any changes.

General Sources

1. The Association of American Geographers
http://www.aag.org
Surf this site of the Association of American Geographers to learn about AAG projects and publications, careers in geography, and information about related organizations.

2. National Geographic Society
http://www.nationalgeographic.com
This site provides links to National Geographic's huge archive of maps, articles, and other documents. Search the site for information about worldwide expeditions of interest to geographers.

3. The New York Times
http://www.nytimes.com
Browsing through the archives of the New York Times will provide you with a wide array of articles and information related to the different subfields of geography.

4. Social Science Internet Resources
http://www.wcsu.ctstateu.edu/library/ss_geography.html
This site is a definitive source for geography-related links to universities, browsers, cartography, associations, and discussion groups.

5. U.S. Geological Survey
http://www.usgs.gov
This site and its many links are replete with information and resources for geographers, from explanations of El Niño, to mapping, to geography education, to water resources. No geographer's resource list would be complete without frequent mention of the USGS.

Geography in a Changing World

6. Alternative Energy Institute
http://www.altenergy.org
The AEI will continue to monitor the transition from today's energy forms to the future in a "surprising journey of twists and turns." This site is the beginning of an incredible journey.

7. Geological Survey of Sweden: Other Geological Surveys
http://www.sgu.se/index_e.htm
This site, providing links to the national geographical surveys of many countries in Europe and elsewhere, including Brazil, South Africa, and the United States, makes for very interesting and informative browsing.

8. Mission to Planet Earth
http://www.earth.nasa.gov
This site will direct you to information about NASA's Mission to Planet Earth program and its Science of the Earth System. Surf here to learn about satellites, El Niño, and even "strategic visions" of interest to geographers.

9. Santa Fe Institute
http://acoma.santafe.edu/sfi/research/
This home page of the Santa Fe Institute—a nonprofit, multidisciplinary research and education center—will lead you to a plethora of valuable links related to its primary goal: to create a new kind of scientific research community, pursuing emerging science. Such links as Evolution of Language, Ecology, and Local Rules for Global Problems are offered.

Human-Environment Relationships

10. The North-South Institute
http://www.nsi-ins.ca/ensi/index.html
Searching this site of the North-South Institute—which works to strengthen international development cooperation and enhance gender and social equity—will help you find information on a variety of development issues.

11. United Nations Environment Programme
http://www.unep.ch
Consult this home page of UNEP for links to critical topics of concern to geographers, including desertification and the impact of trade on the environment. The site will direct you to useful databases and global resource information.

12. World Health Organization
http://www.who.int
This home page of the World Health Organization will provide you with links to a wealth of statistical and analytical information about health in the developing world.

The Region

13. AS at UVA Yellow Pages: Regional Studies
http://xroads.virginia.edu/~YP/regional/regional.html
Those interested in American regional studies will find this site a gold mine. Links to periodicals and other informational resources about the Midwest/Central, Northeast, South, and West regions are provided here.

14. Can Cities Save the Future?
http://www.huduser.org/publications/econdev/habitat/prep2.html
This press release about the second session of the Preparatory Committee for Habitat II is an excellent discussion of the question of global urbanization.

15. IISDnet
http://iisd1.iisd.ca
The International Institute for Sustainable Development, a Canadian organization, presents information through gateways entitled Business and Sustainable Development, Developing Ideas, and Hot Topics. Linkages is its multimedia resource for environment and development policymakers.

16. NewsPage
http://www.individual.com
Individual, Inc., maintains this business-oriented Web site. Geographers will find links to much valuable information about such fields as energy, environmental services, media and communications, and health care.

17. Telecommuting as an Investment: The Big Picture—John Wolf
http://www.svi.org/telework/forums/messages5/48.html
This page deals with the many issues related to telecommuting, including its potential role in reducing environmental pollution. The site discusses such topics as employment law and the impact of telecommuting on businesses and employees.

18. The Urban Environment
http://www.geocities.com/RainForest/Vines/6723/urb/index.html
Global urbanization is discussed fully at this site, which also includes the original 1992 Treaty on Urbanization.

19. Virtual Seminar in Global Political Economy//Global Cities & Social Movements
http://csf.colorado.edu/gpe/gpe95b/resources.html
This Web site is rich in links to subjects of interest in regional studies, such as sustainable cities, megacities, and urban planning. Links to many international nongovernmental organizations are included.

Spatial Interaction and Mapping

20. Edinburgh Geographical Information Systems
http://www.geo.ed.ac.uk/home/gishome.html
This valuable site, hosted by the Department of Geography at the University of Edinburgh, provides information on all aspects of Geographic Information Systems and provides links to other servers worldwide. A GIS reference database as well as a major GIS bibliography is included.

21. GIS Frequently Asked Questions and General Information
http://www.census.gov/ftp/pub/geo/www/faq-index.html
Browse through this site to get answers to FAQs about Geographic Information Systems. It can direct you to general information about GIS as well as guidelines on such specific questions as how to order U.S. Geological Survey maps. Other sources of information are also noted.

22. International Map Trade Association
http://www.maptrade.org
The International Map Trade Association offers this site for those interested in information on maps, geography, and mapping technology. Lists of map retailers and publishers as well as upcoming IMTA conferences and trade shows are noted.

23. PSC Publications
http://www.psc.lsa.umich.edu/pubs/abs/abs94-319.html
Use this site and its links from the Population Studies Center of the University of Michigan for spatial patterns of immigration and discussion of white and black flight from high immigration metropolitan areas in the United States.

24. U.S. Geological Survey
http://www.usgs.gov/research/gis/title.html
This site discusses the uses for Geographic Information Systems and explains how GIS works, addressing such topics as data integration, data modeling, and relating information from different sources.

Population, Resources, and Socioeconomic Development

25. African Studies WWW (U.Penn)
http://www.sas.upenn.edu/African_Studies/AS.html
Access to rich and varied resources that cover such topics as demographics, migration, family planning, and health and nutrition is available at this site.

26. Human Rights and Humanitarian Assistance
http://info.pitt.edu/~ian/resource/human.htm
Through this site, part of the World Wide Web Virtual Library, you can conduct research into a number of human-rights topics in order to gain a greater understanding of the issues affecting indigenous peoples in the modern era.

27. Hypertext and Ethnography
http://www.umanitoba.ca/faculties/arts/anthropology/tutor/aaa_presentation.new.html
This site, presented by Brian Schwimmer of the University of Manitoba, will be of great value to people who are interested in culture and communication. He addresses such topics as multivocality and complex symbolization, among many others.

28. Research and Reference (Library of Congress)
http://lcweb.loc.gov/rr/
This research and reference site of the Library of Congress will lead you to invaluable information on different countries. It provides links to numerous publications, bibliographies, and guides in area studies that can be of great help to geographers.

29. Space Research Institute
http://arc.iki.rssi.ru/Welcome.html
Browse through this home page of Russia's Space Research Institute for information on its Environment Monitoring Information Systems, the IKI Satellite Situation Center, and its Data Archive.

We highly recommend that you review our Web site for expanded information and our other product lines. We are continually updating and adding links to our Web site in order to offer you the most usable and useful information that will support and expand the value of your Annual Editions. You can reach us at: *http://www.dushkin.com/annualeditions/*.

Unit 1

Unit Selections

1. **Rediscovering the Importance of Geography,** Alexander B. Murphy
2. **The Four Traditions of Geography,** William D. Pattison
3. **The American Geographies,** Barry Lopez
4. **Human Domination of Earth's Ecosystems,** Peter M. Vitousek, Harold A. Mooney, Jane Lubchenco, and Jerry M. Melillo
5. **Seas and Soils Emerge as Keys to Climate,** William K. Stevens
6. **Plying a Fabled Waterway: Melting May Open the Northwest Passage,** Tim Appenzeller
7. **Lead in the Inner Cities,** Howard W. Mielke
8. **Uneasy Collaborators: To China, Taiwan Investors Are Both Welcome and Suspect,** Business Week

Key Points to Consider

- Why is geography called an integrating discipline?

- How is geography related to earth science? Give some examples of these relationships.

- What are area studies? Why is the spatial concept so important in geography? What is your definition of geography?

- Why do history and geography rely on each other? How have humans affected the weather? The environment in general?

- How is individual behavior related to our understanding of environment?

- Discuss whether or not change is a good thing. Why is it important to anticipate change?

- What are the prospects for the "two Chinas"?

- What does interconnectedness mean in terms of places? Give examples of how you as an individual interact with people in other places. How are you "connected" to the rest of the world?

- What will the world be like in the year 2010? Tell why you are pessimistic or optimistic about the future. What, if anything, can you do about the future?

 **Links**   www.dushkin.com/online/

6. **Alternative Energy Institute**
   http://www.altenergy.org
7. **Geological Survey of Sweden: Other Geological Surveys**
   http://www.sgu.se/index_e.htm
8. **Mission to Planet Earth**
   http://www.earth.nasa.gov
9. **Santa Fe Institute**
   http://acoma.santafe.edu/sfi/research/

These sites are annotated on pages 6 and 7.

Geography in a Changing World

What is geography? This question has been asked innumerable times, but it has not elicited a universally accepted answer, even from those who are considered to be members of the geography profession. The reason lies in the very nature of geography as it has evolved through time. Geography is an extremely wide-ranging discipline, one that examines appropriate sets of events or circumstances occurring at specific places. Its goal is to answer certain basic questions.

The first question—Where is it?—establishes the location of the subject under investigation. The concept of location is very important in geography, and its meaning extends beyond the common notion of a specific address or the determination of the latitude and longitude of a place. Geographers are more concerned with the relative location of a place and how that place interacts with other places both far and near. Spatial interaction and the determination of the connections between places are important themes in geography.

Once a place is "located," in the geographer's sense of the word, the next question is, Why is it here? For example, why are people concentrated in high numbers on the North China plain, in the Ganges River Valley in India, and along the eastern seaboard in the United States? Conversely, why are there so few people in the Amazon basin and the Central Siberian lowlands? Generally, the geographer wants to find out why particular distribution patterns occur and why these patterns change over time.

The element of time is another extremely important ingredient in the geographical mix. Geography is most concerned with the activities of human beings, and human beings bring about change. As changes occur, new adjustments and modifications are made in the distribution patterns previously established. Patterns change, for instance, as new technology brings about new forms of communication and transportation and as once-desirable locations decline in favor of new ones. For example, people migrate from once-productive regions such as the Sahel when a disaster such as drought visits the land. Geography, then, is greatly concerned with discovering the underlying processes that can explain the transformation of distribution patterns and interaction forms over time. Geography itself is dynamic, adjusting as a discipline to handle new situations in a changing world.

Geography is truly an integrating discipline. The geographer assembles evidence from many sources in order to explain a particular pattern or ongoing process of change. Some of this evidence may even be in the form of concepts or theories borrowed from other disciplines. The first article of this unit stresses the importance of geography as a discipline and proclaims its "rediscovery." The next two articles of this unit provide insight into both the conceptual nature of geography and the development of the discipline over time.

Throughout its history, four main themes have been the focus of research work in geography. These themes or traditions, according to William Pattison in "The Four Traditions of Geography," link geography with earth science, establish it as a field that studies land-human relationships, engage it in area studies, and give it a spatial focus. Although Pattison's article first appeared over 30 years ago, it is still referred to and cited frequently today. Much of the geographical research and analysis engaged in today would fall within one or more of Pattison's traditional areas, but new areas are also opening for geographers. In a particularly thought-provoking essay, the eminent author Barry Lopez discusses local geographies and the importance of a sense of place. In "Human Domination of the Earth's Ecosystems," we see how human activity is significantly altering the globe. This theme continues in the next article, which considers the climatic effect of carbon dioxide accumulation in the atmosphere. Will the Northwest Passage at last be a viable transportation corridor? The next article deals with this topic. Howard Mielke details the problem of lead content in inner-city soils. The prospect of a China-Taiwan economic cooperation is the theme of the final article.

Article 1

POINT OF VIEW

Rediscovering the Importance of Geography

By Alexander B. Murphy

AS AMERICANS STRUGGLE to understand their place in a world characterized by instant global communications, shifting geopolitical relationships, and growing evidence of environmental change, it is not surprising that the venerable discipline of geography is experiencing a renaissance in the United States. More elementary and secondary schools now require courses in geography, and the College Board is adding the subject to its Advanced Placement program. In higher education, students are enrolling in geography courses in unprecedented numbers. Between 1985–86 and 1994–95, the number of bachelor's degrees awarded in geography increased from 3,056 to 4,295. Not coincidentally, more businesses are looking for employees with expertise in geographical analysis, to help them analyze possible new markets or environmental issues.

In light of these developments, institutions of higher education cannot afford simply to ignore geography, as some of them have, or to assume that existing programs are adequate. College administrators should recognize the academic and practical advantages of enhancing their offerings in geography, particularly if they are going to meet the demand for more and better geography instruction in primary and secondary schools. We cannot afford to know so little about the other countries and peoples with which we now interact with such frequency, or about the dramatic environmental changes unfolding around us.

From the 1960s through the 1980s, most academics in the United States considered geography a marginal discipline, although it remained a core subject in most other countries. The familiar academic divide in the United States between the physical sciences, on one hand, and the social sciences and humanities, on the other, left little room for a discipline concerned with how things are organized and relate to one another on the surface of the earth—a concern that necessarily bridges the physical and cultural spheres. Moreover, beginning in the 1960s, the U.S. social-science agenda came to be dominated by pursuit of more-scientific explanations for human phenomena, based on assumptions about global similarities in human institutions, motivations, and actions. Accordingly, regional differences often were seen as idiosyncrasies of declining significance.

Although academic administrators and scholars in other disciplines might have marginalized geography, they could not kill it, for any attempt to make sense of the world must be based on some understanding of the changing human and physical patterns that shape its evolution—be they shifting vegetation zones or expanding economic contacts across international boundaries. Hence, some U.S. colleges and universities continued to teach geography, and the discipline was often in the background of many policy issues—for example, the need to assess the risks associated with foreign investment in various parts of the world.

By the late 1980s, Americans' general ignorance of geography had become too widespread to ignore. Newspapers regularly published reports of surveys demonstrating that many Americans could not identify major countries or oceans on a map. The real problem, of course, was not the inability to answer simple questions that might be asked on *Jeopardy!*; instead, it was what that inability demonstrated about our collective understanding of the globe.

Geography's renaissance in the United States is due to the growing recognition that physical and human processes such as soil erosion and ethnic unrest are inextricably tied to their geographical context. To understand modern Iraq, it is not enough to know who is in power and how the political system functions. We also need to know something about the country's ethnic groups and their settlement patterns, the different physical environments and resources within the country, and its ties to surrounding countries and trading partners.

Those matters are sometimes addressed by practitioners of other disciplines, of course, but they are rarely central to the analysis. Instead, generalizations are often made at the level of the state, and little attention is given to spatial patterns and practices that play out on local levels or across international boundaries. Such preoccupations help to explain why many scholars were caught off guard by the explosion of ethnic unrest in Eastern Europe following the fall of the Iron Curtain.

Similarly, comprehending the dynamics of El Niño requires more than knowledge of the behavior of ocean and air currents; it is also important to understand how those currents are situated with respect to land masses and how they relate to other climatic patterns, some of which have been altered by the burning of fossil fuels and other human activities. And any attempt to understand the nature and extent of humans' impact on the environment requires consideration of the relationship between human and physical contributions to environmental change. The factories and cars in a city produce smog, but surrounding mountains may trap it, increasing air pollution significantly.

TODAY, academics in fields including history, economics, and conservation biology are turning to geographers for help with some of their concerns. Paul Krugman, a noted economist at the Massachusetts Institute of Technology, for example, has turned conventional wisdom on its head by pointing out the role of historically rooted regional inequities in how international trade is structured.

Geographers work on issues ranging from climate change to ethnic conflict to urban sprawl. What unites their work is its focus on the shifting organization and character of the earth's surface. Geographers examine changing patterns of vegetation to study global warming; they analyze where ethnic groups live in Bosnia to help understand the pros and cons of competing administrative solutions to the civil war there; they map AIDS cases in Africa to learn how to reduce the spread of the disease.

Geography is reclaiming attention because it addresses such questions in their relevant spatial and environmental contexts. A growing number of scholars in other disciplines are realizing that it is a mistake to treat all places as if they were essentially the same (think of the assumptions in most economic models), or to undertake research on the environment that does not include consideration of the relationships between human and physical processes in particular regions.

Still, the challenges to the discipline are great. Only a small number of primary- and secondary-school teachers have enough training in geography to offer students an exciting introduction to the subject. At the college level, many geography departments are small; they are absent altogether at some high-profile universities.

Perhaps the greatest challenge is to overcome the public's view of geography as a simple exercise in place-name recognition. Much of geography's power lies in the insights it sheds on the nature and meaning of the evolving spatial arrangements and landscapes that make up our world. The importance of those insights should not be underestimated at a time of changing political boundaries, accelerated human alteration of the environment, and rapidly shifting patterns of human interaction.

Alexander B. Murphy is a professor and head of the geography department at the University of Oregon, and a vice-president of the American Geographical Society.

Article 2

The Four Traditions of Geography

William D. Pattison

Late Summer, 1990

To Readers of the *Journal of Geography:*

I am honored to be introducing, for a return to the pages of the *Journal* after more than 25 years, "The Four Traditions of Geography," an article which circulated widely, in this country and others, long after its initial appearance—in reprint, in xerographic copy, and in translation. A second round of life at a level of general interest even approaching that of the first may be too much to expect, but I want you to know in any event that I presented the paper in the beginning as my gift to the geographic community, not as a personal property, and that I re-offer it now in the same spirit.

In my judgment, the article continues to deserve serious attention—perhaps especially so, let me add, among persons aware of the specific problem it was intended to resolve. The background for the paper was my experience as first director of the High School Geography Project (1961–63)—not all of that experience but only the part that found me listening, during numerous conference sessions and associated interviews, to academic geographers as they responded to the project's invitation to locate "basic ideas" representative of them all. I came away with the conclusion that I had been witnessing not a search for consensus but rather a blind struggle for supremacy among honest persons of contrary intellectual commitment. In their dialogue, two or more different terms had been used, often unknowingly, with a single reference, and no less disturbingly, a single term had been used, again often unknowingly, with two or more different references. The article was my attempt to stabilize the discourse. I was proposing a basic nomenclature (with explicitly associated ideas) that would, I trusted, permit the development of mutual comprehension and confront all parties concerned with the pluralism inherent in geographic thought.

This intention alone could not have justified my turning to the NCGE as a forum, of course. The fact is that from the onset of my discomfiting realization I had looked forward to larger consequences of a kind consistent with NCGE goals. As finally formulated, my wish was that the article would serve "to greatly expedite the task of maintaining an alliance between professional geography and pedagogical geography and at the same time to promote communication with laymen" (see my fourth paragraph). I must tell you that I have doubts, in 1990, about the acceptability of my word choice, in saying "professional," "pedagogical," and "layman" in this context, but the message otherwise is as expressive of my hope now as it was then.

I can report to you that twice since its appearance in the *Journal,* my interpretation has received more or less official acceptance—both times, as it happens, at the expense of the earth science tradition. The first occasion was Edward Taaffe's delivery of his presidential address at the 1973 meeting of the Association of American Geographers (see *Annals AAG,* March 1974, pp. 1–16). Taaffe's working-through of aspects of an interrelations among the spatial, area studies, and man-land traditions is by far the most thoughtful and thorough of any of which I am aware. Rather than fault him for omission of the fourth tradition, I compliment him on the grace with which he set it aside in conformity to a meta-epistemology of the American university which decrees the integrity of the social sciences as a consortium in their own right. He was sacrificing such holistic claims as geography might be able to muster for a freedom to argue the case for geography as a social science.

The second occasion was the publication in 1984 of *Guidelines for Geographic Education: Elementary and Secondary Schools,* authored by a committee jointly representing the AAG and the NCGE. Thanks to a recently published letter (see *Journal of Geography,* March-April 1990, pp. 85–86), we know that, of five themes commended to teachers in this source,

The committee lifted the human environmental interaction theme directly from Pattison. The themes of place and location are based on Pattison's spatial or geometric geography, and the theme of region comes from Pattison's area studies or regional geography.

Having thus drawn on my spatial, area studies, and man-land traditions for four of the five themes, the committee could have found the remaining theme, movement, there too—in the spatial tradition (see my sixth paragraph). However that may be, they did not avail themselves of the earth science tradition, their reasons being readily surmised. Peculiar to the elementary and secondary schools is a curriculum category framed as much by theory of citizenship as by theory of knowledge: the social studies. With admiration, I see already in the committee members' adoption of the theme idea a strategy for assimilation of their program to the established repertoire of social studies practice. I see in their exclusion of the earth science tradition an intelligent respect for social studies' purpose.

Here's to the future of education in geography: may it prosper as never before.

W. D. P., 1990

2. Four Traditions of Geography

Reprinted from the Journal of Geography, 1964, pp. 211–216.

In 1905, one year after professional geography in this country achieved full social identity through the founding of the Association of American Geographers, William Morris Davis responded to a familiar suspicion that geography is simply an undisciplined "omnium-gatherum" by describing an approach that as he saw it imparts a "geographical quality" to some knowledge and accounts for the absence of the quality elsewhere.[1] Davis spoke as president of the AAG. He set an example that was followed by more than one president of that organization. An enduring official concern led the AAG to publish, in 1939 and in 1959, monographs exclusively devoted to a critical review of definitions and their implications.[2]

Every one of the well-known definitions of geography advanced since the founding of the AAG has had its measure of success. Tending to displace one another by turns, each definition has said something true of geography.[3] But from the vantage point of 1964, one can see that each one has also failed. All of them adopted in one way or another a monistic view, a singleness of preference, certain to omit if not to alienate numerous professionals who were in good conscience continuing to participate creatively in the broad geographic enterprise.

The thesis of the present paper is that the work of American geographers, although not conforming to the restrictions implied by any one of these definitions, has exhibited a broad consistency, and that this essential unity has been attributable to a small number of distinct but affiliated traditions, operant as binders in the minds of members of the profession. These traditions are all of great age and have passed into American geography as parts of a general legacy of Western thought. They are shared today by geographers of other nations.

There are four traditions whose identification provides an alternative to the competing monistic definitions that have been the geographer's lot. The resulting pluralistic basis for judgment promises, by full accommodation of what geographers do and by plain-spoken representation thereof, to greatly expedite the task of maintaining an alliance between professional geography and pedagogical geography and at the same time to promote communication with laymen. The following discussion treats the traditions in this order: (1) a spatial tradition, (2) an area studies tradition, (3) a man-land tradition and (4) an earth science tradition.

Spatial Tradition

Entrenched in Western thought is a belief in the importance of spatial analysis, of the act of separating from the happenings of experience such aspects as distance, form, direction and position. It was not until the 17th century that philosophers concentrated attention on these aspects by asking whether or not they were properties of things-in-themselves. Later, when the 18th century writings of Immanuel Kant had become generally circulated, the notion of space as a category including all of these aspects came into widespread use. However, it is evident that particular spatial questions were the subject of highly organized answering attempts long before the time of any of these cogitations. To confirm this point, one need only be reminded of the compilation of elaborate records concerning the location of things in ancient Greece. These were records of sailing distances, of coastlines and of landmarks that grew until they formed the raw material for the great *Geographia* of Claudius Ptolemy in the 2nd century A.D.

A review of American professional geography from the time of its formal organization shows that the spatial tradition of thought had made a deep penetration from the very beginning. For Davis, for Henry Gannett and for most if not all of the 44 other men of the original AAG, the determination and display of spatial aspects of reality through mapping were of undoubted importance, whether contemporary definitions of geography happened to acknowledge this fact or not. One can go further and, by probing beneath the art of mapping, recognize in the behavior of geographers of that time an active interest in the true essentials of the spatial tradition—*geometry* and *movement*. One can trace a basic favoring of movement as a subject of study from the turn-of-the-century work of Emory R. Johnson, writing as professor of transportation at the University of Pennsylvania, through the highly influential theoretical and substantive work of Edward L. Ullman during the past 20 years and thence to an article by a younger geographer on railroad freight traffic in the U.S. and Canada in the *Annals* of the AAG for September 1963.[4]

One can trace a deep attachment to geometry, or positioning-and-layout, from articles on boundaries and population densities in early 20th century volumes of the *Bulletin of the American Geographical Society,* through a controversial pronouncement by Joseph Schaefer in 1953 that granted geographical legitimacy only to studies of spatial patterns[5] and so onward to a recent *Annals* report on electronic scanning of cropland patterns in Pennsylvania.[6]

One might inquire, is discussion of the spatial tradition, after the manner of the remarks just made, likely to bring people within geography closer to an understanding of one another and people outside geography closer to an understanding of geographers? There seem to be at least two reasons for being hopeful. First, an appreciation of this tradition allows one to see a bond of fellowship uniting the elementary school teacher, who attempts the most rudimentary instruction in directions and mapping, with the contemporary research geographer, who dedicates himself to an exploration of central-place theory. One cannot only open the eyes of many teachers to the potentialities of their own instruction, through proper exposition of the spatial tradition, but one can also "hang a bell" on research quantifiers in geography, who are often thought to have wandered so far in their intellectual adventures as to have become lost from the rest. Looking outside geography, one may anticipate benefits from the readiness of countless persons to associate the name "geography" with maps. Latent within this readiness is a willingness to recognize as geography, too, what maps are about—and that is the geometry of and the movement of what is mapped.

Area Studies Tradition

The area studies tradition, like the spatial tradition, is quite strikingly represented in classical antiquity by a practitioner to whose surviving work we can point. He is Strabo, celebrated for his *Geography* which is a massive production addressed to the statesmen of Augustan Rome and intended to sum up and regularize knowledge not of the location of places and associated cartographic facts, as in the somewhat later case of Ptolemy, but of the nature of places, their character and their differentiation. Strabo exhibits interesting attributes of the area-studies tradition that can hardly be overemphasized. They are a pronounced tendency toward subscription primarily to literary standards, an almost omnivorous appetite for information and a self-conscious companionship with history.

It is an extreme good fortune to have in the ranks of modern American geography the scholar Richard Hartshorne, who has pondered the meaning of the area-studies tradition with a legal acuteness that few persons would challenge. In his *Nature of Geography,* his 1939 monograph already cited,[7] he scrutinizes exhaustively the implications of the "interesting attributes" identified in connection with Strabo, even though his concern is with quite other and much later authors, largely German. The major literary problem of unities or wholes he considers from every angle. The Gargantuan appetite for miscellaneous information he accepts and rationalizes. The companionship between area studies and history he clarifies by appraising the so-called idiographic content of both and by affirming the tie of both to what he and Sauer have called "naively given reality."

The area-studies tradition (otherwise known as the chorographic tradition) tended to be excluded from early American professional geography. Today it is beset by certain champions of the spatial tradition who would have one believe that somehow the area-studies way of organizing knowledge is only a subdepartment of spatialism. Still, area-studies as a method of presentation lives and prospers in its own right. One can turn today for reassurance on this score to practically any issue of the *Geographical Review*, just as earlier readers could turn at the opening of the century to that magazine's forerunner.

What is gained by singling out this tradition? It helps toward restoring the faith of many teachers who, being accustomed to administering learning in the area-studies style, have begun to wonder if by doing so they really were keeping in touch with professional geography. (Their doubts are owed all too much to the obscuring effect of technical words attributable to the very professionals who have been intent, ironically, upon protecting that tradition.) Among persons outside the classroom the geographer stands to gain greatly in intelligibility. The title "area-studies" itself carries an understood message in the United States today wherever there is contact with the usages of the academic community. The purpose of characterizing a place, be it neighborhood or nation-state, is readily grasped. Furthermore, recognition of the right of a geographer to be unspecialized may be expected to be forthcoming from people generally, if application for such recognition is made on the merits of this tradition, explicitly.

Man-Land Tradition
That geographers are much given to exploring man-land questions is especially evident to anyone who examines geographic output, not only in this country but also abroad. O. H. K. Spate, taking an international view, has felt justified by his observations in nominating as the most significant ancient precursor of today's geography neither Ptolemy nor Strabo nor writers typified in their outlook by the geographies of either of these two men, but rather Hippocrates, Greek physician of the 5th century B.C. who left to posterity an extended essay, *On Airs, Waters and Places*.[8] In this work made up of reflections on human health and conditions of external nature, the questions asked are such as to confine thought almost altogether to presumed influence passing from the latter to the former, questions largely about the effects of winds, drinking water and seasonal changes upon man. Understandable though this uni-directional concern may have been for Hippocrates as medical commentator, and defensible as may be the attraction that this same approach held for students of the condition of man for many, many centuries thereafter, one can only regret that this narrowed version of the man-land tradition, combining all too easily with social Darwinism of the late 19th century, practically overpowered American professional geography in the first generation of its history.[9] The premises of this version governed scores of studies by American geographers in interpreting the rise and fall of nations, the strategy of battles and the construction of public improvements. Eventually this special bias, known as environmentalism, came to be confused with the whole of the man-land tradition in the minds of many people. One can see now, looking back to the years after the ascendancy of environmentalism, that although the spatial tradition was asserting itself with varying degrees of forwardness, and that although the area-studies tradition was also making itself felt, perhaps the most interesting chapters in the story of American professional geography were being written by academicians who were reacting against environmentalism while deliberately remaining within the broad man-land tradition. The rise of culture historians during the last 30 years has meant the dropping of a curtain of culture between land and man, through which it is asserted all influence must pass. Furthermore work of both culture historians and other geographers has exhibited a reversal of the direction of the effects in Hippocrates, man appearing as an independent agent, and the land as a sufferer from action. This trend as presented in published research has reached a high point in the collection of papers titled *Man's Role in Changing the Face of the Earth*. Finally, books and articles can be called to mind that have addressed themselves to the most difficult task of all, a balanced tracing out of interaction between man and environment. Some chapters in the book mentioned above undertake just this. In fact the separateness of this approach is discerned only with difficulty in many places; however, its significance as a general research design that rises above environmentalism, while refusing to abandon the man-land tradition, cannot be mistaken.

The NCGE seems to have associated itself with the man-land tradition, from the time of founding to the present day, more than with any other tradition, although all four of the traditions are amply represented in its official magazine, *The Journal of Geography* and in the proceedings of its annual meetings. This apparent preference on the part of the NCGE members *for defining geography in terms of the man-land tradition* is strong evidence of the appeal that man-land ideas, separately stated, have for persons whose main job is teaching. It should be noted, too, that this inclination reflects a proven acceptance by the general public of learning that centers on resource use and conservation.

Earth Science Tradition
The earth science tradition, embracing study of the earth, the waters of the earth, the atmosphere surrounding the earth and the association between earth and sun, confronts one with a paradox. On the one hand one is assured by professional geographers that their participation in this tradition has declined precipitously in the course of the past few decades, while on the other one knows that college departments of geography across the nation rely substantially, for justification of their role in general education, upon curricular content springing directly from this tradition. From all the reasons that combine to account for this state of affairs, one may, by selecting only two, go far toward achieving an understanding of this tradition. First, there is the fact that American college geography, growing out of departments of geology in many crucial instances, was at one time greatly overweighted in favor of earth science, thus rendering the field unusually liable to a sense of loss as better balance came into being. (This one-time disproportion found reciprocate support for many years in the narrowed, environmentalistic interpretation of the man-land tradition.) Second, here alone in earth science does one encounter subject matter in the normal sense of the term as one reviews geographic traditions. The spatial tradition abstracts certain aspects of reality; area studies is distinguished by a point of view; the man-land tradition dwells upon relationships; but earth science is identifiable through concrete objects. Historians, sociologists and other academicians tend not only to accept but also to ask for help from this part of geography. They readily appreciate earth science as something physically associated with their subjects of study, yet generally beyond their competence to treat. From this appreciation comes strength for geography-as-earth-science in the curriculum.

Only by granting full stature to the earth science tradition can one make sense out of the oft-repeated addage, "Geography is the mother of sciences." This is the tradition that emerged in ancient Greece, most clearly in the work of Aristotle, as a wide-ranging study of natural processes in and near the surface of the earth. This is the tradition that was rejuvenated by Varenius in the 17th century as "Geographia Generalis." This is the tradition that has been subjected to subdivision as the development of science has approached the present day, yielding mineralogy, paleontology, glaciology, meterology and other specialized fields of learning.

Readers who are acquainted with American junior high schools may want to make a challenge at this point, being aware that a current revival of earth sciences is being sponsored in those schools by the field of geology. Belatedly, geography has joined in support of this revival.[10] It may be said that in this connection and in others, American professional geography may have faltered in its adherence to the earth science tradition but not given it up.

In describing geography, there would appear to be some advantages attached to isolating this final tradition. Separation improves the geographer's chances of successfully explaining to educators why geography has extreme difficulty in accommodating itself to social studies programs. Again, separate attention allows one to make understanding contact with members of the American public for whom surrounding nature is known as the geographic environment. And finally, specific reference to the geographer's earth science tradition brings into the open the basis of what is, almost without a doubt, morally the most significant concept in the entire geographic heritage, that of the earth as a unity, the single common habitat of man.

An Overview

The four traditions though distinct in logic are joined in action. One can say of geography that it pursues concurrently all four of them. Taking the traditions in varying combinations, the geographer can explain the conventional divisions of the field. Human or cultural geography turns out to consist of the first three traditions applied to human societies; physical geography, it becomes evident, is the fourth tradition prosecuted under constraints from the first and second traditions. Going further, one can uncover the meanings of "systematic geography," "regional geography," "urban geography," "industrial geography," etc.

It is to be hoped that through a widened willingness to conceive of and discuss the field in terms of these traditions, geography will be better able to secure the inner unity and outer intelligibility to which reference was made at the opening of this paper, and that thereby the effectiveness of geography's contribution to American education and to the general American welfare will be appreciably increased.

Notes

1. William Morris Davis, "An Inductive Study of the Content of Geography," *Bulletin of the American Geographical Society,* Vol. 38, No. 1 (1906), 71.
2. Richard Hartshorne, *The Nature of Geography,* Association of American Geographers (1939), and idem., *Perspective on the Nature of Geography,* Association of American Geographers (1959).
3. The essentials of several of these definitions appear in Barry N. Floyd, "Putting Geography in Its Place," *The Journal of Geography,* Vol. 62, No. 3 (March, 1963). 117–120.
4. William H. Wallace, "Freight Traffic Functions of Anglo-American Railroads," *Annals of the Association of American Geographers,* Vol. 53, No. 3 (September, 1963), 312–331.
5. Fred K. Schaefer, "Exceptionalism in Geography: A Methodological Examination," *Annals of the Association of American Geographers,* Vol. 43, No. 3 (September, 1953), 226–249.
6. James P. Latham, "Methodology for an Instrumental Geographic Analysis," *Annals of the Association of American Geographers,* Vol. 53, No. 2 (June, 1963), 194–209.
7. Hartshorne's 1959 monograph, *Perspective on the Nature of Geography,* was also cited earlier. In this later work, he responds to dissents from geographers whose preferred primary commitment lies outside the area studies tradition.
8. O. H. K. Spate, "Quantity and Quality in Geography," *Annals of the Association of American Geographers,* Vol. 50, No. 4 (December, 1960), 379.
9. Evidence of this dominance may be found in Davis's 1905 declaration: "Any statement is of geographical quality if it contains . . . some relation between an element of inorganic control and one of organic response" (Davis, *loc. cit.*).
10. Geography is represented on both the Steering Committee and Advisory Board of the Earth Science Curriculum Project, potentially the most influential organization acting on behalf of earth science in the schools.

The American Geographies

Americans are fast becoming strangers in a strange land, where one roiling river, one scarred patch of desert, is as good as another. America the beautiful exists— a select few still know it intimately—but many of us are settling for a homogenized national geography.

Barry Lopez

Barry Lopez has written The Rediscovery of North America *(Vintage), and his most recent book is* Field Notes *(Knopf).*

It has become commonplace to observe that Americans know little of the geography of their country, that they are innocent of it as a landscape of rivers, mountains, and towns. They do not know, supposedly, the location of the Delaware Water Gap, the Olympic Mountains, or the Piedmont Plateau; and, the indictment continues, they have little conception of the way the individual components of this landscape are imperiled, from a human perspective, by modern farming practices or industrial pollution.

I do not know how true this is, but it is easy to believe that it is truer than most of us would wish. A recent Gallup Organization and National Geographic Society survey found Americans woefully ignorant of world geography. Three out of four couldn't locate the Persian Gulf. The implication was that we knew no more about our own homeland, and that this ignorance undermined the integrity of our political processes and the efficiency of our business enterprises.

As Americans, we profess a sincere and fierce love for the American landscape, for our rolling prairies, freeflowing rivers, and "purple mountains' majesty"; but it is hard to imagine, actually, where this particular landscape is. It is not just that a nostalgic landscape has passed away—Mark Twain's Mississippi is now dammed from Minnesota to Missouri and the prairies have all been sold and fenced. It is that it has always been a romantic's landscape. In the attenuated form in which it is presented on television today, in magazine articles and in calendar photographs, the essential wildness of the American landscape is reduced to attractive scenery. We look out on a familiar, memorized landscape that portends adventure and promises enrichment. There are no distracting people in it and few artifacts of human life. The animals are all beautiful, diligent, one might even say well-

---

To truly understand geography requires not only time but a kind of local expertise, an intimacy with place few of us ever develop.

---

behaved. Nature's unruliness, the power of rivers and skies to intimidate, and any evidence of disastrous human land management practices are all but invisible. It is, in short, a magnificent garden, a colonial vision of paradise imposed on a real place that is, at best, only selectively known.

The real American landscape is a face of almost incomprehensible depth and complexity. If one were to sit for a few days, for example, among the ponderosa pine forests and black lava fields of the Cascade Mountains in western Oregon, inhaling the pines' sweet balm on an evening breeze from some point on the barren rock, and then were to step off to the Olympic Peninsula in Washington, to those rain forests with sphagnum moss floors soft as fleece underfoot and Douglas firs too big around for five people to hug, and then head south to walk the ephemeral creeks and sun-blistered playas of the Mojave Desert in southern California, one would be reeling under the sensations. The contrast is not only one of plants and soils, a different array say, of brilliantly colored beetles. The shock to the senses comes from a different shape to the silence, a difference in the very quality of light, in the weight of the air. And this relatively short journey down the West Coast would still leave the traveler with all that lay to the east to explore—the anomalous sand hills of Nebraska, the heat and frog voices of Okefenokee Swamp, the fetch of Chesapeake Bay, the hardwood copses and black bears of the Ozark Mountains.

No one of these places, of course, can be entirely fathomed, biologically or aesthetically. They are mysteries upon which we impose names. Enchantments. We tick the names off glibly but lovingly. We mean no disrespect. Our genuine desire, though we might be skeptical about the time it would take and uncertain of its practical value to us, is to actually know these places. As deeply ingrained in the American psyche as the desire to conquer and control the land is the desire to sojourn in it, to sail up and down Pamlico Sound, to paddle a

3. American Geographies

canoe through Minnesota's boundary waters, to walk on the desert of the Great Salt Lake, to camp in the stony hardwood valleys of Vermont.

To do this well, to really come to an understanding of a specific American geography, requires not only time but a kind of local expertise, an intimacy with place few of us ever develop. There is no way around the former requirement: If you want to know you must take the time. It is not in books. A specific geographical understanding, however, can be sought out and borrowed. It resides with men and women more or less sworn to a place, who abide there, who have a feel for the soil and history, for the turn of leaves and night sounds. Often they are glad to take the outlander in tow.

These local geniuses of American landscape, in my experience, are people in whom geography thrives. They are the antithesis of geographical ignorance. Rarely known outside their own communities, they often seem, at the first encounter, unremarkable and anonymous. They may not be able to recall the name of a particular wildflower—or they may have given it a name known only to them. They might have forgotten the precise circumstances of a local historical event. Or they can't say for certain when the last of the Canada geese passed through in the fall, or can't differentiate between two kinds of trout in the same creek. Like all of us, they have fallen prey to the fallacies of memory and are burdened with ignorance; but they are nearly flawless in the respect they bear these places they love. Their knowledge is intimate rather than encyclopedic, human but not necessarily scholarly. It rings with the concrete details of experience.

America, I believe, teems with such people. The paradox here, between a faulty grasp of geographical knowledge for which Americans are indicted and the intimate, apparently contradictory familiarity of a group of largely anonymous people, is not solely a matter of confused scale. (The local landscape is easier to know than a national geography.) And it is not simply ironic. The paradox is dark. To be succinct: The politics and advertising that seek a national audience must project a national geography; to be broadly useful that geography must, inevitably, be generalized and it is often romantic. It is therefore frequently misleading and imprecise. The same holds true with the entertainment industry, but here the problem might be clearer. The same films, magazines, and television features that honor an imaginary American landscape also tout the worth of the anonymous men and women who interpret it. Their affinity for the land is lauded, their local allegiance admired. But the rigor of their local geographies, taken as a whole, contradicts a patriotic, national vision of unspoiled, untroubled land. These men and women are ultimately forgotten, along with the details of the landscapes they speak for, in the face of more pressing national matters. It is the chilling nature of modern society to find an ignorance of geography, local or national, as excusable as an ignorance of hand tools; and to find the commitment of people to their home places only momentarily entertaining. And finally naive.

If one were to pass time among Basawara people in the Kalahari Desert, or with Kreen-Akrora in the Amazon Basin, or with Pitjantjatjara Aborigines in Australia, the most salient impression they might leave is of an absolutely stunning knowledge of their local geography—geology, hydrology, biology, and weather. In short, the extensive particulars of their intercourse with it.

In 40,000 years of human history, it has only been in the last few hundred years or so that a people could afford to ignore their local geographies as completely as we do and still survive. Technological innovations from refrigerated trucks to artificial fertilizers, from sophisticated cost accounting to mass air transportation, have utterly changed concepts of season, distance, soil productivity, and the real cost of drawing sustenance from the land. It is now possible for a resident of Boston to bite into a fresh strawberry in the dead of winter; for someone in San Francisco to travel to Atlanta in a few hours with no worry of how formidable might be crossing of the Great Basin Desert or the Mississippi River; for an absentee farmer to gain a tax advantage from a farm that leaches poisons into its water table and on which crops are left to rot. The Pitjantjatjara might shake their heads in bewilderment and bemusement, not because they are primitive or ignorant people, not because they have no sense of irony or are incapable of marveling, but because they have not (many would say not yet) realized a world in which such manipulation of the land—surmounting the imperatives of distance it imposes, for example, or turning the large-scale destruction of forests and arable land in wealth—is desirable or plausible.

In the years I have traveled through America, in cars and on horseback, on foot and by raft, I have repeatedly been brought to a sudden state of awe by some gracile or savage movement of animal, some odd wrapping of tree's foliage by the wind, an unimpeded run of dew-laden prairie stretching to a horizon flat as a coin where a pin-dot sun pales the dawn sky pink. I know these things are beyond intellection, that they are the vivid edges of a world that includes but also transcends the human world. In memory, when I dwell on these things, I know that in a truly national literature there should be odes to the Triassic reds of the Colorado Plateau, to the sharp and ghostly light of the Florida Keys, to the aeolian soils of southern Minnesota, and the Palouse in Washington, though the modern mind abjures the literary potential of such subjects. (If the sand and flood water farmers of Arizona and New Mexico were to take the black loams of Louisiana in their hands they would be flabbergasted, and that is the beginning of literature.) I know there should be eloquent evocations of the cobbled beaches of Maine, the plutonic walls of the Sierra Nevada, the orange canyons of the Kaibab Plateau. I have no doubt, in fact, that there are. They are as numerous and diverse as the eyes and fingers that ponder the country—it is that only a handful of them are known. The great majority are to be found in drawers and boxes, in the letters and private journals of millions of workaday people who have regarded their encounters with the land as an engagement bordering on the spiritual, as being fundamentally linked to their state of health.

One cannot acknowledge the extent and the history of this kind of testimony without being forced to the realization that something strange, if not dangerous, is afoot. Year by year, the number of people with firsthand experience in the land dwindles. Rural populations continue to shift to the cities. The family farm is in a state of demise, and government and industry continue to apply pressure on the native peoples of North America to sever their ties with the land. In the wake of this loss of personal and local knowledge from which a real geography is derived, the knowledge on which a country must ultimately stand, has [be]come something hard to define but I think sinister and unsettling—the packaging and marketing of land as a form of entertainment. An incipient industry, capitalizing on the nostalgia Americans feel for the imagined virgin landscapes of their fathers, and on a desire for adventure, now offers people a convenient though sometimes incomplete or even spurious geography as an inducement to purchase a unique experience. But the line between authentic experience and a superficial exposure to the elements of experience is blurred. And the real landscape, in all its complexity, is distorted even further in the public imagination. No longer innately mysterious and dignified, a ground from which experience grows, it becomes a curiously generic backdrop on which experience is imposed.

In theme parks the profound, subtle, and protracted experience of running a river is reduced to a loud, quick, safe equivalence, a pleasant distraction. People only able to venture into the countryside on annual vacations are, increasingly, schooled in the belief that wild land will, and should, provide thrills and exceptional scenery on a timely basis. If it does not, something is wrong, either with the land itself or possibly with the company outfitting the trip.

People in America, then, face a convoluted situation. The land itself, vast and differentiated, defies the notion of a national geography. If applied at all it must be applied lightly and it must grow out of the concrete detail of local geographies. Yet Americans are daily presented with, and have become accustomed to talking about, a homogenized national geography. One that seems to operate independently of the land, a collection of objects rather than a continuous bolt of fabric. It appears in advertisements, as a background in movies, and in patriotic calendars. The suggestion is that there can be national geography because the constituent parts are interchangeable and can be treated as commodities. In day-to-day affairs, in other words, one place serves as well as another to convey one's point. On reflection, this is an appalling condescension and a terrible imprecision, the very antithesis of knowledge. The idea that either the Green River in Utah or the Salmon River in Idaho will do, or that the valleys of Kentucky and West Virginia are virtually interchangeable, is not just misleading. For people still dependent on the soil for their sustenance, or for people whose memories tie them to those places, it betrays a numbing casualness, utilitarian, expedient, and commercial frame of mind. It heralds a society in which it is no longer necessary for human beings to know where they live, except as those places are described and fixed by numbers. The truly difficult and lifelong task of discovering where one lives is finally disdained.

If a society forgets or no longer cares where it lives, then anyone with the political power and the will to do so can manipulate the landscape to conform to certain social ideals or nostalgic visions. People may hardly notice that anything has happened, or assume that whatever happens—a mountain stripped of timber and eroding into its creeks—is for the common good. The more superficial a society's knowledge of the real dimensions of the land it occupies becomes, the more vulnerable the land is to exploitation, to manipulation for short-term gain. The land, virtually powerless before political and commercial entities, finds itself finally with no defenders. It finds itself bereft of intimates with indispensable, concrete knowledge. (Oddly, or perhaps not oddly, while American society continues to value local knowledge as a quaint part of its heritage, it continues to cut such people off from any real political power. This is as true for small farmers and illiterate cowboys as it is for American Indians, native Hawaiians, and Eskimos.)

The intense pressure of imagery in America, and the manipulation of images necessary to a society with specific goals, means the land will inevitably be treated like a commodity; and voices that tend to contradict the proffered image will, one way or another, be silenced or discredited by those in power. This is not new to America; the promulgation in America of a false or imposed geography has been the case from the beginning. All local geographies, as they were defined by hundreds of separate, independent native traditions, were denied in the beginning in favor of an imported and unifying vision of America's natural history. The country, the landscape itself, was eventually defined according to dictates of Progress like Manifest Destiny, and laws like the Homestead Act which reflected a poor understanding of the physical lay of the land.

When I was growing up in southern California, I formed the rudiments of a local geography—eucalyptus trees, February rains, Santa Ana winds. I lost much of it when my family moved to New York City, a move typical of the modern, peripatetic style of American life, responding to the exigencies of divorce and employment. As a boy I felt a hunger to know the American landscape that was extreme; when I was finally able to travel on my own, I did so. Eventually I visited most of the United States, living for brief periods of time in Arizona, Indiana, Alabama, Georgia, Wyoming, New Jersey, and Montana before settling 20 years ago in western Oregon.

The astonishing level of my ignorance confronted me everywhere I went. I knew early on that the country could not be held together in a few phrases, that its geography was magnificent and incomprehensible, that a man or woman could devote a lifetime to its elucidation and still feel in the end that he had but sailed many thousands of miles over the surface of the ocean. So I came into the habit of traversing landscapes I wanted to know with local tutors and reading what had previously been written about, and in, those places. I came to value exceedingly novels and essays and works of nonfiction that connected human enterprise to real and specific places, and I grew to be mildly distrustful of work that occurred in no particular place, work so cerebral and detached as to be refutable only in an argument of ideas.

These sojourns in various corners of the country infused me, somewhat to my surprise on thinking about it, with a great sense of hope. Whatever despair I had come to feel at a waning sense of the real land and the emergence of false geographies—elements of the land being manipulated, for example, to create erroneous but useful patterns in advertising—was dispelled by the depth of a single person's local knowledge, by the serenity that seemed to come with that intelligence. Any harm that might be done by people who cared nothing for the land, to whom it was not innately worthy but only something ultimately for sale, I thought, would one day have to meet this kind of integrity, people with the same dignity and transcendence as the land they occupied. So when I traveled, when I rolled my sleeping bag out on the shores of the Beaufort Sea, or in the high pastures of the Absaroka

Range in Wyoming, or at the bottom of the Grand Canyon, I absorbed those particular testaments to life, the indigenous color and songbird song, the smell of sun-bleached rock, damp earth, and wild honey, with some crude appreciation of the singular magnificence of each of those places. And the reassurance I felt expanded in the knowledge that there were, and would likely always be, people speaking out whenever they felt the dignity of the Earth imperiled in those places.

The promulgation of false geographies, which threaten the fundamental notion of what it means to live somewhere, is a current with a stable and perhaps growing countercurrent. People living in New York City are familiar with the stone basements, the cratonic geology, of that island and have a feeling for birds migrating through in the fall, their sequence and number. They do not find the city alien but human, its attenuated natural history merely different from that of rural Georgia or Kansas. I find the countermeasure, too, among Eskimos who cannot read but who might engage you for days on the subtleties of sea-ice topography. And among men and women who, though they have followed in the footsteps of their parents, have come to the conclusion that they cannot farm or fish or log in the way their ancestors did; the finite boundaries to this sort of wealth have appeared in their lifetime. Or among young men and women who have taken several decades of book-learned agronomy, zoology, silviculture and horticulture, ecology, ethnobotany, and fluvial geomorphology and turned it into a new kind of local knowledge, who have taken up residence in a place and sought, both because of and in spite of their education, to develop a deep intimacy with it. Or they have gone to work, idealistically, for the National Park Service or the fish and wildlife services or for a private institution like the Nature Conservancy. They are people to whom the land is more than politics and economics. These are people for whom the land is alive. It feeds them, directly, and that is how and why they learn its geography.

In the end, then, if one begins among the blue crabs of Chesapeake Bay and wanders for several years, down through the Smoky Mountains and back to the bluegrass hills, along the drainages of the Ohio and into the hill country of Missouri, where in summer a chorus of cicadas might drown out human conversation, then up the Missouri itself, reading on the way the entries of Meriwether Lewis and William Clark and musing on the demise of the plains grizzly and the sturgeon, crosses west into the drainage of the Platte and spends the evenings with Gene Weltfish's *The Lost Universe,* her book about the Pawnee who once thrived there, then drops south to the Palo Duro Canyon and the irrigated farms of the Llano Estacado in Texas, turns west across the Sangre de Cristo, southernmost of the Rocky Mountain ranges, and moves north and west up onto the slickrock mesas of Utah, those browns and oranges, the ocherous hues reverberating in the deep canyons, then goes north, swinging west to the insular ranges that sit like battleships in the pelagic space of Nevada, camps at the steaming edge of the sulfur springs in the Black Rock desert, where alkaline pans are glazed with a ferocious light, a heat to melt iron, then crosses the northern Sierra Nevada, waist-deep in summer snow in the passes, to descend to the valley of the Sacramento, and rises through groves of the elephantine redwoods in the Coast Range, to arrive at Cape Mendocino, before Balboa's Pacific, cormorants and gulls, gray whales headed north for Unimak Pass in the Aleutians, the winds crashing down on you, facing the ocean over the blue ocean that gives the scene its true vastness, making this crossing, having been so often astonished at the line and the color of the land, the ingenious lives of its plants and animals, the varieties of its darknesses, the intensity of the stars overhead, you would be ashamed to discover, then, in yourself, any capacity to focus on ravages in the land that left you unsettled. You would have seen so much, breathtaking, startling, and outsize, that you might not be able for a long time to break the spell, the sense, especially finishing your journey in the West, that the land had not been as rearranged or quite as compromised as you had first imagined.

After you had slept some nights on the beach, however, with that finite line of the ocean before you and the land stretching out behind you, the wind first battering then cradling you, you would be compelled by memory, obligated by your own involvement, to speak of what left you troubled. To find the rivers dammed and shrunken, the soil washed away, the land fenced, a tracery of pipes and wires and roads laid down everywhere and animals, cutting the eye off repeatedly and confining it—you had expected this. It troubles you no more than your despair over the ruthlessness, the insensitivity, the impetuousness of modern life. What underlies this obvious change, however, is a less noticeable pattern of disruption: acidic lakes, the skies empty of birds, fouled beaches, the poisonous slags of industry, the sun burning like a molten coin in ruined air.

It is a tenet of certain ideologies that man is responsible for all that is ugly, that everything nature creates is beautiful. Nature's darkness goes partly unreported, of course, and human brilliance is often perversely ignored. What is true is that man has a power, literally beyond his comprehension, to destroy. The lethality of some of what he manufactures, the incompetence with which he stores it or seeks to dispose of it, the cavalier way in which he employs in his daily living substances that threaten his health, the leniency of the courts in these matters (as though products as well as people enjoyed the protection of the Fifth Amendment), and the treatment of open land, rivers, and the atmosphere as if, in some medieval way they could still be regarded as disposal sinks of infinite capacity, would make you wonder, standing face to in the wind at Cape Mendocino, if we weren't bent on an errant of madness.

The geographies of North America, the myriad small landscapes that make up the national fabric, are threatened— by ignorance of what makes them unique, by utilitarian attitudes, by failure to include them in the moral universe, and by brutal disregard. A testament of minor voices can clear away an ignorance of any place, can inform us of its special qualities; but no voice, by merely telling a story, can cause the poisonous wastes that saturate some parts of the land to decompose, to evaporate. This responsibility falls ultimately to the national community, a vague and fragile entity to be sure, but one that, in America, can be ferocious in exerting its will.

Geography, the formal way in which we grapple with this areal mystery, is finally knowledge that calls up something in the land we recognize and respond to. It gives us a sense of place and a sense of community. Both are indispensable to a state of well-being, an individual's and a country's.

One afternoon on the Siuslaw River in the Coast Range of Oregon, in January, I hooked a steelhead, a sea-run trout, that told me, through the muscles of my hands and arms and shoulders, something of the nature of the thing I was calling "the Siuslaw River." Years ago I had stood under a pecan tree in Upson Country, Georgia, idly eating the nuts, when slowly it occurred to me that these nuts would taste different from pecans growing somewhere up in South Carolina. I didn't need a sharp sense of taste to know this, only to pay attention at a level no one had ever told me was necessary. One November dawn, long before the sun rose, I began a vigil at the Dumont Dunes in the Mojave Desert in California, which I kept until a few minutes after the sun broke the horizon. During that time I named to myself the colors by which the sky changed and by which the sand itself flowed like a rising tide through grays and silvers and blues into yellows, pinks, washed duns, and fallow beiges.

It is through the power of observation, the gifts of eye and ear, of tongue and nose and finger, that a place first rises up in our mind; afterward, it is memory that carries the place, that allows it to grow in depth and complexity. For as long as our records go back we have held these two things dear, landscape and memory. Each infuses us with a different kind of life. The one feeds us, figuratively and literally. The other protects us from lies and tyranny. To keep landscapes intact and the memory of them, our history in them, alive, seems as imperative a task in modern time as finding the extent to which individual expression can be accommodated, before it threatens to destroy the fabric of society.

If I were now to visit another country, I would ask my local companion, before I saw any museum or library, any factory or fabled town, to walk me in the country of his or her youth, to tell me the names of things and how, traditionally, they have been fitted together in a community. I would ask for the stories, the voice of memory over the land. I would ask about the history of storms there, the age of the trees, the winter color of the hills. Only then would I ask to see the museum. I would want first the sense of a real place, to know that I was not inhabiting an idea. I would want to know the lay of the land first, the real geography, and take some measure of the love of it in my companion before [having] stood before the painting or read works of scholarship. I would want to have something real and remembered against which I might hope to measure their truth.

Human Domination of Earth's Ecosystems

Peter M. Vitousek, Harold A. Mooney, Jane Lubchenco, Jerry M. Melillo

Human alteration of Earth is substantial and growing. Between one-third and one-half of the land surface has been transformed by human action; the carbon dioxide concentration in the atmosphere has increased by nearly 30 percent since the beginning of the Industrial Revolution; more atmospheric nitrogen is fixed by humanity than by all natural terrestrial sources combined; more than half of all accessible surface fresh water is put to use by humanity; and about one-quarter of the bird species on Earth have been driven to extinction. By these and other standards, it is clear that we live on a human-dominated planet.

All organisms modify their environment, and humans are no exception. As the human population has grown and the power of technology has expanded, the scope and nature of this modification has changed drastically. Until recently, the term "human-dominated ecosystems" would have elicited images of agricultural fields, pastures, or urban landscapes; now it applies with greater or lesser force to all of Earth. Many ecosystems are dominated directly by humanity, and no ecosystem on Earth's surface is free of pervasive human influence.

This article provides an overview of human effects on Earth's ecosystems. It is not intended as a litany of environmental disasters, though some disastrous situations are described; nor is it intended either to downplay or to celebrate environmental successes, of which there have been many. Rather, we explore how large humanity looms as a presence on the globe—how, even on the grandest scale, most aspects of the structure and functioning of Earth's ecosystems cannot be understood without accounting for the strong, often dominant influence of humanity.

We view human alterations to the Earth system as operating through the interacting processes summarized in Fig. 1. The growth of the human population, and growth in the resource base used by humanity, is maintained by a suite of human enterprises such as agriculture, industry, fishing, and international commerce. These enterprises transform the land surface (through cropping, forestry, and urbanization), alter the major biogeochemical cycles, and add or remove species and genetically distinct populations in most of Earth's ecosystems. Many of these changes are substantial and reasonably well quantified; all are ongoing. These relatively well-documented changes in turn entrain further alterations to the functioning of the Earth system, most notably by driving global climatic change (*1*) and causing irreversible losses of biological diversity (*2*).

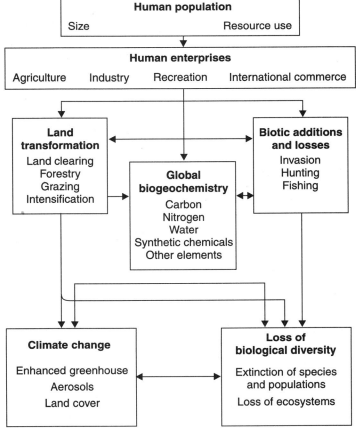

Fig. 1 A conceptual model illustrating humanity's direct and indirect effects on the Earth system [modified from (*56*)].

P. M. Vitousek and H. A. Mooney are in the Department of Biological Sciences, Stanford University, Stanford, CA 94305, USA. J. Lubchenco is in the Department of Zoology, Oregon State University, Corvallis, OR 97331, USA. J. M. Melillo is at the U.S. Office of Science and Technology Policy, Old Executive Office Building, Room 443, Washington, DC 20502, USA.

Land Transformation

The use of land to yield goods and services represents the most substantial human alteration of the Earth system. Human use of land alters the structure and functioning of ecosystems, and it alters how ecosystems interact with the atmosphere, with aquatic systems, and with surrounding land. Moreover, land transformation interacts strongly with most other components of global environmental change.

The measurement of land transformation on a global scale is challenging; changes can be measured more or less straightforwardly at a given site, but it is difficult to aggregate these changes regionally and globally. In contrast to analyses of human alteration of the global carbon cycle, we cannot install instruments on a tropical mountain to collect evidence of land transformation. Remote sensing is a most useful technique, but only recently has there been a serious scientific effort to use high-resolution civilian satellite imagery to evaluate even the more visible forms of land transformation, such as deforestation, on continental to global scales (3).

Land transformation encompasses a wide variety of activities that vary substantially in their intensity and consequences. At one extreme, 10 to 15% of Earth's land surface is occupied by row-crop agriculture or by urban-industrial areas, and another 6 to 8% has been converted to pastureland (4); these systems are wholly changed by human activity. At the other extreme, every terrestrial ecosystem is affected by increased atmospheric carbon dioxide (CO_2), and most ecosystems have a history of hunting and other low-intensity resource extraction. Between these extremes lie grassland and semiarid ecosystems that are grazed (and sometimes degraded) by domestic animals, and forests and woodlands from which wood products have been harvested; together, these represent the majority of Earth's vegetated surface.

The variety of human effects on land makes any attempt to summarize land transformations globally a matter of semantics as well as substantial uncertainty. Estimates of the fraction of land transformed or degraded by humanity (or its corollary, the fraction of the land's biological production that is used or dominated) fall in the range of 39 to 50% (5) (Fig. 2). These numbers have large uncertainties, but the fact that they are large is not at all uncertain. Moreover, if anything these estimates understate the global impact of land transformation, in that land that has not been transformed often has been divided into fragments by human alteration of the surrounding areas. This fragmentation affects the species composition and functioning of otherwise little modified ecosystems (6).

Overall, land transformation represents the primary driving force in the loss of biological diversity worldwide. Moreover, the effects of land transformation extend far beyond the boundaries of transformed lands. Land transformation can affect climate directly at local and even regional scales. It contributes ~20% to current anthropogenic CO_2 emissions, and more substantially to the increasing concentrations of the greenhouse gases methane and nitrous oxide; fires associated with it alter the reactive chemistry of the troposphere, bringing elevated carbon monoxide concentrations and episodes of urban-like photochemical air pollution to remote tropical areas of Africa and South America; and it causes runoff of sediment and nutrients that drive substantial changes in stream, lake, estuarine, and coral reef ecosystems (7–10).

The central importance of land transformation is well recognized within the community of researchers concerned with global environmental change. Several research programs are focused on aspects of it (9, 11); recent and substantial progress toward understanding these aspects has been made (3), and much more progress can be anticipated. Understanding land transformation is a difficult challenge; it requires integrating the social, economic, and cultural causes of land transformation with evaluations of its biophysical nature and consequences. This interdisciplinary approach is essential to predicting the course, and to any hope of affecting the consequences, of human-caused land transformation.

Oceans

Human alterations of marine ecosystems are more difficult to quantify than those of terrestrial ecosystems, but several kinds of information suggest that they are substantial. The human population is concentrated near coasts—about 60% within 100 km—and the oceans' productive coastal margins have been affected strongly by humanity. Coastal wetlands that mediate interactions between land and sea have been altered over large areas; for example, approximately 50% of mangrove ecosystems globally have been transformed or destroyed by human activity (12). Moreover, a recent analysis suggested that although humans use about 8% of the primary production of the oceans, that fraction grows to more than 25% for upwelling areas and to 35% for temperate continental shelf systems (13).

Many of the fisheries that capture marine productivity are focused on top predators, whose removal can alter marine ecosystems out of proportion to their abundance. Moreover, many such fisheries have proved to be unsustainable, at least at our present level of knowledge and control. As of 1995, 22% of recognized marine fisheries were overexploited or already depleted, and 44% more were at their limit of exploitation (14) (Figs. 2 and 3). The consequences of fisheries are not restricted to their target organisms; commercial marine fisheries around the world discard 27 million tons of nontarget animals annually, a quantity nearly one-third as large as total landings (15). Moreover, the dredges and trawls used in some fisheries damage habitats substantially as they are dragged along the sea floor.

A recent increase in the frequency, extent, and duration of harmful algal blooms in coastal areas (16) suggests that human activity has affected the base as well as the top of marine food chains. Harmful algal blooms

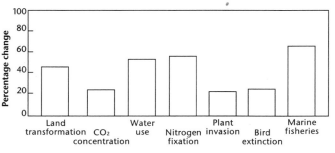

Fig. 2. Human dominance or alteration of several major components of the Earth system, expressed as (from left to right) percentage of the land surface transformed (5); percentage of the current atmospheric CO_2 concentration that results from human action (17); percentage of accessible surface fresh water used (20); percentage of terrestrial N fixation that is human-caused (28); percentage of plant species in Canada that humanity has introduced from elsewhere (48); percentage of bird species on Earth that have become extinct in the past two millennia, almost all of them as a consequence of human activity (42); and percentage of major marine fisheries that are fully exploited, overexploited, or depleted (14).

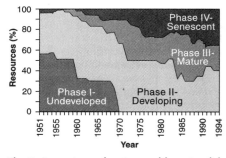

Fig. 3. Percentage of major world marine fish resources in different phases of development, 1951 to 1994 [from (57)]. Undeveloped = a low and relatively constant level of catches; developing = rapidly increasing catches; mature = a high and plateauing level of catches; senescent = catches declining from higher levels.

4. Human Domination of Earth's Ecosystems

are sudden increases in the abundance of marine phytoplankton that produce harmful structures or chemicals. Some but not all of these phytoplankton are strongly pigmented (red or brown tides). Algal blooms usually are correlated with changes in temperature, nutrients, or salinity; nutrients in coastal waters, in particular, are much modified by human activity. Algal blooms can cause extensive fish kills through toxins and by causing anoxia; they also lead to paralytic shellfish poisoning and amnesic shellfish poisoning in humans. Although the existence of harmful algal blooms has long been recognized, they have spread widely in the past two decades (16).

Alterations of the Biogeochemical Cycles

Carbon. Life on Earth is based on carbon, and the CO_2 in the atmosphere is the primary resource for photosynthesis. Humanity adds CO_2 to the atmosphere by mining and burning fossil fuels, the residue of life from the distant past, and by converting forests and grasslands to agricultural and other low-biomass ecosystems. The net result of both activities is that organic carbon from rocks, organisms, and soils is released into the atmosphere as CO_2.

The modern increase in CO_2 represents the clearest and best documented signal of human alteration of the Earth system. Thanks to the foresight of Roger Revelle, Charles Keeling, and others who initiated careful and systematic measurements of atmospheric CO_2 in 1957 and sustained them through budget crises and changes in scientific fashions, we have observed the concentration of CO_2 as it has increased steadily from 315 ppm to 362 ppm. Analysis of air bubbles extracted from the Antarctic and Greenland ice caps extends the record back much further; the CO_2 concentration was more or less stable near 280 ppm for thousands of years until about 1800, and has increased exponentially since then (17).

There is no doubt that this increase has been driven by human activity, today primarily by fossil fuel combustion. The sources of CO_2 can be traced isotopically; before the period of extensive nuclear testing in the atmosphere, carbon depleted in ^{14}C was a specific tracer of CO_2 derived from fossil fuel combustion, whereas carbon depleted in ^{13}C characterized CO_2 from both fossil fuels and land transformation. Direct measurements in the atmosphere, and analyses of carbon isotopes in tree rings, show that both ^{13}C and ^{14}C in CO_2 were diluted in the atmosphere relative to ^{12}C as the CO_2 concentration in the atmosphere increased.

Fossil fuel combustion now adds 5.5 ± 0.5 billion metric tons of CO_2-C to the atmosphere annually, mostly in economically developed regions of the temperate zone (18) (Fig. 4). The annual accumulation of CO_2-C has averaged 3.2 ± 0.2 billion metric tons recently (17). The other major terms in the atmospheric carbon balance are net ocean-atmosphere flux, net release of carbon during land transformation, and net storage in terrestrial biomass and soil organic matter. All of these terms are smaller and less certain than fossil fuel combustion or annual atmospheric accumulation; they represent rich areas of current research, analysis, and sometimes contention.

The human-caused increase in atmospheric CO_2 already represents nearly a 30% change relative to the pre-industrial era (Fig. 2), and CO_2 will continue to increase for the foreseeable future. Increased CO_2 represents the most important human enhancement to the greenhouse effect; the consensus of the climate research community is that it probably already affects climate detectably and will drive substantial climate change in the next century (1). The direct effects of increased CO_2 on plants and ecosystems may be even more important. The growth of most plants is enhanced by elevated CO_2, but to very different extents; the tissue chemistry

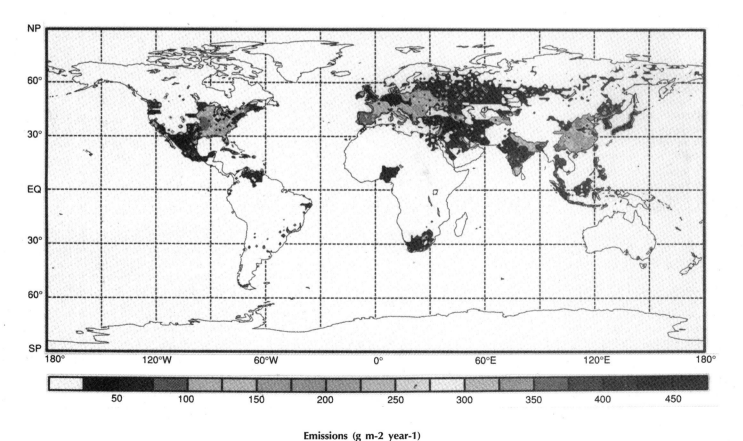

Fig. 4 Geographical distribution of fossil fuel sources of CO_2 as of 1990. The global mean is 12.2 g m^{-2} year^{-1}; most emissions occur in economically developed regions of the north temperate zone. EQ, equator; NP, North Pole; SP, South Pole. [Prepared by A.S. Denning, from information in (18)]

of plants that respond to CO_2 is altered in ways that decrease food quality for animals and microbes; and the water use efficiency of plants and ecosystems generally is increased. The fact that increased CO_2 affects species differentially means that it is likely to drive substantial changes in the species composition and dynamics of all terrestrial ecosystems (19).

Water. Water is essential to all life. Its movement by gravity, and through evaporation and condensation, contributes to driving Earth's biogeochemical cycles and to controlling its climate. Very little of the water on Earth is directly usable by humans; most is either saline or frozen. Globally, humanity now uses more than half of the runoff water that is fresh and reasonably accessible, with about 70% of this use in agriculture (20) (Fig. 2). To meet increasing demands for the limited supply of fresh water, humanity has extensively altered river systems through diversions and impoundments. In the United States only 2% of the rivers run unimpeded, and by the end of this century the flow of about two-thirds of all of Earth's rivers will be regulated (21). At present, as much as 6% of Earth's river runoff is evaporated as a consequence of human manipulations (22). Major rivers, including the Colorado, the Nile, and the Ganges, are used so extensively that little water reaches the sea. Massive inland water bodies, including the Aral Sea and Lake Chad, have been greatly reduced in extent by water diversions for agriculture. Reduction in the volume of the Aral Sea resulted in the demise of native fishes and the loss of other biota; the loss of a major fishery; exposure of the salt-laden sea bottom, thereby providing a major source of windblown dust; the production of a drier and more continental local climate and a decrease in water quality in the general region; and an increase in human diseases (23).

Impounding and impeding the flow of rivers provides reservoirs of water that can be used for energy generation as well as for agriculture. Waterways also are managed for transport, for flood control, and for the dilution of chemical wastes. Together, these activities have altered Earth's freshwater ecosystems profoundly, to a greater extent than terrestrial ecosystems have been altered. The construction of dams affects biotic habitats indirectly as well; the damming of the Danube River, for example, has altered the silica chemistry of the entire Black Sea. The large number of operational dams (36,000) in the world, in conjunction with the many that are planned, ensure that humanity's effects on aquatic biological systems will continue (24). Where surface water is sparse or over-exploited, humans use groundwater—and in many areas the groundwater that is drawn upon is nonrenewable, or fossil, water (25). For example, three-quarters of the water supply of Saudi Arabia currently comes from fossil water (26).

Alterations to the hydrological cycle can affect regional climate. Irrigation increases atmospheric humidity in semiarid areas, often increasing precipitation and thunderstorm frequency (27). In contrast, land transformation from forest to agriculture or pasture increases albedo and decreases surface roughness; simulations suggest that the net effect of this transformation is to increase temperature and decrease precipitation regionally (7, 26).

Conflicts arising from the global use of water will be exacerbated in the years ahead, with a growing human population and with the stresses that global changes will impose on water quality and availability. Of all of the environmental security issues facing nations, an adequate supply of clean water will be the most important.

Nitrogen. Nitrogen (N) is unique among the major elements required for life, in that its cycle includes a vast atmospheric reservoir (N_2) that must be fixed (combined with carbon, hydrogen, or oxygen) before it can be used by most organisms. The supply of this fixed N controls (at least in part) the productivity, carbon storage, and species composition of many ecosystems. Before the extensive human alteration of the N cycle, 90 to 130 million metric tons of N (Tg N) were fixed biologically on land each year; rates of biological fixation in marine systems are less certain, but perhaps as much was fixed there (28).

Human activity has altered the global cycle of N substantially by fixing N_2—deliberately for fertilizer and inadvertently during fossil fuel combustion. Industrial fixation of N fertilizer increased from <10 Tg/year in 1950 to 80 Tg/year in 1990; after a brief dip caused by economic dislocations in the former Soviet Union, it is expected to increase to >135 Tg/year by 2030 (29). Cultivation of soybeans, alfalfa, and other legume crops that fix N symbiotically enhances fixation by another ~40 Tg/year, and fossil fuel combustion puts >20 Tg/year of reactive N into the atmosphere globally—some by fixing N_2, more from the mobilization of N in the fuel. Overall, human activity adds at least as much fixed N to terrestrial ecosystems as do all natural sources combined (Fig. 2), and it mobilizes >50 Tg/year more during land transformation (28, 30).

Alteration of the N cycle has multiple consequences. In the atmosphere, these include (i) an increasing concentration of the greenhouse gas nitrous oxide globally; (ii) substantial increases in fluxes of reactive N gases (two-thirds or more of both nitric oxide and ammonia emissions globally are human-caused); and (iii) a substantial contribution to acid rain and to the photochemical smog that afflicts urban and agricultural areas throughout the world (31). Reactive N that is emitted to the atmosphere is deposited downwind, where it can influence the dynamics of recipient ecosystems. In regions where fixed N was in short supply, added N generally increases productivity and C storage within ecosystems, and ultimately increases losses of N and cations from soils, in a set of processes termed "N saturation" (32). Where added N increases the productivity of ecosystems, usually it also decreases their biological diversity (33).

Human-fixed N also can move from agriculture, from sewage systems, and from N-saturated terrestrial systems to streams, rivers, groundwater, and ultimately the oceans. Fluxes of N through streams and rivers have increased markedly as human alteration of the N cycle has accelerated; river nitrate is highly correlated with the human population of river basins and with the sum of human-caused N inputs to those basins (8). Increases in river N drive the eutrophication of most estuaries, causing blooms of nuisance and even toxic algae, and threatening the sustainability of marine fisheries (16, 34).

Other cycles. The cycles of carbon, water, and nitrogen are not alone in being altered by human activity. Humanity is also the largest source of oxidized sulfur gases in the atmosphere; these affect regional air quality, biogeochemistry, and climate. Moreover, mining and mobilization of phosphorus and of many metals exceed their natural fluxes; some of the metals that are concentrated and mobilized are highly toxic (including lead, cadmium, and mercury) (35). Beyond any doubt, humanity is a major biogeochemical force on Earth.

Synthetic organic chemicals. Synthetic organic chemicals have brought humanity many beneficial services. However, many are toxic to humans and other species, and some are hazardous in concentrations as low as 1 part per billion. Many chemicals persist in the environment for decades; some are both toxic and persistent. Long-lived organochlorine compounds provide the clearest examples of environmental consequences of persistent compounds. Insecticides such as DDT and its relatives, and industrial compounds like polychlorinated biphenyls (PCBs), were used widely in North America in the 1950s and 1960s. They were transported globally, accumulated in organisms, and magnified in concentration through food chains; they devastated populations of some predators (notably falcons and eagles) and entered parts of the human food supply in concentrations higher than was prudent. Domestic use of these compounds was phased out in the 1970s in the United States and Canada, and their concentrations declined thereafter. However, PCBs in particular remain readily detectable in many organisms, sometimes approaching thresholds of public health concern (36). They will continue to circulate through organisms for many decades.

Synthetic chemicals need not be toxic to cause environmental problems. The fact that the persistent and volatile chlorofluorocarbons (CFCs) are wholly nontoxic contributed to their widespread use as refrigerants and even aerosol propellants. The subsequent discovery that CFCs drive the breakdown of stratospheric ozone, and especially the later discovery of the Antarctic ozone hole and their role in it, represent great surprises in global environmental science (37). Moreover, the response of the international political system to those discoveries is the best extant illustration that global environmental change can be dealt with effectively (38).

Particular compounds that pose serious health and environmental threats can be and often have been phased out (although PCB production is growing in Asia). Nonetheless, each year the chemical industry produces more than 100 million tons of organic chemicals representing some 70,000 different compounds, with about 1000 new ones being added annually (39). Only a small fraction of the many chemicals produced and released into the environment are tested adequately for health hazards or environmental impact (40).

Biotic Changes

Human modification of Earth's biological resources—its species and genetically distinct populations—is substantial and growing. Extinction is a natural process, but the current rate of loss of genetic variability, of populations, and of species is far above background rates; it is ongoing; and it represents a wholly irreversible global change. At the same time, human transport of species around Earth is homogenizing Earth's biota, introducing many species into new areas where they can disrupt both natural and human systems.

Losses. Rates of extinction are difficult to determine globally, in part because the majority of species on Earth have not yet been identified. Nevertheless, recent calculations suggest that rates of species extinction are now on the order of 100 to 1000 times those before humanity's dominance of Earth (41). For particular well-known groups, rates of loss are even greater; as many as one-quarter of Earth's bird species have been driven to extinction by human activities over the past two millennia, particularly on oceanic islands (42) (Fig. 2). At present, 11% of the remaining birds, 18% of the mammals, 5% of fish, and 8% of plant species on Earth are threatened with extinction (43). There has been a disproportionate loss of large mammal species because of hunting; these species played a dominant role in many ecosystems, and their loss has resulted in a fundamental change in the dynamics of those systems (44), one that could lead to further extinctions. The largest organisms in marine systems have been affected similarly, by fishing and whaling. Land transformation is the single most important cause of extinction, and current rates of land transformation eventually will drive many more species to extinction, although with a time lag that masks the true dimensions of the crisis (45). Moreover, the effects of other components of global environmental change—of altered carbon and nitrogen cycles, and of anthropogenic climate change—are just beginning.

As high as they are, these losses of species understate the magnitude of loss of genetic variation. The loss to land transformation of locally adapted populations within species, and of genetic material within populations, is a human-caused change that reduces the resilience of species and ecosystems while precluding human use of the library of natural products and genetic material that they represent (46).

Although conservation efforts focused on individual endangered species have yielded some successes, they are expensive—and the protection or restoration of whole ecosystems often represents the most effective way to sustain genetic, population, and species diversity. Moreover, ecosystems themselves may play important roles in both natural and human-dominated landscapes. For example, mangrove ecosystems protect coastal areas from erosion and provide nurseries for offshore fisheries, but they are threatened by land transformation in many areas.

Invasions. In addition to extinction, humanity has caused a rearrangement of Earth's biotic systems, through the mixing of floras and faunas that had long been isolated geographically. The magnitude of transport of species, termed "biological invasion," is enormous (47); invading species are present almost everywhere. On many islands, more than half of the plant species are nonindigenous, and in many continental areas the figure is 20% or more (48) (Fig. 2).

As with extinction, biological invasion occurs naturally—and as with extinction, human activity has accelerated its rate by orders of magnitude. Land transformation interacts strongly with biological invasion, in that human-altered ecosystems generally provide the primary foci for invasions, while in some cases land transformation itself is driven by biological invasions (49). International commerce is also a primary cause of the breakdown of biogeographic barriers; trade in live organisms is massive and global, and many other organisms are inadvertently taken along for the ride. In freshwater systems, the combination of upstream land transformation, altered hydrology, and numerous deliberate and accidental species introductions has led to particularly widespread invasion, in continental as well as island ecosystems (50).

In some regions, invasions are becoming more frequent. For example, in the San Francisco Bay of California, an average of one new species has been established every 36 weeks since 1850, every 24 weeks since 1970, and every 12 weeks for the last decade (51). Some introduced species quickly become invasive over large areas (for example, the Asian clam in the San Francisco Bay), whereas others become widespread only after a lag of decades, or even over a century (52).

Many biological invasions are effectively irreversible; once replicating biological material is released into the environment and becomes successful there, calling it back is difficult and expensive at best. Moreover, some species introductions have consequences. Some degrade human health and that of other species; after all, most infectious diseases are invaders over most of their range. Others have caused economic losses amounting to billions of dollars; the recent invasion of North America by the zebra mussel is a well-publicized example. Some disrupt ecosystem processes, altering the structure and functioning of whole ecosystems. Finally, some invasions drive losses in the biological diversity of native species and populations; after land transformation, they are the next most important cause of extinction (53).

Conclusions

The global consequences of human activity are not something to face in the future—as Fig. 2 illustrates, they are with us now. All of these changes are ongoing, and in many cases accelerating; many of them were entrained long before their importance was recognized. Moreover, all of these seemingly disparate phenomena trace to a single cause—the growing scale of the human enterprise. The rates, scales, kinds, and combinations of changes occurring now are fundamentally different from those at any other time in history; we are changing Earth more rapidly than we are understanding it. We live on a human-dominated planet—and the momentum of human population growth, together with the imperative for further economic development in most of the world, ensures that our dominance will increase.

The papers in this special section summarize our knowledge of and provide specific policy recommendations concerning major human-dominated ecosystems. In addition, we suggest that the rate and extent of human alteration of Earth should affect how we think about Earth. It is clear that we control much of Earth, and that our activities affect the rest. In a very real sense, the world is in our hands—and how we handle it will determine its composition and dynamics, and our fate.

Recognition of the global consequences of the human enterprise suggests three complementary directions. First, we can work to reduce the rate at which we alter the Earth system. Humans and human-dominated systems may be able to adapt to slower change, and ecosystems and the species they support may cope more effectively with the changes we impose, if those changes are slow. Our footprint on the planet (54) might then be stabilized at a point where enough space and resources remain to sustain most of the other species on Earth, for their sake and our own. Reducing the rate of growth in human effects on Earth involves slowing human population growth and using resources as efficiently as is practical. Often it is the waste products and by-products of human activity that drive global environmental change.

Second, we can accelerate our efforts to understand Earth's ecosystems and how they interact with the numerous components of human-caused global change. Ecological research is inherently complex and demanding: It requires measurement and monitoring of populations and ecosystems; experimental studies to elucidate the regulation of ecological processes; the development, testing, and validation of regional and global models; and integration with a broad range of biological, earth, atmospheric, and marine sciences. The challenge of understanding a human-dominated planet further requires that the human dimensions of global change—the social, economic, cultural, and other drivers of human actions—be included within our analyses.

Finally, humanity's dominance of Earth means that we cannot escape responsibility for managing the planet. Our activities are causing rapid, novel, and substantial changes to Earth's ecosystems. Maintaining populations, species, and ecosystems in the face of those changes, and maintaining the flow of goods and services they provide humanity (55), will require active management for the foreseeable future. There is no clearer illustration of the extent of human dominance of Earth than the fact that maintaining the diversity of "wild" species and the functioning of "wild" ecosystems will require increasing human involvement.

REFERENCES AND NOTES

1. Intergovernmental Panel on Climate Change, *Climate Change 1995* (Cambridge Univ. Press, Cambridge, 1996), pp. 9–49.
2. United Nations Environment Program, *Global Biodiversity Assessment*, V. H. Heywood, Ed. (Cambridge Univ. Press, Cambridge, 1995).
3. D. Skole and C. J. Tucker, *Science* **260**, 1905 (1993).
4. J. S. Olson, J. A. Watts, L. J. Allison, *Carbon in Live Vegetation of Major World Ecosystems* (Office of Energy Research, U.S. Department of Energy, Washington, DC, 1983).
5. P. M. Vitousek, P. R. Ehrlich, A. H. Ehrlich, P. A. Matson, *Bioscience* **36**, 368 (1986); R. W. Kates, B. L. Turner, W. C. Clark, in (35), pp. 1–17; G. C. Daily, *Science* **269**, 350 (1995).
6. D. A. Saunders, R. J. Hobbs, C. R. Margules, *Conserv. Biol.* **5**, 18 (1991).
7. J. Shukla, C. Nobre, P. Sellers, *Science* **247**, 1322 (1990).
8. R. W. Howarth et al., *Biogeochemistry* **35**, 75 (1996).
9. W. B. Meyer and B. L. Turner II, *Changes in Land Use and Land Cover: A Global Perspective* (Cambridge Univ. Press, Cambridge, 1994).
10. S. R. Carpenter, S. G. Fisher, N. B. Grimm, J. F. Kitchell, *Annu. Rev. Ecol. Syst.* **23**, 119 (1992); S. V. Smith and R. W. Buddemeier, *ibid.*, p. 89; J. M. Melillo, I. C. Prentice, G. D. Farquhar, E.-D. Schulze, O. E. Sala, in (1), pp. 449–481.
11. R. Leemans and G. Zuidema, *Trends Ecol. Evol.* **10**, 76 (1995).
12. World Resources Institute, *World Resources 1996–1997* (Oxford Univ. Press, New York, 1996).
13. D. Pauly and V. Christensen, *Nature* **374**, 257 (1995).
14. Food and Agricultural Organization (FAO), *FAO Fisheries Tech. Pap. 335* (1994).
15. D. L. Alverson, M. H. Freeberg. S. A. Murawski, J. G. Pope, *FAO Fisheries Tech. Pap. 339* (1994).
16. G. M. Hallegraeff, *Phycologia* **32**, 79 (1993).
17. D. S. Schimel et al., in *Climate Change 1994: Radiative Forcing of Climate Change*, J. T. Houghton et al., Eds. (Cambridge Univ. Press, Cambridge, 1995), pp. 39–71.
18. R. J. Andres, G. Marland, I. Y. Fung, E. Matthews, *Global Biogeochem. Cycles* **10**, 419 (1996).
19. G. W. Koch and H. A. Mooney, *Carbon Dioxide and Terrestrial Ecosystems* (Academic Press, San Diego, CA, 1996); C. Körner and F. A. Bazzaz, *Carbon Dioxide, Populations, and Communities* (Academic Press, San Diego, CA, 1996).
20. S. L. Postel, G. C. Daily, P. R. Ehrlich, *Science* **271**, 785 (1996).
21. J. N. Abramovitz, *Imperiled Waters, Impoverished Future: The Decline of Freshwater Ecosystems* (Worldwatch Institute, Washington, DC, 1996).
22. M. I. L'vovich and G. F. White, in (35), pp. 235–252; M. Dynesius and C. Nilsson, *Science* **266**, 753 (1994).
23. P. Micklin, *Science* **241**, 1170 (1988), V. Kotlyakov, *Environment* **33**, 4 (1991).
24. C. Humborg, V. Ittekkot, A. Cociasu, B. Bodungen, *Nature* **386**, 385 (1997).
25. P. H. Gleick, Ed., *Water in Crisis* (Oxford Univ. Press, New York, 1993).
26. V. Gornitz, C. Rosenzweig, D. Hillel, *Global Planet. Change* **14**, 147 (1997).
27. P. C. Milly and K. A. Dunne, *J. Clim.* **7**, 506 (1994).
28. J. N. Galloway, W. H. Schlesinger, H. Levy II, A. Michaels, J. L. Schnoor, *Global Biogeochem. Cycles* **9**, 235 (1995).
29. J. N. Galloway, H. Levy II, P. S. Kasibhatla, *Ambio* **23**, 120 (1994).
30. V. Smil, in (35), pp. 423–436.
31. P. M. Vitousek et al., *Ecol. Appl.*, in press.
32. J. D. Aber, J. M. Melillo, K. J. Nadelhoffer, J. Pastor, R. D. Boone, *ibid.* **1**, 303 (1991).
33. D. Tilman, *Ecol. Monogr.* **57**, 189 (1987).
34. S. W. Nixon et al., *Biogeochemistry* **35**, 141 (1996).
35. B. L. Turner II et al., Eds., *The Earth As Transformed by Human Action* (Cambridge Univ. Press, Cambridge, 1990).
36. C. A. Stow, S. R. Carpenter, C. P. Madenjian, L. A. Eby, L. J. Jackson, *Bioscience* **45**, 752 (1995).
37. F. S. Rowland, *Am. Sci.* **77**, 36 (1989); S. Solomon, *Nature* **347**, 347 (1990).
38. M. K. Tolba et al., Eds., *The World Environment 1972–1992* (Chapman & Hall, London, 1992).
39. S. Postel, *Defusing the Toxics Threat: Controlling Pesticides and Industrial Waste* (Worldwatch Institute, Washington, DC, 1987).
40. United Nations Environment Program (UNEP). *Saving Our Planet—Challenges and Hopes* (UNEP, Nairobi, 1992).
41. J. H. Lawton and R. M. May, Eds., *Extinction Rates* (Oxford Univ. Press, Oxford, 1995); S. L. Pimm, G. J. Russell, J. L. Gittleman, T. Brooks, *Science* **269**, 347 (1995).
42. S. L. Olson, in *Conservation for the Twenty-First Century*, D. Western and M. C. Pearl, Eds. (Oxford Univ. Press, Oxford, 1989), p. 50; D. W. Steadman, *Science* **267**, 1123 (1995).
43. R. Barbault and S. Sastrapradja. in (2), pp. 193–274.
44. R. Dirzo and A. Miranda, in *Plant-Animal Interactions*, P. W. Price, T. M. Lewinsohn, W. Fernandes, W. W. Benson, Eds. (Wiley Interscience, New York, 1991), p. 273.
45. D. Tilman, R. M. May, C. Lehman, M. A. Nowak, *Nature* **371**, 65 (1994).
46. H. A. Mooney, J. Lubchenco, R. Dirzo, O. E. Sala, in (2), pp. 279–325.
47. C. Elton, *The Ecology of Invasions by Animals and Plants* (Methuen, London, 1958); J. A. Drake et al., Eds., *Biological Invasions. A Global Perspective* (Wiley, Chichester, UK, 1989).
48. M. Rejmanek and J. Randall, *Madrono* **41**, 161 (1994).
49. C. M. D'Antonio and P. M. Vitousek, *Annu. Rev. Ecol. Syst.* **23**, 63 (1992).
50. D. M. Lodge, *Trends Ecol. Evol.* **8**, 133 (1993).
51. A. N. Cohen and J. T. Carlton, *Biological Study: Nonindigenous Aquatic Species in a United States Estuary. A Case Study of the Biological Invasions of the San Francisco Bay and Delta* (U.S. Fish and Wildlife Service, Washington, DC, 1995).
52. I. Kowarik, in *Plant Invasions—General Aspects and Special Problems*, P. Pysek, K. Prach, M. Rejmánek, M. Wade, Eds. (SPB Academic, Amsterdam, 1995), p. 15.
53. P. M. Vitousek, C. M. D'Antonio, L. L. Loope, R. Westbrooks, *Am. Sci.* **84**, 468 (1996).
54. W. E. Rees and M. Wackernagel, in *Investing in Natural Capital: The Ecological Economics Approach to Sustainability*, A. M. Jansson, M. Hammer, C. Folke, R. Costanza, Eds. (Island, Washington, DC, 1994).
55. G. C. Daily, Ed., *Nature's Services* (Island, Washington, DC, 1997).
56. J. Lubchenco et al., *Ecology*, **72**, 371 (1991), P. M. Vitousek, *ibid.* **75**, 1861 (1994).
57. S. M. Garcia and R. Grainger. *FAO Fisheries Tech. Pap.* 359 (1996).
58. We thank G. C. Daily, C. B. Field, S. Hobbie, D. Gordon, P.A. Matson, and R. L. Naylor for constructive comments on this paper, A. S. Denning and S. M. Garcia for assistance with illustrations, and C. Nakashima and B. Lilley for preparing text and figures for publication.

Seas and Soils Emerge as Keys to Climate

By WILLIAM K. STEVENS

Over the last 150 years, the burning of coal, oil and natural gas has released some 270 billion tons of carbon into the air in the form of heat-trapping carbon dioxide. If all that had stayed in the atmosphere, much of the substantial global warming predicted for a century from now might have already taken place.

Luckily, more than half the carbon dioxide emitted by fossil-fuel burning is absorbed from the air by the oceans, by plants (which use it in converting sunlight to chemical energy) and by soils. Fewer than half the emissions, scientists believe, remain airborne to warm the earth.

All of this is part of one of nature's grand and endlessly complicated global recycling networks. In it, carbon is constantly exchanged among the air, the terrestrial biosphere, the oceans and the solid rock of the earth at varying rates on time scales ranging from hours to millions of years.

Now, as the economic and political difficulty of reducing carbon dioxide emissions at the source becomes ever more obvious, a broad-based move is under way to manipulate the carbon cycle so that it will remove more of the heat-trapping gases from the air. It is an ambitious idea, but on closer inspection one surrounded by questions, potential risks and obstacles.

In one approach, researchers are investigating ways to remove carbon chemically from industrial and automotive emissions before they are released to the atmosphere, then to sequester it at some other way station in the natural carbon cycle—by injecting the removed carbon back into underground reservoirs or into the ocean, for instance.

The United States Department of Energy has expanded research into carbon-removal technologies at eight of its national laboratories. But this option, the department said in a recent study, is "truly radical" and most experts believe it is some distance from providing a practical alternative.

More promising in the near term, many scientists believe, are prospects for increasing the absorption of atmospheric carbon dioxide by what is called the terrestrial biosphere: that is, the whole complex of living things on land, mainly plants and various tiny inhabitants and detritus of the soil.

Every time a plant sprouts and grows to maturity, it absorbs carbon dioxide from the atmosphere, breaks it down chemically and uses carbon as construction material for roots, stem, branches, flowers and leaves. When the plant dies, some of the carbon goes back into the atmosphere, but some is also released into the soil as the dead plant decays. Decomposing animals, including people, further add to the carbon reservoir in the soil. Globally, soil contains about five times as much carbon as vegetation.

The expansion of these carbon reservoirs, or "sinks," in vegetation and soil has been sanctioned as one way industrialized countries might meet their obligations under the 1997 Kyoto Protocol, which is aimed at reducing emissions of carbon dioxide and other heat-trapping greenhouse gases.

Lately, intense argument has developed over the degree to which countries like the United States, which may have great potential for expanding its carbon sinks, ought to be allowed to meet their emissions reduction targets under the protocol by means like planting trees, improving forest management and conserving soils rather than reducing carbon dioxide emissions from cars, factories and power generating stations.

Delegates from more than 100 countries are to grapple with the issue when they meet in November in The Hague for what is widely seen as a make-or-break effort to make the Kyoto Protocol operational.

The rules governing carbon sinks are among the many sets of yet-to-be adopted regulations necessary to put the protocol into effect, and only a few countries have ratified the agreement pending action on them. The United States, where the protocol has encountered strong opposition in Congress on several grounds, is one of those holding back.

In preparation for the critical November meeting, scientists of the Intergovernmental Panel on Climate Change, created by the United Nations more than a decade ago to advise the world's governments, issued a report on the carbon issue at a meeting last week in Montreal. The analysis suggested that expansion of terrestrial carbon sinks in the industrialized countries could potentially more than enable them to meet their entire combined emissions reduction quota under the protocol.

Stakes are high in a debate on how to fight global warming.

But experts associated with the report questioned how much of that potential could be achieved, given the difficulty of actually expanding the world's carbon sinks in practice.

"Judging the technical potential is easy," said Dr. Robert T. Watson, chairman of the intergovernmental panel, who is also the chief atmospheric scientist and environmental adviser of the World Bank. "The real question is, what is the likelihood of realizing that technical potential?"

Under the Kyoto Protocol, the rich industrial countries as a group are obligated to reduce their greenhouse gas emissions 5 percent below 1990 levels by about 2010. The reductions are widely seen as a first step in eventually stabilizing atmospheric concentrations. If emissions are not reduced, prominent scientists say, the average global temperature will rise by about 3.5 degrees Fahrenheit by the year 2100. (By comparison, the world is now 5 to 9 degrees warmer than in the depths of the last ice age 18,000 to 20,000 years ago.)

This much warming, the experts say, would bring rising seas, more severe droughts, rainstorms, heat waves and floods, along with broad shifts in climatic and agricultural zones that would benefit some regions but seriously harm others.

Environmentalists and some parties to the Kyoto Protocol, especially the European Union, strenuously object to allowing any country to meet all or most of its emissions reduction goal by expanding carbon sinks.

"Sucking carbon out of the atmosphere is not the same as preventing it permanently from leaving the exhaust pipe or power plant," said Jennifer Morgan, director of the World Wildlife Fund's climate change campaign.

Environmentalists argue that since forests can always be cut down as well as planted, and soils can always be plowed up as well as conserved, the terrestrial biosphere is not nearly as secure a carbon reservoir as the underground reservoirs where coal, oil and gas are contained. So, they say, why remove them from that secure spot in the first place?

They also fear that reliance on sinks to meet the Kyoto targets could delay the adoption of measures to reduce fossil fuel use and replace it with other energy technologies.

In any event, scientists say, carbon sinks will eventually probably reach a saturation point, at which they will be unable to absorb any more carbon dioxide. The dominant view therefore remains that if the world is serious about reducing and stabilizing the atmospheric concentrations, emissions must be cut at the source by burning fossil fuels more efficiently, switching from relatively high-carbon fuels like coal to lower-carbon natural gas, and replacing fossil fuels altogether by alternatives like hydrogen fuel cells.

Given the slow progress in that direction, though, proponents of the carbon sink idea argue that it could be a valuable interim measure.

In many ways, however, human activity has been shrinking carbon sinks rather than expanding them. Deforestation and farming have obliterated great stretches of forest and grassland that once absorbed carbon dioxide. Farming has also chopped up soils in ways that cause them to give off carbon dioxide rather than to sequester it.

In fact, said the scientific report issued in Montreal, human use of the terrestrial biosphere, including the burning of wood, has released some 135 billion tons of carbon into the air since 1850—about half as much as has been emitted by fossil fuel burning.

The net result of all this is that overall atmospheric concentrations of carbon dioxide are about 30 percent higher now than at the start of the industrial revolution.

A de facto counterattack has been under way for some time in the Northern Hemisphere, where the replanting of forests on deforested lands has made headway in recent decades. One provision of the Kyoto Protocol allows for industrialized countries to gain credit toward their reduction targets for tree planting since 1990—but also to incur debits for any deforestation since then. To reap meaningful credit, a country would have to show a net gain.

The protocol also allows countries to gain credit for reductions through a variety of other ways of expanding terrestrial carbon stocks should the parties to the agreement so decide. These include, for instance, expanding carbon stores in soil through no-till agriculture, in which no plowing takes place and seeds are sown in the carbon-rich detritus of the previous year's harvest.

Other possibilities include conversion of cropland into grassland, fertilization of pastureland and forest management practices designed to increase carbon stocks—for example, lengthening the period between harvests of a forest's timber, or promoting faster tree growth and thicker stands of trees, plantation-style.

Estimates suggest that these kinds of measures might allow some countries to achieve large cuts in net carbon dioxide emissions. The potential of such measures in the United States, Canada and Russia is great, according to a recent calculation made for the World Wildlife Fund by Kevin Gurney, an atmospheric researcher at Colorado State University who studies the carbon cycle.

Mr. Gurney estimated, for example, that without planting any trees, the measures would potentially enable the United States to remove about 150 million tons of carbon a year from the atmosphere. This exceeds its Kyoto emissions reduction target by tens of millions of tons.

"They're pretty big," Mr. Gurney said of the estimates.

The Montreal report estimated that in theory, measures other than tree-planting by the industrialized countries could remove nearly 290 million more tons of carbon a year from the atmosphere by about 2010 than are removed by such measures now. By comparison, the Kyoto Protocol would require the rich countries to cut emissions by about 200 million tons a year below 1990 levels by about 2010.

The report also raised caveats about how much of this estimated potential might be achieved. Even with "an ambitious policy agenda," it said, the estimates were "likely to be on the high side."

Far smaller gains could be achieved by the rich countries by planting trees and slowing deforestation, the report concluded. It said that a 20 percent increase in forest planting would probably offset no more than 3 million tons of carbon emissions a year, while slowing deforestation by 20 percent could offset about 18 million tons.

Even if the potential for carbon sequestration can be realized to a substantial degree, many headaches still present themselves. For instance, forest growth is spurred—and terrestrial carbon stocks are increased—by a warming climate, and by rising levels of carbon dioxide and of atmospheric nitrogen deposited on forests as a result of industrial emissions.

In assessing a country's carbon stocks for purposes of assigning credit, how can these effects of human activity be separated from the growth in carbon stocks brought about by tree-planting efforts alone? It is almost impossible to make such a separation, Dr. Watson said.

Another difficulty for some kinds of carbon sequestration efforts is that of monitoring and accounting for carbon stocks. Although there are well-established means of measuring the increase of carbon in a stand of growing trees, other sinks may present problems. One way to get around the difficulty in the case of agricultural soils might be to set up a business-as-usual control plot, in which conventional farming methods are used, against which to measure the change in carbon stocks as a result of no-till farming.

Then there are ecological concerns. For instance, managers of carbon-sequestration projects might be tempted to replace naturally functioning forests with tree plantations, which in biological terms are comparatively poor and sterile, with little of the biological richness and variety of the wild woods. Likewise, managers might be tempted to cut down mature forests so as to plant new, faster-growing ones that sequester carbon at a higher rate.

It is on questions of detail like these that the success or failure of the Kyoto Protocol may ultimately hinge.

Plying a fabled waterway

Melting may open the Northwest Passage

BY TIM APPENZELLER

John Franklin should have waited. The Northwest Passage, the ice-choked Arctic shortcut from Europe to Asia, had defeated generations of explorers, but the Englishman's 1845 attempt to find a route was the most spectacular failure. Pack ice trapped his two ships in the maze of channels north of Canada, and Franklin and his 128 men were never seen again, despite many searches. It wasn't much easier decades later, when Norwegian explorer Roald Amundsen became the first to conquer the passage. When he finally reached the Pacific in 1906, he had spent three years—including two icebound winters—making the crossing.

These days, the fierce, fabled Northwest Passage can be a cakewalk. In the summers of 1998 and 1999, "there was not that much ice at all," says Andy Maillet, superintendent of the Canadian Coast Guard's Arctic Icebreaking Program. "You could almost do the whole passage in open water."

A great thaw has set in at the top of the world. The crust of floating ice that covers the Arctic Ocean almost from edge to edge is shrinking, and it has thinned by almost half since the late 1950s. No one knows exactly what's causing the melting, although global warming is an obvious suspect. And no one knows how far it will go, although a few scientists are predicting that within decades the Arctic Ocean could be completely ice free in summer, which conjures visions of tourist ships sailing to the North Pole. Already, the effects of the thaw are unmistakable along the southern edge of the Arctic: Polar bears are reportedly losing weight as their icy hunting grounds liquefy, and small flotillas of ships are starting to negotiate the once impassable Northwest Passage.

Cold facts. Arctic researchers knew the ice was dwindling, but even so the numbers that emerged late last year from two major studies were startling. One, based on satellite measurements, showed that the area of "multiyear ice"—the thick, hard ice that covers much of the Arctic and lasts through the summer—had shrunk by 14 percent, or an area larger than France, in the last 20 years. The second study examined data on ice thickness gathered by nuclear submarines traveling under the floes. Over 40 years, from the late 1950s to the late 1990s, the ice had thinned from an average of 10 feet to less than 6 feet thick.

At that rate, the ice would be gone in another 50 years—a total polar meltdown. Because the ice is already floating, the sea level would not rise, any more than a cold drink overflows as the ice cubes melt. But the effects on global climate could be drastic, although it's unclear whether the melting would push the planet's thermostat up or down. By turning reflective ice into dark open water, the melting might create a huge solar collector at the pole, speeding up global warming. Or it might send a chill, at least in Europe, if meltwater flooding into the North Atlantic blocked the

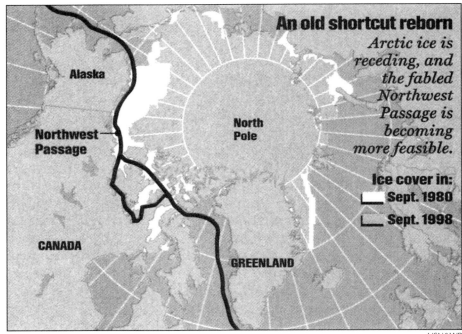

An old shortcut reborn
Arctic ice is receding, and the fabled Northwest Passage is becoming more feasible.

Ice cover in:
■ Sept. 1980
■ Sept. 1998

USN&WR

Sources: Ola Johannessen, Nansen Environmental and Remote Sensing Center; Britannica.com

6. Plying a Fabled Waterway

warm currents that now bring mild winters to England and Scandinavia.

But many climate specialists think that today's rapid melting could slow down. To the extent that the receding ice is caused by the accumulation of heat-trapping "greenhouse" gases, it will likely continue, but most doubt that this phenomenon accounts fully for the thaw. "The ice is melting too fast," says Ola Johannessen of the Nansen Environmental and Remote Sensing Center at the University of Bergen in Norway, who led the team that measured the shrinking multiyear ice. He points instead to a nest of storms in the North Atlantic that has been pumping warm air and water into the Arctic and whipping up winds that drive ice southward. The persistent weather pattern may be one extreme of a natural climate cycle that takes decades to flip back and forth. If this natural cycle has in fact been accelerating the meltdown, the ice might stabilize when the cycle reverses.

Still, the melting so far "is certainly a warning that something is happening," says Gary Maykut of the University of Washington, who was part of the team that studied the sub measurements. Another warning comes from the Hudson Bay's polar bears, which are about 15 percent lighter than two decades ago and have fewer young. Ian Stirling of the Canadian Wildlife Service says the bears are slimming down because the bay's seasonal ice is breaking up two to three weeks earlier each year, cutting into their prime seal-hunting season. If warming continues, biologists say, walruses could suffer as the sea ice where they rest and sleep retreats northward, away from the coastal waters where they feed.

Shipping news. Meanwhile a new kind of activity is stirring the Arctic waters: shipping. Last summer some of the first commercial shipments sailed the Northwest Passage from ocean to ocean. A pair of icebreakers escorted a floating dry dock through the passage from Siberia to Bermuda, and a set of barges plied the route from the western Arctic on their way to the Great Lakes. A Finnish company is now building ice-resistant oil tankers to ply the waters north of Siberia—the so-called Northeast Passage, where the ice has also dwindled over the past decade. And growing numbers of ecotourism ships are braving Arctic waters.

It's just the beginning, says political scientist Rob Huebert of the University of Calgary, who notes that a ship sailing between Europe and Asia could save 2,500 miles by choosing the Northwest Passage over the Panama Canal. "I'm quite convinced that we are going to see an increase in commercial shipping," he says. But he notes that shipping will stir up a festering diplomatic issue. While the United States and most other countries view the passage as international waters, Canada considers it to be internal waters, subject to Canadian environmental and navigation standards.

In the past, says Huebert, "the ice [did] freeze the American-Canadian dispute." And for now, it still does. But the course of climate change doesn't always run smooth: After two ice-free years, the Northwest Passage has reverted to something of its frosty old self this year. "There's still a fair amount of ice in the passage today," says Maillet. Indeed, even today the explorer Franklin would have no guarantees of smooth passage.

Lead in the Inner Cities

Policies to reduce children's exposure to lead may be overlooking a major source of lead in the environment

Howard W. Mielke

In the middle of the 1970s, U.S. health officials identified what some called a "silent epidemic." They were referring to childhood lead poisoning, a problem that is easily overlooked and underappreciated. Of all of the metal-poisoning episodes to date, none has come close in sheer numbers. Toxicology textbooks mention cadmium poisoning in Japan in the 1950s and methyl mercury poisoning in both Japan in the 1950s and Iraq in 1972. Some hundreds of deaths were attributed to each of these events. But most textbooks fail to mention lead poisoning, in spite of the fact that since the 1920s millions of American children have been quietly poisoned by lead, and thousands of deaths are attributed to this over the long term.

Although childhood lead exposure has diminished over the past 20 or so years, the problem has by no means been solved. Rather, the demographics have shifted. Some groups, mainly minority and poor children living in the inner city, suffer from high rates of lead poisoning. Over 50 percent (some studies place this figure at around 70 percent) of children living in the inner cities of New Orleans and Philadelphia have blood lead levels above the current guideline of 10 micrograms per deciliter (ug/dl). In contrast, in the concrete jungle of Manhattan, where very little of the soil is exposed and almost all apartments and housing contain lead-based paints, between 5 and 7 percent of children under the age of 6 have been reported to have blood-lead levels of 10 ug/dl or higher. Interestingly, just across the river in Brooklyn, where yards containing soil are common, the percentage of affected children is several times higher.

Exposure remains such a problem that early in the 1990s, the U.S. Centers for Disease Control and Prevention (CDC) in Atlanta called lead poisoning "one of the most common pediatric health problems in the United States today," but added that the problem was "entirely preventable." Effective prevention, however, assumes an accurate identification of the environmental reservoirs of lead.

Current policies to reduce lead exposure are based on the assumption that the greatest lead hazard comes from lead-based paints. Poorly maintained paints decay and release lead on their surfaces in the form of dust. In addition, lead tastes sweet, and young children may be tempted to eat leaded paint chips as though they were candy. The health consequences of this can be severe. Lead is a neurotoxin that can be especially dangerous to the developing nervous systems of infants and young children. To deal with this problem, most lead paints have been removed from the market, some lead paints are being stripped

---

Howard Mielke is a professor at the College of Pharmacy of Xavier University in New Orleans. He served as a Peace Corps Volunteer in Malawi, Africa, before obtaining his M.S. in biology and his Ph.D. in geography at the University of Michigan. He began his urban lead research in 1971 while teaching at the University of California, Los Angeles, and continued these studies at the University of Maryland, Baltimore County, and later at Macalester College in Minnesota. Mielke serves as the program director, and is a principal investigator of a multimedia study of metals in the urban and rural environment, for the Substance Specific Applied Research Program as part of a cooperative agreement with the Minority Health Professions Foundation/Agency for Toxic Substances and Disease Registry. He is a corresponding member of the Working Group on Geoscience and Health, Commission on Geological Sciences for Environmental Planning of the International Union of Geological Sciences. Address: Xavier University of Louisiana, College of Pharmacy, 7325 Palmetto Street, New Orleans, LA 70125. Internet: hmielke@mail.xula.edu.

7. Lead in the Inner Cities

Figure 1. Lead additives were used in gasoline for over 50 years until they were banned in 1986. Although lead is no longer allowed in most gasoline products, the legacy of its use remains embedded in soils along roadways and in U.S. cities. Where there is a long history of traffic congestion, such as in the inner cities of many urban areas, lead accumulations are especially high in the soils. Children who live in these areas and play in these soils are particularly vulnerable to lead poisoning. Some simple and relatively inexpensive methods can be used, says the author, to reduce children's exposure to this environmental toxin. (Except where noted, photographs are courtesy of the author.)

off of walls, and parents are instructed to guard their children from eating paint flakes.

But such policies deal with only part of the lead hazard. Work done in my laboratory demonstrates that there exist additional sources of lead in the environment that pose as great, if not greater, threats to children. To be sure, lead paint is a major contributor of the lead in the environment. Lead was used in residential paint between 1884 and 1978 and remains on the walls of many older buildings. But paint is neither the most abundant nor the most accessible source of lead. The common problem is lead dust. For most children the issue is whether there is a source of lead dust in the environment. Research in my laboratory and others shows that in predictable locations of many cities, the soil is a giant reservoir of tiny particles of lead. This means that many children face their greatest risk for exposure in the yards around their houses and, to a lesser extent, in the open spaces such as public playgrounds in which they play. My colleagues and I believe that an accurate and complete appreciation of the distribution of lead in the environment can help shape policies that more effectively protect the health of children.

Sources of Lead

Lead is versatile and formulated into many products. Some products, such as common lead-acid batteries used in cars, trucks, boats,

1 ❖ GEOGRAPHY IN A CHANGING WORLD

Figure 2. Bare soils in play areas, especially those near busy roads, constitute one of the most common vectors by which children become poisoned. Children play in these soils, put their hands in their mouths, or touch objects that do, and ingest the lead. Children are most vulnerable to the toxic effects of lead, which include damage to the nervous system, learning impairments and behavioral problems.

motorcycles and the like, are sealed and if appropriately recycled, should not be the cause of poisoning of ordinary citizens. Other products allow lead to be released into the pathway of human exposure. Lead solder was used to seal seams in the canning industry until it was voluntarily withdrawn, first from baby food containers and then all canning facilities in the early 1980s. The same canning solder is used in other countries, and imported canned food continues to be tainted with lead. Some brightly colored ceramic plates and cups as well as leaded crystal may, in the presence of acidic foods (tomatoes, pineapple, wine etc.), release lead and contaminate the food. Lead-based paint was banned for household use in 1978, but lead is still an ingredient of "specialty" paints. Leaded gasoline was banned in 1986, although lead additives are still in use in racing fuels (up to 6 grams per gallon), boat fuels, farm tractors and personal watercraft fuels despite the fact that they are not required in any of these applications. Alternative octane boosters are available.

Lead acetate, or "sugar of lead," is water-soluble and one of the most bioavailable forms of lead. It is an ingredient in some hair-coloring cosmetic products. The Food and Drug Administration allows up to 6,000 parts per million of lead acetate in cosmetics. Several brands of slow-acting hair-coloring cosmetics are used daily by a sizable number of people with graying hair in many homes both in the U.S. and abroad. When users pour the cosmetic into their bare hands to rub into their hair, they become conveyers of a very toxic substance. Some lead acetate may be spilled on the sink, and it is indistinguishable from drops of water. On the hands it can be easily transferred to other items such as toothbrushes, faucet handles, combs and dental floss. It can be absorbed through the skin and shows up in sweat and saliva, but not in blood, as does ingested lead. Many plastics and vinyl products contain lead as a stabilizer or coloring agent. Products become a hazard if they deteriorate into fine dust particles or otherwise directly release lead onto hands (or paws), from which it is transferred into the mouths of unsuspecting creatures, including people.

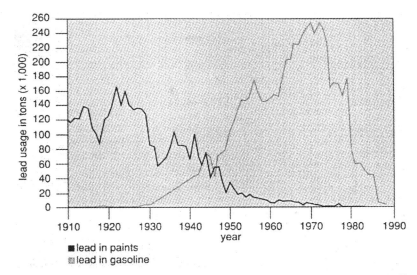

Figure 3. Lead use in paint peaked in the U.S. in the mid-1920s and then gradually died out, just as the use of leaded gasoline was on the rise. The use of leaded gasoline reached its peak in the 1970s and then declined until Congress banned it in 1986. There is a common perception that lead-based paints alone account for the amount of lead in the environment. In reality, both sources of lead contribute to the problem.

Lead in paint and gasoline together accounts for most of the lead now in the human environment. In terms of raw tonnage, the amount of lead in gasoline over only the 57 years of its use from 1929 to 1986 roughly equals all of the lead in paints in 94 years of lead-paint production, from 1884 to 1978. The peak use of lead-based paint came during the 1920s when the U.S. economy was largely agrarian and rural. Most lead paints still exist as a thin mass on the walls and structures of older buildings. Deteriorated or sanded and scraped paint contributes to lead dust accumulation in the soil.

In contrast, the use of leaded gasoline peaked early in the 1970s, a time during which the U.S. economy had become industrial and urban and reliant on automobiles for transportation. About 75 percent of gasoline lead was emitted from exhaust pipes in the form of a fine lead dust (the remaining 25 percent of the lead ended up in the oil or was trapped on the internal surfaces of the engine and exhaust system). It is estimated that the use of 5.9 million metric tons of lead in gasoline left a residue of 4 to 5 million metric tons in the environment. From these facts, we expected to find that lead would be disproportionately concentrated along roadways with the highest traffic flows—those running through cities.

Figure 4. Car exhaust spewed lead particles into the air when leaded gasolines were used. Soils in areas with a long history of traffic congestion, such as the inner cities, are heavily contaminated with lead. Lead resists movement and does not decompose, so it remains a permanent feature of the environment until people take measures to deal with it.

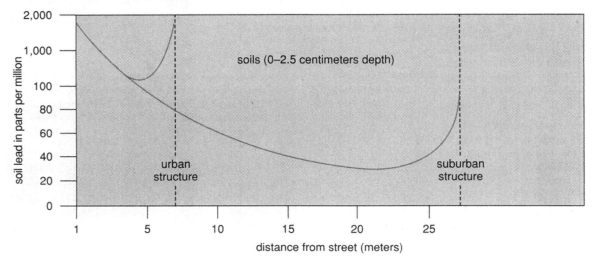

Figure 5. Two different-sized lead particles are spewed into the air when leaded gasoline is used. The trajectory of a typical particle is indicated by the arrows (*top*). Larger, heavier particles settle near the street. Smaller particles are carried by the wind until they meet a barrier, such as a tree or a house, to which they stick. Eventually, they are washed into the soil by rain. Lighter particles can be carried a greater distance and are eventually scavenged by precipitation and then fall to the ground. Assuming unpainted homes, the graph (*bottom*) generalizes the quantity of lead in the soil around an inner-city home situated about 7 meters from a heavily traveled road versus around a home more than 25 meters from the road.

Health Consequences

Lead poisoning was known in colonial times and has roots in antiquity. Clinical symptoms of lead poisoning were observed in the work force of trades that handled lead. Benjamin Franklin noticed in 1786 that some of his lead type–setters at the print shop suffered from a debilitating weakness of the hands called dropsy. He observed that those workers who suffered from dropsy failed to wash their hands before eating their sandwiches, and attributed the disease to their slovenly and unclean habits. When he learned of a 60-year-old report noting similar health problems, he commented, "and you will observe with concern how long a useful truth may be known and exist before it is generally received and practiced on."

The clinical problems of lead poisoning were described by the Greeks and were so prevalent during Roman times that some attribute the fall of Rome to lead poisoning of the aristocracy. It was considered a scourge (Saturnine poisoning) in Europe in the Middle Ages, when people frequently sweetened sour wine with lead acetate.

In other words, lead poisoning is not new, but because of the industrial way of life the poisoning has shifted to children, and the number of its victims has skyrocketed. Consider the magnitude of the man-made products that use and release lead into global environmental circulation. Jerome Nriagu of the University of Michigan School of Public Health estimates the total amount of lead produced from mines to be about 260 million metric tons in the past 3,000 years. And all of this mining has left its imprint in the geological record. Lead is found trapped in datable layers of the glaciers of Greenland, peat-bog cores of Switzerland and sediment cores of the Mississippi Gulf Coast. Although this geochemical perspective provides a framework for the history of the global quantity of mining and environmental emissions over three millennia, it does not assist us in appreciating the nature and degree of the local insult that lead has had on our urbanized society. The impact—particularly on children was not fully appreciated until the past few decades.

Re-evaluations

Until the 1970s in the United States, the health guidelines for lead exposures were essentially the same for children as for adults. In 1979 guidelines for both the amount of lead and the differences in vulnerability between children and adults were revised when Herbert Needleman at Harvard Medical School reported that children's cognitive abilities were affected by much lower lead exposures than previously recognized.

At about the same time that Needleman was doing his work, my colleague Rufus Chaney and I were conducting a study at the U.S. Department of Agriculture in Beltsville, Maryland, of the lead content of garden soils of Baltimore, Maryland. While we were performing the field studies, supervisors and other people frequently reminded me that the major source of lead in soil is lead-based paint. When we pored over the results of our analytical determinations, we came up with a problem of statistics. The distribution of the numerical results did not follow a normal curve. The statisticians I initially worked with wanted to truncate the results so that they conformed to the normal distribution.

In essence, the highest numbers were being removed from the database to satisfy the requirements of the parametric statistical models being used at that time. My brother, Paul Mielke, is a professor of statistics at Colorado State University, and I called on him for help. Paul and Ken Berry, also at Colorado State University, had recently completed their work on a non-parametric statistical test, Multi-Response Permutation Procedures, which provided a method for testing the kind of data that I had.

Paul had an appropriate nonparametric statistical model but no data base to analyze; I had the data base with no model to evaluate it. Serendipity played a role in our initial realization about the extraordinary accumulation of lead in urban environments. The research results were surprising. The pattern of lead did not match what we expected to find if paint alone was the major source of lead in Baltimore soils.

Our data did not support the lead-based paint hypothesis. Our observation was that in the inner city, where the soil-lead concentrations are highest, Baltimore had mostly unpainted brick buildings, since its inner city underwent a huge fire in 1904. On reconstruction, the new building codes required the use of fireproof (hence, brick) construction materials.

The sites of old housing constructed with painted wood siding are located in outlying parts of the city, where the lead content of garden soils is lowest. The highest lead-containing garden soils appeared to be associated with the inner-city location, not the presence of painted wood structures. The geographic pattern was extremely strong. Because of this simple observation, I sought an alternative source of lead.

Leaded gasoline fit the requirements for that alternative. We learned that gaso-

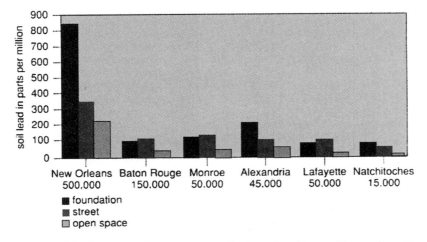

Figure 6. Soil-lead concentrations are greatest in the soils of large cities, such as New Orleans, and lower in the soils of smaller cities and towns, such as Baton Rouge, Monroe, Alexandria, Lafayette and Natchitoches. Soil surrounding the foundations of houses and buildings is more heavily contaminated than are soils around the street or in open areas. (The 1990 population of each city is given below the city's name.)

line was a huge source of lead, and we predicted that it was exhausted in a pattern that corresponded to the flow and congestion of traffic. The environment of the city was undergoing lead accumulations because of the daily commuter traffic flows that strongly characterize the industrial way of urban life. We suspected that the lead particles released with gasoline exhaust could travel through the air until they hit a barrier, such as the side of a house or apartment building. We also imagined that these lead particles would be washed down the sides of the buildings and into the soil anytime it rained. Based on this scenario, we would expect the greatest accumulations of lead to be found in the soil surrounding the foundations of buildings. We also thought it was possible that lead accumulated in soil was a major contributor to the childhood lead problem of Baltimore.

Our findings were published in 1983. Did the automobile act as a toxic-substance delivery system in all large cities? Critics of our publication argued that Baltimore is an unusual city because of all of the heavy industries there. They asserted that industrial emissions, and not leaded gasoline, accounted for the soil-lead accumulation problem in Baltimore. In 1979 I moved to Macalester College in St. Paul, Minnesota. I received a small grant to study St. Paul soils. The family of one of my students, Handy Wade of Boston, funded the purchase of analytical equipment to determine the lead content of soil samples. By 1984, the results were available.

We found that St. Paul and Minneapolis, where, in contrast to Baltimore, there is no heavy industry, had the same inner-city soil-lead accumulation as first observed in Baltimore. Even the quantities were similar. We also studied small towns and large cities and found that city size was a more important factor than age of city.

The soil-lead concentration in old communities of large cities is 10 to 100 times greater than comparable neighborhoods of smaller cities. In addition, soil-lead concentrations diminish with distance from a city center.

For example, in Baltimore, the highest garden-soil contamination is so tightly clustered around the city center that the probability that this distribution could be due to chance is 1 in 10^{23}. The concentration of lead in the soil of the Twin Cities (Minneapolis and St. Paul) is 10 times greater than that in adjacent suburbs that have old housing, where lead-based paint concentrations are as high. In 1988, I joined the faculty of Xavier University in New Orleans. I repeated the soil-lead studies conducted in Minnesota and obtained similar results. Soil-lead concentrations were higher in congested high-traffic, inner-city regions of New Orleans, as compared with its older suburbs and with smaller towns. Although the age of the houses in these communities is similar, the traffic flows vary widely. In New Orleans, with a population of about 500,000 people, the daily traffic count at major intersections in the inner-city averaged about 100,000 vehicles in 1970. In contrast, in a small town, such as Thibodaux, with its population

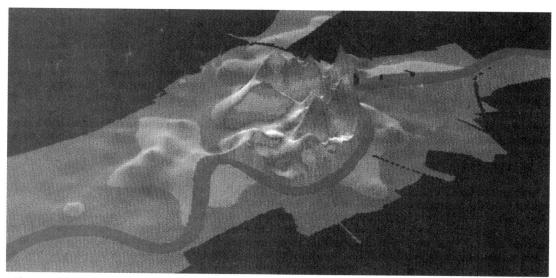

Figure 7. Three-dimensional surface plot shows that New Orleans soil-lead concentrations peak in the inner city, where traffic is heaviest, and then decrease exponentially toward the outlying parts of the city. (Image by James Pepersack, courtesy of the author.)

of about 14,000, the busiest intersections averaged only about 10,000 vehicles a day in 1970. These data strongly suggested to us that proximity to a high-traffic route is a better predictor of soil-lead concentrations than is the age of the buildings in the area or the amount of lead-based paint in the buildings. We further speculated that the soil of larger cities is more concentrated in lead than is the soil of smaller cities because of the greater volume of traffic in larger cities and because each vehicle remains inside the larger city longer than it would inside a smaller city.

In New Orleans, we tested our hypothesis that soil-lead accumulations do indeed follow traffic volumes with a comparison of large versus small cities. To do this, my colleagues and I calculated the quantities of lead emitted within one-half mile of major intersections of the inner-city regions of New Orleans and Thibodaux using the average daily vehicle traffic records during the peak lead-use years. Both of these communities have older housing with lead-based paint located around the community core, but their traffic flows differ by a factor of 10.

Next, we estimated the lead emissions in these two communities, basing our calculations on available records of vehicular use of gasoline during the years in which lead gasoline was in use, which we applied to the traffic volumes in the two communities. From our calculations, we determined that during the peak years, 5.15 metric tons of lead was emitted annually in New Orleans, as opposed to 0.45 metric ton in Thibodaux—less than one-tenth of the New Orleans concentrations. The

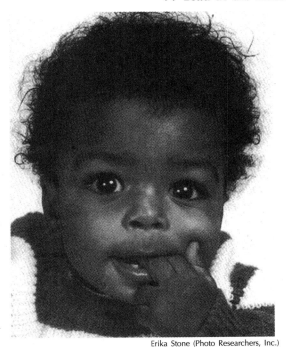

Figure 8. Young children frequently put objects and hands in their mouths, by which they ingest lead-contaminated dust and soil.

distinctive pattern of lead that remains in cities has created "urban metal islands." A consequence of the accumulation of lead in urban communities is that a widespread environmental health hazard is now imposed on a large segment of society.

The fact that the lead is not distributed uniformly throughout the city provides strong support for our hypothesis. Within the cities we studied in Maryland, Minnesota and Louisiana, the greatest amount of lead can be found in the central part of the city where the

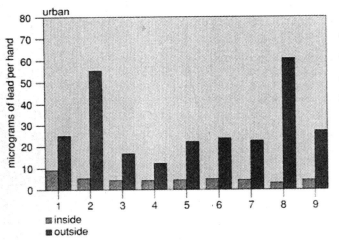
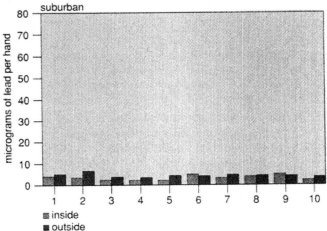

Figure 9. Measuring the lead on children's hands reveals that children pick up more lead when they play outdoors than indoors. This graph reflects data collected at private day-care centers at residential dwellings. The children of inner-city New Orleans (*left graph*) playing outdoors encounter as much as ten times the lead on their hands as children who play outdoors in non-inner city and suburban neighborhoods (*right graph*) around New Orleans.

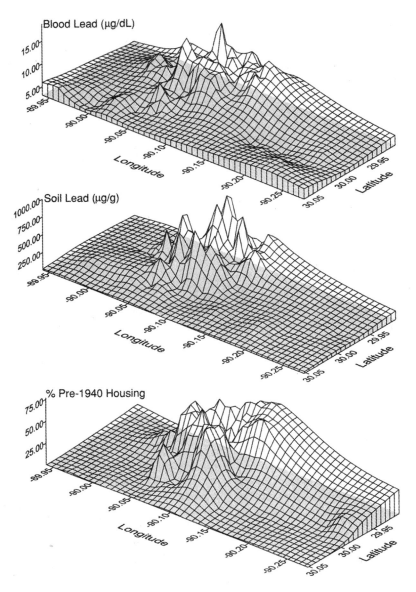

Figure 10. Three-dimensional surface plots, matched by census tracts, relate blood-lead levels (a), soil-lead quantities (b) and percentage of houses built before 1940 in New Orleans. Peaks for soil-lead and blood-lead levels are visibly similar. The percent of pre-1940 housing, which corresponds to the period of peak use of lead-based paint, does not match as closely to blood-lead levels as do soil-lead levels. Soil lead was found to be more predictive of children's exposure than was the age of housing. (Image by Christopher Gonzales, courtesy of the author.)

highway networks concentrate traffic. In effect, urban highways can be considered inadvertent dispersal systems of lead in densely populated areas surrounding the city center. Some people may argue that since lead has been removed from gasoline, it no longer poses a threat to children's health. But the clays and organic matter in soil weakly bind lead, and our research findings give us reason to believe that lead remains in the soil of the inner cities for a long time. Our results refute the commonly held notion that lead concentrations are highest in *all* areas with older housing.

Children at Risk

Lead is a neurotoxin that can be especially harmful to the developing brains and nervous systems of young people. High doses can be fatal. Chronic exposures to as little as 10 micrograms of lead per deciliter of blood can permanently impair the brain's functioning, thus limiting a child's intellectual and social development. It can also result in behavioral problems, which are disruptive to other children in the classroom.

The survey conducted in the 1970s showed that nine of eleven children in the United States had lead concentrations of 10 ug/dL or higher in their blood. By that time, lead-based paint production had been curtailed. Since that time, lead has been removed by mandate from gasoline and eliminated from food processing, so that a survey conducted late in the 1980s indicated that only one out of eleven children was exposed to dangerously high levels of lead. Surveys show, however, that the children affected now are disproportionately African-Americans living in the inner cities of major U.S. cities.

The Environmental Protection Agency (EPA) has estimated that more than 12 million children living within U.S. urban environments are exposed to risk from 10 million metric tons of lead residues resulting from the use of leaded gasoline and lead-based paint. A child can safely tolerate an intake of only 6 micrograms of lead per day. The mass of lead in America's urban environments that is potentially available to children is about 19 orders of magnitude (10^{19}) greater than safe levels. Clearly, there is an unacceptable amount of lead *potentially* available to children. In the laboratory we have attempted to measure how much lead is *actually* available to the child.

Children frequently put their hands in their mouths, and this is the most common route of exposure to lead. To get an idea of how much lead children routinely ingest, Latonia Viverette, then on staff at Xavier, and I measured the amount of lead on children's hands in New Orleans.

Figure 11. Sandblasting and other traditional sanding methods for removing lead-based paint from buildings release fine lead-dust particles into the environment. Measures to limit lead contamination of soil include using modern sanders that capture lead dust and alternative methods to remove or encapsulate paint.

We visited several day-care centers in various parts of the city. Children's hands were wiped after they had played indoors only and then a second time after they had played outside. Our general conclusion was that children have several times more lead on their hands after playing outdoors than they do after playing indoors. The amount of lead varies directly with amount of lead measured in the outdoor soil. For private day-care centers located in ordinary homes with soil in the yards, the amount of lead picked up by children varied directly with the amount of lead in the yard soil. The amount of lead in the soil also related to the part of the city in which the day-care center was located. In the inner city, the children picked up the most lead, whereas in day-care facilities in the outer city, children were least exposed. We also observed that soil was absent from public Head Start day-care centers. The soil in outside play areas was completely covered with rubberized matting or other playground covering and did not contain bare soil. Even in the inner city, children at public day-care centers did not pick up appreciable amounts of lead on their hands.

In collaboration with Dianne Dugas and her staff at the Department of Environmental Epidemiology and Toxicology in the Louisiana Office of Public Health, we then looked at the age of homes and soil lead to determine how they related to children's blood-lead levels. Correlating the age of housing with blood-lead levels yields mixed results. In 96 percent of the areas where we found low levels of lead in children's blood, a majority of the houses were newer, having been built after 1940, when the use of lead-based paint had been reduced. By contrast, in areas where children had high levels of lead in their blood, 51 percent of the houses in these areas were built after 1940, and 49 percent were built before. Newer housing is a good predictor of low blood-lead levels, whereas old housing is a poor predictor for the group of greatest interest, children with high blood-lead levels.

The strongest correlation linked soil-lead concentrations and blood-lead concentrations, as we had expected. The association between soil lead and blood lead was 12 orders of magnitude (10^{12}) stronger than the association between the age of housing and blood lead levels. Soil-lead concentrations are more predictive of childhood lead exposure than is the age of housing. In New Orleans, housing tracts with low levels of lead in the soil are very strongly associated with low blood-lead concentrations, whereas areas high in soil lead are likewise associated with high blood-lead concentrations.

Figure 12. Artificial surfaces, such as this wooden deck, can cover over already-contaminated soils, thus limiting a child's potential exposure to toxic soils. The child in this household recovered rapidly from lead poisoning after this deck was installed. Other relatively inexpensive landscaping measures can likewise safeguard children from exposure to environmental lead.

Prevention

During a routine check-up in 1983, my daughter Beverly, then about two, was found to have a blood-lead level of about 10 to 15 ug/dL. I became haunted by the fact that my daughter and all other children were being poisoned, a finding that was reinforced by the 1978—83 national survey results. Given what I knew, I spent several days observing her activities and taking soil and paint samples. Her licensed daycare home was located several hundred yards from a major freeway. The state's licensing inspection focused on the condition of the inside facilities. The sandbox where the children played daily contained a mixture of soil and sand, was located next to the house and was surrounded by soil that contained over 500 parts per million of lead. I replaced the sand and laid down indoor-outdoor carpet on the bare soil around the sandbox. A month after I made these changes, Beverly's blood-lead levels dropped by half. My daughter's exposure riveted my attention to our nation's lead problem.

This experience and the results from our work in Minnesota prompted the formation of the Minnesota Lead Coalition. Patrick Reagan of the Midwest Environmental Education and Research Association became a major co-worker with the coalition. The lead coalition lobbied the Minnesota legislature to ban leaded gasoline. The effort failed because, unknown to us, additives to gasoline, as stated in the Federal Air Pollution Regulations, are the exclusive jurisdiction of the federal government. The only way that lead could be banned from gasoline was by EPA regulation or congressional action. Senator David Durenberger was on the subcommittee that conducted hearings to consider new lead regulations.

The Senate hearings took place in June of 1984, and I was invited to make a presentation on behalf of the Minnesota Lead Coalition. Our statement played a prominent role in the deliberations. Senator Durenberger apparently expected us to fully agree with and testify in support of the Senate bill that delayed banning lead in gasoline until the mid-1990s. We could not support this provision and instead testified in favor of the House version of the bill, which banned lead on January 1, 1986. Senator Durenberger stopped the hearing to publicly reprimand me for not testifying in support of the Senate bill. Nevertheless, the next speaker, from Ashland Oil Company, threw down his prepared statement and agreed with us about the January 1, 1986 date. He said that his company had tooled up to meet the demand for unleaded fuel and was now paying a large monthly sum to store their unused lead-free fuel refining capabilities. The rapid phase-down became law on January 1, 1986. Unfortunately, the automobile had already spread the toxin. It played a role as a toxic-substance delivery system.

It is important to note that during recent decades public actions for lead exposure prevention have concentrated on screening children for elevated blood lead and treating children on a case-by-case basis for lead poisoning. This is secondary exposure prevention because a victim must be identified before treatment can proceed. Treatment invariably focused on limiting exposure to lead-based paint, usually by removing it. But the treatment often resulted in the release of lead dust, which exacerbated the lead poisoning.

There have, however, been phenomenal gains in the reduction of children's blood-lead levels for the whole population of U.S.

children, most of the gains taking place even before the federal lead-based paint intervention programs were in place. Although the gains have been remarkable, nevertheless, the poisoning problem continues to plague our society.

Our studies have shown that for most urban areas, decades of leaded paint and gasoline use have resulted in the accumulation of lead in the soil. The major conclusion from our New Orleans study is that a balanced prevention program must include activities to limit exposure to contaminated soil.

The Federal Residential Lead-Based Paint Hazard Reduction Act of 1992 includes soil and dust as sources of lead. Nevertheless, in practice, these vectors are not adequately acknowledged as the dangers they are. For example, Patrick Reagan of the Minnesota Environmental Research Association has documented that only 9 of the 26 member countries of the Organization for Economic Cooperation and Development regulate lead in soil, as opposed to 17 members that regulate lead in paint.

This is certainly true in the United States, where there exist more programs to limit exposure to lead paint than to leaded soil. In recently submitted rules, the EPA proposes 2,000 parts per million as an acceptable standard for soil lead. For children, this is not a protective standard. In Sweden, the standard soil-lead concentration for play areas is 80 parts per million. As guest scientist to Stockholm, I saw first hand the thoroughness of the implication of the standard. The attitude of the people of Stockholm is that a city safe for children is a city safe for everyone.

The U.S. Department of Housing and Urban Development (HUD) proposes less stringent regulation for lead in soil (500-1,000 parts per million) than for lead in paint (600 parts per million). These proposals come in spite of the fact that HUD among other federal agencies concluded that soil and dust are important sources of lead. HUD notes that "for infants and young children...* surface dust and soil are important pathways...." The Agency for Toxic Substances and Disease Registry (ATSDR) specifically states that the two major sources of lead are paint and soil. And a recent CDC report states that "lead-based paint and lead-contaminated dusts and soil remain the primary sources...." The EPA reports that "the three major sources of elevated blood lead are lead-based paint, urban soil and dust ... and lead in drinking water."

There are more than 20 other government reports that recognize the major hazards posed by leaded soil and dust. When, and only when, the sources and pathways of lead contamination are explicitly recognized as entwined will measures to prevent childhood exposure improve the situation.

Recognizing that soil in play areas is a major source of environmental lead provides an obvious course for prevention: Parents can provide a clean play area for their children and they can teach their children to wash their hands after playing outside. Communities might consider joining together to resurface play areas.

In Minneapolis and St. Paul a pilot project to reduce children's exposure to lead was implemented during the summer, when children spend the most time outdoors. In the target communities, children had their blood-lead concentrations measured at the beginning of the project in May or early June and at the end of the project in late August and September. Adults in this community were taught how to reduce the children's contact with bare soil. At the end of the trial, 96 percent of the children in the target community either remained the same or experienced *decreases* of blood lead, and 4 percent experienced modest rises. In contrast, children within a control community had their blood-lead levels measured, but no further education was provided. At the end of the trial 53 percent of these children experienced an *increase* in blood-lead levels, while 47 percent either remained the same or experienced decreases. Eighteen percent of the children in the control group had blood-lead levels of 40 ug/dL or higher, which necessitated clinical intervention. This contrasted sharply with the target group, where none of the children required clinical intervention.

A second study was conducted in Minnesota during the winter to evaluate the changes in blood lead when the ground is covered with snow and ice. The target group received special house-cleaning assistance, which the control group did not receive. Nevertheless, at the end of the winter both groups of children displayed the same degree of reductions in blood lead, relative to levels at the beginning of the winter, regardless of the cleaning treatment in their houses. This result further reinforces the hypothesis that outdoor soils are a major reservoir of lead.

The community of Trail, British Columbia, the site of the largest lead-zinc smelter in North America, provides an example of another community that successfully dealt with the problem of toxic levels of lead in its soil. Initially community leaders feared that alerting the public to the potential hazards of the soil would drive people out of the community. In spite of those fears, a primary-prevention program was put in place. Parents and child-care providers were educated about the problem, and contaminated soils were covered with plants or wood chips. New play boxes containing clean sand were put in play areas as an alternative to bare soil. Service organizations such as the Kiwanis Club, the Knights of Columbus and the Rotary Club joined to carry out landscaping projects on properties where parents could not afford to do it themselves. Because of these efforts, the levels of lead in children's blood dropped significantly 2 to 4 ug/dL (in three out of five years) or remained unchanged (during the other two years). These measures—particularly the new landscaping—had the added benefit of improving the community's overall appearance. Ultimately, the city even attracted an influx of new residents. Now all residents are routinely provided information about the lead problem and are educated about preventive measures.

The quantity of lead in New Orleans inner-city soils is about the same as that of the lead-smelter community of Trail. There is reason to believe that the success achieved by Trail could be repeated in other urban areas with contaminated soil. To reduce lead exposure of New Orleans children, I envision a program similar to Trail's. The major stakeholders include public officials, community organizations, public health and health-care providers (including pharmacists who have daily and direct contact with customers), parents, teachers, church leaders, painters, homeowners and landlords. The overall goal would be to prevent children from exposure to lead dust and contaminated soil. The projects must include interior dust control combined with measures to prevent exposure to exterior soil hazards.

Lead paint can be prevented from entering the environment by preventing its unsafe removal from both interior and exterior surfaces of houses. Alternatively, surfaces coated with unsafe paints can be covered over with a physical barrier, such as wall board. Some New Orleans painters have already agreed to use wet scraping and modern sanding equipment that captures the paint dust, rather than releasing it to the environment.

To reduce exposure from soil-based contaminants, the safest measure would be to prevent children from playing in bare soil, especially soil in play areas next to residential dwellings in the inner city. Soils should also be covered and maintained with a tight and vigorous grass or alternative surface, such as rubberized matting, bark, gravel, decking or cement. New Orleans has an enormous supply of river sediment that originates from farmlands upstream, which can be used to cover over the currently contaminated soils. In addition, lead-safe play areas, such as sandboxes, can be erected. These measures can be paid for by a tax on gasoline. A program to relandscape children's play areas would also provide useful employment to city residents.

Ultimately, successful prevention programs can be achieved only through collaborative efforts that combine public education with community-based primary-prevention measures. These efforts must focus on the ideal of clean environments as a means to healthy communities. It took nearly 10 decades for lead to accumulate to its current levels in urban areas. With judicious planning, the problem can be resolved in much less time.

Bibliography

ATSDR. 1988. *The Nature and Extent of Lead Poisoning in Children in the United States: A Report to Congress.* Atlanta: U.S. Public Health Service.

Brody, D. J., J. L. Pirkle, R. A. Kramer, K. M. Flegal, T. D. Matte, E. W. Gunter and D. C. Paschal. 1994. Blood lead levels in the US population: Phase 1 of the third national health and nutritional examination survey, NHANES III 1988 to 1991. *Journal of the American Medical Association* 272:277–283.

Centers for Disease Control. 1991. *Preventing Lead Poisoning in Young Children.* Atlanta: Centers for Disease Control.

Charney, E., J. Sayre and M. Coulter. 1980. Increased lead absorption in inner city children: Where does the lead come from? *Pediatrics* 65:226–231.

Hammond, P. B., C. S. Clark, P. S. Gartside, O. Berger, A. Walker and L. W. Michael. 1980. Fecal lead excretion in young children as related to sources of lead in their environment. *International Archives of Occupational and Environmental Health* 46:191–202.

Hilts, S. R. 1996. A cooperative approach to risk management in an active lead/zinc smelter community. *Environmental Geochemistry and Health* 18: 17–24.

Hilts, S. R., S. E. Bock, T. L. Oke, C. L. Yats and R. A. Copes. 1998. Effect of interventions on children's blood lead levels. *Environmental Health Perspectives* 106:79–83.

Mielke, H. W., J. C. Anderson, K. J. Berry, P. W. Mielke and R. L. Chaney. 1983. Lead concentrations in inner city soils as a factor in the child lead problem. *American Journal of Public Health* 73:1366–1369.

Mielke, H. W., B. Blake, S. Burroughs and N. Hassinger. 1984. Urban lead levels in Minneapolis: The case of the Hmong children. *Environmental Research* 34:64–76.

Mielke, H. W., S. Burroughs, S. Wade, T. Yarrow and P. W. Mielke. 1984–1985. Urban lead in Minnesota: Soil transects of four cities. *Journal of Minnesota Academy of Science* 50:19–24.

Mielke, H. W., J. L. Adams, P. L. Reagan and P. W. Mielke, 1989. Soil-dust lead and childhood lead exposure as a function of city size and community traffic flow: The case for lead abatement in Minnesota. In "Lead in Soil," ed. B. E. Davies and B. G. Wixson. *Environmental Geochemistry and Health Supplement 9*, 253–271.

Mielke, H. W., J. E. Adams, B. Huff, J. Pepersack, P. L. Reagan, D. Stoppel and P. W. Mielke, Jr. 1992. Dust control as a means of reducing inner-city childhood Pb exposure. *Trace Substances and Environmental Health* 25:121–128.

Mielke, H. W. 1993. Lead dust contaminated USA cities: Comparison of Louisiana and Minnesota. *Applied Geochemistry* (Supplementary Issue 2):257–261.

Mielke, H. W. 1994. Lead in New Orleans soils: New images of an urban environment. *Environmental Geochemistry and Health* 16:123–128.

Mielke, H. W., D. Dugas, P. W. Mielke, K. S. Smith, S. L. Smith and C. R. Gonzales. 1997. Associations between soil lead and childhood blood lead in urban New Orleans and rural Lafourche Parishes of Louisiana USA. *Environmental Health Perspectives* 105:950–954.

Mielke, H. W., and P. L. Reagan. 1998. Soil is an important pathway of human lead exposure. *Environmental Health Perspectives* 106 (Suppl. 1):217–229.

Pirkle, J. L., D. J. Brody, E. W. Gunter, R. A. Kramer, D. C. Paschal, K. M. Flegal and T. D. Matt. 1994. The decline in blood lead levels in the United States: The national health and nutritional examination surveys NHANES. *Journal of the American Medical Association* 272:284–291.

Sayre, J. W., E. Charney, J. Vostal and I. B. Pless. 1974. House and hand dust as a potential source of childhood lead exposure. *American Journal of the Diseases of Children* 127:167–170.

Statement of the Ethyl Corporation. 1984. S 2609—A Bill to Amend the Clean Air Act with Regard to Mobile Source Emission Control. Hearings before the Committee on Environment and Public Works, U.S. Senate, 98th Contress 2nd Session, June 22.

Viverette, L., H. W. Mielke, M. Brisco, A. Dixon, J. Schaefer and K. Pierre. 1996. Environmental health in minority and other underserved populations: benign methods for identifying lead hazards at day care centres of new Orleans. *Environmental Geochemistry and Health* 18:41–45.

UNEASY COLLABORATORS

To China, Taiwan investors are both welcome and suspect

Huang Chun-che flips on the police-issue siren in his Nissan Cefiro and steps on the gas. Huang has just finished blasting clay pigeons at the skeet-shooting range he manages, a place where he and other Taiwanese businessmen living in the southern Chinese city of Dongguan go to let off steam. Weaving from lane to lane, Huang gripes that the tangled traffic is making him late for the next appointment. But it is a snarl he helped create. In the last 10 years, Huang and others like him have transformed this once sleepy city into a veritable Taiwan Town.

> **TAIWAN'S DEEP-POCKET COMMITMENT TO CHINA**
> - Taiwanese companies have invested $40 billion in China's economy
> - 250,000 Taiwanese are employed in China
> - Taiwan-linked companies in China now number 40,000
> - Taiwanese companies are responsible for 12% of China's total exports

Located in the middle of Guangdong province's industrial-export belt, Dongguan is home to some 40,000 Taiwanese and 3,000 of their factories. Their imprint is everywhere—from the luxury villas to the flashy karaoke bars. Nearby, what will be China's first Taiwan-run school is nearing completion. Every other day, it seems, another delegation of Taiwanese comes to meet their Dongguan counterparts.

Such collaboration has flourished across China. Some 250,000 Taiwanese living on the mainland run companies that are responsible for 12% of China's exports. Taiwan investors have pumped an estimated $40 billion into the mainland economy. Their real contribution may be even greater, since many disguise their activities through Hong Kong front companies to avoid antagonizing Taipei, which is leery about becoming too cozy with China.

NEW WAVE. Moreover, Taiwan investors are moving well beyond the small-time activities of cowboy capitalists. The first wave of Taiwan businessmen made low-end clothing, shoes, and furniture. The new wave, by contrast, is dominated by Taiwan multinationals with high-tech production lines and deals with China's most powerful people. For example, Winston Wang, son of Taiwanese petrochemical tycoon Wang Yung-ching, plans to build a semiconductor plant in Shanghai, according to Taiwan sources. His prospective partner is none other than Jiang Mianheng, son of Chinese President Jiang Zemin.

But the March election of Taiwan President Chen Shuibian is sending jitters through China's Taiwanese community here. In the past, Chen has backed Taiwan's independence from China, a matter of grave concern in Beijing. Taiwan investors fret that a diplomatic downturn could derail their mainland projects. The June disappearance of General Pan His-hsien has heightened anxieties. The former chief of personnel at Taiwan's National Security Bureau went missing after taking up a senior position with a Taiwan electronics company in Dongguan. Taiwanese believe he was nabbed by China's State Security Ministry. Beijing has not confirmed the reports.

The fear is that Beijing may be trying to use Taiwan businesses on the mainland—especially companies run by the handful of executives who advise Chen on economic policy—to extract concessions from his government. Who better to push Chen away from independence, the theory goes, than those with the most to lose in China?

Some Taiwanese warn that pressure from Beijing could prompt them to curb expansion plans. But most investors are simply hunkering down for the long haul, hoping that common sense will prevail. "We're afraid," acknowledges C. M. Wu, managing director of Acer Peripherals (Suzhou) Inc., which manufactures CD-ROM drives, keyboards, monitors, and other equipment. "But we still want to expand."

The truth is, neither Taiwan nor China can afford to disengage. Today's biggest investors in China are the likes of Taiwan computer equipment companies Acer and Delta Electronics. Their factories are turning China into a major export platform for everything from monitors to motherboards. For Taiwan companies facing rising costs at home, China offers a nearly limitless pool of cheap labor and engineering talent. For Beijing, the Taiwanese provide capital and jobs at a time when bloated state enterprises are laying off millions. They also bring new technology and management systems, which Chinese industry needs as it prepares for competition upon joining the World Trade Organization.

The hope is that the burgeoning economic relationship will lead to a lasting political thaw. "Both governments and the people win" with Taiwanese investment, says H. D. Yeh, chairman of Dongguan Primax Electronic Products Ltd., a global producer of computer mice. "We hope that through this economic cooperation, both governments can keep the peace."

Beijing is rolling out incentives to keep Taiwan companies happy. Last December, the government made it easier for them to get local loans and strengthened legal protections. Local governments, meanwhile, are slashing taxes and land-use fees. Guangdong's government has opened special information offices for Taiwan investors and launched a Web site where business people can post complaints and suggestions.

SINGLED OUT. The trouble is that even as China beckons with one hand, it intimidates with the other. Many Taiwan factory owners complain of being targeted in crackdowns by Chinese officials over labor standards and customs rules. It's hard to tell whether Taiwanese are being singled out. But it does seem clear that companies most visibly associated with President Chen are most likely to be the victims of cross-Strait politics. "Business leaders who support Taiwan independence and have a bad influence, but still hope to benefit from China's strong economy, are not welcome," says Jiang Changfang, a Guangdong official responsible for Taiwan affairs. Chinese officials also have warned business in Hong Kong not to help Taiwan companies that back independence.

Evergreen Group Chairman Chang Yung-fa is one Chen adviser who seems to have run into trouble. Taiwan businessmen in Guangdong say customs officials are holding up Evergreen containers from Hong Kong. As a result, they say, Evergreen has lost business. An Evergreen official in Taiwan denies that.

And in what some analysts viewed as a snub, Chinese officials refused to see Acer Group Chairman Stan Shih, another Chen adviser, when he visited Beijing in April for a computer show. Acer officials in Taipei play down the incident, but company execs in China worry that Beijing could block new investments. In the past, Acer avoided cross-Strait politics, says J. T. Wang, the Taiwan-based chairman of Acer Sertek Inc., the group's sales and marketing arm. "But now politics comes to us." That means, Wang says, that Acer must assure Beijing it opposes independence and backs its policy that Taiwan is part of China.

Some Taiwan executives, however, are irritated at their own government. After being urged by the administration of previous President Lee Teng-hui to go slow in China, many had high hopes that Chen's election would lead to a friendlier climate. During the campaign, Chen vowed to quickly lift decades-old restrictions on direct transportation, communication, and postal links. For now, though, many constraints remain. Taipei still bans investment in finance and transportation as well as in products it considers technologically advanced, such as PCs or motherboards containing anything more powerful than a 486-class microprocessor. Taipei also caps mainland investment at $50 million. "We do have a national security concern," says Tsai Ing-wen, of the Mainland Affairs Council in Taipei, which oversees cross-Strait relations.

Not surprisingly, Taiwan investors have devised ways to skirt the rules. Many route investment through companies in Hong Kong, the U.S., or Japan. Several Guangdong-based companies admit they simply don't tell Taipei how much they invest in China or disclose what products they make.

Many Taiwanese just hope the controversy will blow over. There have been many stormy episodes in cross-Strait relations, and each time Beijing has refrained from seriously harming commerce. The less optimistic view is that the time is coming for Beijing to tighten the screws. With Hong Kong and Macau returned to the motherland, President Jiang and other leaders have made Taiwan a top priority. Now that some of Taiwan's premier companies have big money at stake, Beijing has valuable leverage. Taiwan industry has long learned to thrive despite political ambiguity. But by putting Taiwan investment at risk, Beijing may be warning that the era of carefree neutrality is coming to an end.

By Dexter Roberts in Dongguan, with Alysha Webb in Suzhou, and Mark L. Clifford and Brent Hannon in Taipei.

Unit 2

Unit Selections

9. **Global Warming: The Contrarian View,** Willliam K. Stevens
10. **The Great Climate Flip-Flop,** William H. Calvin
11. **Breaking the Global-Warming Gridlock,** Daniel Sarewitz and Roger Pielke Jr.
12. **Beyond the Valley of the Dammed,** Bruce Barcott
13. **A River Runs Through It,** Andrew Murr
14. **There Goes the Neighborhood,** Fen Montaigne
15. **The Race to Save Open Space,** Weston Kosova
16. **Past and Present Land Use and Land Cover in the USA,** William B. Meyer
17. **Operation Desert Sprawl,** William Fulton and Paul Shigley
18. **Turning the Tide: A Coastal Refuge Holds Its Ground as Sprawl Spills Into Southern Maine,** Richard M. Stapleton
19. **A Greener, or Browner, Mexico?** *The Economist*

Key Points to Consider

❖ What are the long-range implications of atmospheric pollution? Explain the greenhouse effect.

❖ How can the problem of regional transfer of pollutants be solved?

❖ The manufacture of goods needed by humans produces pollutants that degrade the environment. How can this dilemma be solved?

❖ Where in the world are there serious problems of desertification and drought? Why are these areas increasing in size?

❖ What will be the major forms of energy in the 21st century?

❖ How are you as an individual related to the land? Does urban sprawl concern you? Explain.

❖ Can humankind do anything to ensure the protection of the environment? Describe.

❖ What is your attitude toward the environment?

❖ What is your view on global warming?

 **Links**   www.dushkin.com/online/

10. **The North-South Institute**
    http://www.nsi-ins.ca/ensi/index.html
11. **United Nations Environment Programme**
    http://www.unep.ch
12. **World Health Organization**
    http://www.who.int

These sites are annotated on pages 6 and 7.

Human-Environment Relationships

The home of humankind is Earth's surface and the thin layer of atmosphere enveloping it. Here the human populace has struggled over time to change the physical setting and to create the telltale signs of occupation. Humankind has greatly modified Earth's surface to suit its purposes. At the same time, we have been greatly influenced by the very environment that we have worked to change.

This basic relationship of humans and land is important in geography and, in unit 1, William Pattison identified it as one of the four traditions of geography. Geographers observe, study, and analyze the ways in which human occupants of Earth have interacted with the physical environment. This unit presents a number of articles that illustrate the theme of human-environment relationships. In some cases, the association of humans and the physical world has been mutually beneficial; in others, environmental degradation has been the result.

At the present time, the potential for major modifications of Earth's surface and atmosphere is greater than at any other time in history. It is crucial that the environmental consequences of these modifications be clearly understood before such efforts are undertaken.

The first three articles in this unit deal with opposing viewpoints regarding the issue of global warming. "The Great Climate Flip-Flop" suggests that there may be global cooling following the global warming trend and offers ways to prevent it. Two articles follow that deal with new attitudes toward dams in the United States. The next two articles from *Audubon* address urban sprawl. William Meyer then addresses changing land use in the United States. "Operation Desert Sprawl" follows with an analysis of Las Vegas's reliance on transportation systems and the availability of fresh water. The next article reviews attemps to save coastal Maine wilderness from urban development. Finally, the question of NAFTA's impact on economics and the environment is covered.

This unit provides a small sample of the many ways in which humans interact with the environment. The outcomes of these interactions may be positive or negative. They may enhance the position of humankind and protect the environment, or they may do just the opposite. We human beings are the guardians of the physical world. We have it in our power to protect, to neglect, or to destroy.

Global Warming: The Contrarian View

By WILLIAM K. STEVENS

Over the years, skeptics have tried to cast doubt on the idea of global warming by noting that measurements taken by earth satellites since 1979 have found little or no temperature rise in large parts of the upper atmosphere. The satellites' all-encompassing coverage yields more reliable results than temperature samplings showing a century-long warming trend at the earth's surface, they argued.

In January, a special study by the National Research Council, the research arm of the National Academy of Sciences, declared that the "apparent disparity" between the two sets of measurements over the 20-year history of the satellite measurements "in no way invalidates the conclusion that surface temperature has been rising." The surface warming "is undoubtedly real," the study panel said.

But the dissenters are a long way from conceding the debate, and they have seized on other aspects of the panel's report in an effort to bolster their case.

To be sure, according to interviews with some prominent skeptics, there is now wide agreement among them that the average surface temperature of the earth has indeed risen.

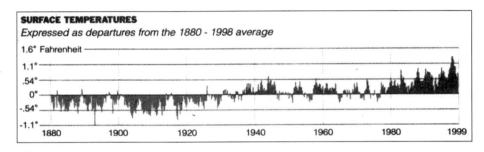

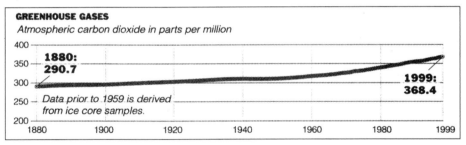

"I don't think we're arguing over whether there's any global warming," said Dr. William M. Gray, an atmospheric scientist at Colorado State University, known for his annual predictions of Atlantic hurricane activities as well as his staunch, long-time dissent on global climate change. "The question is, 'What is the cause of it?'"

On that issue, and on the remaining big question of how the climate might change in the future, skeptics continue to differ sharply with the dominant view among climate experts.

The dominant view is that the surface warming is at least partly attributable to emissions of heat-trapping waste industrial gases like carbon dioxide, a product of the burning of fossil fuels like coal, oil and natural gas. A United Nations scientific panel has predicted that unless these greenhouse gas emissions are reduced, the earth's average surface temperature will rise by some 2 to 6 degrees Fahrenheit over the next century, with a best estimate of about 3.5 degrees, compared with a rise of 5 to 9 degrees since the

Two sides, two data sets

In support of their position, skeptics in the debate over climate change cite an apparent disparity between two sets of temperature data.

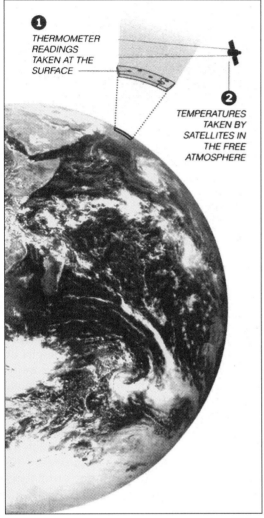

1. THERMOMETER READINGS TAKEN AT THE SURFACE
2. TEMPERATURES TAKEN BY SATELLITES IN THE FREE ATMOSPHERE

Sources: National Research Council; Carbon Dioxide Information Analysis Center

depths of the last ice age 18,000 to 20,000 years ago. This warming, the panel said, would touch off widespread disruptions in climate and weather and cause the global sea level to rise and flood many places.

Dr. Gray and others challenge all of this. To them, the observed surface warming of about 1 degree over the last century—with an especially sharp rise in the last quarter century—is mostly or wholly natural, and there is no significant human influence on global climate. They also adhere firmly to their long-held opinion that any future warming will be inconsequential or modest at most, and that its effects will largely be beneficial.

In some ways, though, adversaries in the debate are not so far apart. For instance, some dissenters say that future warming caused by greenhouse gases will be near the low end of the range predicted by the United Nations scientific panel. And most adherents of the dominant view readily acknowledge that the size of the human contribution to global warming is not yet known.

The thrust-and-parry of the climate debate goes on nevertheless.

With the National Research Council panel's conclusion that the surface warming is real, "one of the key arguments of the contrarians has evaporated," said Dr. Michael Oppenheimer, an atmospheric scientist with Environmental Defense, formerly the Environmental Defense Fund.

But those on the other side of the argument see things differently.

To them, the most important finding of the panel is its validation of satellite readings showing less warming, and maybe none, in parts of the upper atmosphere, said Dr. S. Fred Singer, an independent atmospheric scientist who is an outspoken dissenter.

For him and other climate dissenters, this disparity is key, in that it does not show up in computer models scientists use to predict future trends. The fact that these models apparently missed the difference in warming between the surface and the upper air, the skeptics say, casts doubt on their reliability over all.

Experts on all sides of the debate acknowledge that the climate models are imperfect, and even proponents of their use say their results should be interpreted cautiously.

9. Global Warming

The surface warming is real; humanity's role is at issue.

A further problem is raised by the divergent temperatures at the surface and the upper air, said Dr. Richard S. Lindzen, an atmospheric scientist at the Massachusetts Institute of Technology, who is a foremost skeptic. "Both are right," But, he said, increasing levels of greenhouse gases should warm the entire troposphere (the lower 6 to 10 miles of the atmosphere). That they have not, he said, suggests that "what's happening at the surface is not related to the greenhouse effect."

Skeptics also argue that the lower temperatures measured by the satellites are confirmed by instruments borne aloft by weather balloons. But over a longer period, going back 40 years, there is no discrepancy between surface readings and those obtained by the balloons, said Dr. John Michael Wallace, an atmospheric scientist at the University of Washington in Seattle, who was chairman of the research council study panel.

Although the climate debate has usually been portrayed as a polarized argument between believers and contrarians, there is actually a broad spectrum of views among scientists. And while the views of skeptics display some common themes, there are many degrees of dissent, many permutations and combinations of individual opinion. The views of some have changed materially over the years, while others have expressed essentially the same basic opinions all along.

One whose views have evolved is Dr. Wallace, who describes himself as "more skeptical than most people" but "fairly open to arguments on both sides" of the debate. He says the especially sharp surface warm-

2 ❖ HUMAN-ENVIRONMENT RELATIONSHIPS

❶ TEMPERATURE CHANGES AT OR NEAR THE EARTH'S SURFACE, 1979–1998

CLIMATE DEBATE:

THE DOMINANT VIEW
Surface warming is at least partly attributable to emissions of heat-trapping gases, like carbon dioxide, produced by the burning of fossil fuels.

THE SKEPTICS
Although the surface does appear to be warming, the cause is wholly or mostly natural.

Thermometers at land stations and aboard ships show a general global warming trend. The data are subject to distortion, but a recent study by the National Research Council says, nevertheless, that the surface warming is "undoubtedly real."

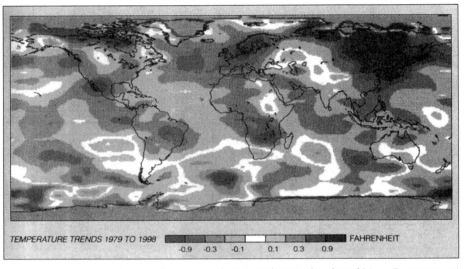

The New York Times; Images by NASA (Earth); National Academy of Science (Temperature maps)

ing trend of the 1990's has "pulled me in a mainstream direction." While he once believed the warming observed in recent decades was just natural variation in climate, he said, he is now perhaps 80 percent sure that it has been induced by human activity—"but that's still a long way from being willing to stake my reputation on it."

A decade ago, Dr. Wallace said, many skeptics questioned whether there even was a surface warming trend, in part because what now appears to be a century-long trend had been interrupted in the 1950's, 1960's and early 1970's, and it had not yet resumed all that markedly. But the surge in the 1980's and 1990's changed the picture substantially.

Today, Dr. Wallace said, few appear to doubt that the earth's surface has warmed. One prominent dissenter on the greenhouse question, Dr. Robert Balling, a climatologist at Arizona State University, says, "the surface temperatures appear to be rising, no doubt," and other skeptics agree.

There also appears to be general agreement that atmospheric concentrations of greenhouse gases are rising. At 360 parts per million, up from 315 parts in the late 1950's, the concentration of carbon dioxide is nearly 30 percent higher than before the Industrial Revolution, and the highest in the last 420,000 years.

Mainstream scientists, citing recent studies, suggest that the relatively rapid warming of the last 25 years cannot be explained without the greenhouse effect. Over that period, according to federal scientists, the average surface temperature rose at a rate equivalent to about 3.5 degrees per century—substantially more than the rise for the last century as a whole, and about what is predicted by computer models for the 21st century.

But many skeptics, including Dr. Gray and Dr. Singer, maintain that the warming of the past 25 years can be explained by natural causes, most likely changes in the circulation of heat-bearing ocean waters. In fact, Dr. Gray says he expects that over the next few decades, the warming will end and there will be a resumption of the cooling of the 1950's and 1960's.

At bottom, people on all sides of the debate agree, the question of the warming's cause has not yet been definitively answered. In December, a group of 11 experts on the question looked at the status of the continuing quest to detect the greenhouse signal amid the "noise" of the climate's natural variability.

The lead author of the study was Dr. Tim P. Barnett, a climatologist at the Scripps Institution of Oceanography in La Jolla, Calif. Dr. Barnett, who has long worked on detecting the greenhouse signal, describes himself as a "hard-nosed" skeptic on that particular issue, even though he believes that global warming in the long run will be a serious problem.

The study by Dr. Barnett and others, published in The Bulletin of the American Meteorological Society, concluded that the "most probable cause" of the observed warming had been a combination of natural and human-made factors. But they said scientists had not yet been able to separate the greenhouse signal from the natural climate fluctuations. This state of affairs, they wrote, "is not satisfactory."

Two big questions complicate efforts to predict the course of the earth's climate over the next century: how sensitive is the climate system, inherently, to the warming effect of greenhouse gases? And how much will atmospheric levels of the gases rise over coming decades?

9. Global Warming

❷ TEMPERATURE CHANGES IN THE FREE ATMOSPHERE, UP TO 5 MILES, 1979–1998

CLIMATE DEBATE:
THE SKEPTICS
Computer models on which projections of future warming are based failed to reflect the surface-satellite disparity, casting doubt on the projections' accuracy. They say future warming will be small to modest, and largely beneficial.

THE DOMINANT VIEW
Scientists adhere to the computer models which say the average global surface temperature will rise by 2 to 6 degrees Fahrenheit over the next century if greenhouse gas emissions are not reduced.

Instruments aboard satellites show little warming, and even some cooling, in the lower to middle part of the free atmosphere, mainly in the tropics. Although scientists cannot fully explain the disparity, some say that over a longer period it might disappear.

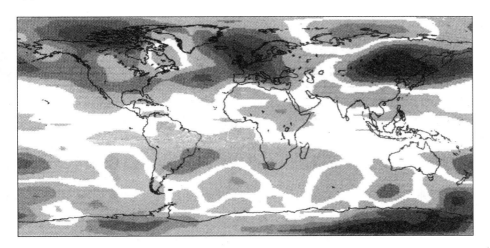

The mainstream view, based on computerized simulations of the climate system, is that a doubling of greenhouse gas concentrations would produce a warming of about 3 to 8 degrees. But Dr. Lindzen and Dr. Gray, pointing to what they consider the models' problems with the physics of the atmosphere, say they overestimate possible warming. It "will be extremely little," Dr. Gray said. How little? Dr. Lindzen pegs it at about half a degree to a bit less than 2 degrees, if atmospheric carbon dioxide doubles.

Other factors in the climate system modify the response to heat-trapping gases, and the United Nations panel's analysis included these to arrive at its projection of a 2- to 6-degree rise in the average global surface temperature by 2100. One factor is various possible levels of future carbon dioxide emissions. Dr. Singer, saying that improving energy efficiency will have a big impact on emissions, predicts a warming of less than 1 degree by 2100.

Dr. Balling projects a warming just shy of 1 degree for the next 50 years, not out of line with the United Nations panel's lower boundary. Another skeptic, Dr. Patrick J. Michaels, a climatologist at the University of Virginia, similarly forecasts a greenhouse warming rise of 2.3 degrees over the next century. As is now the case, he says, the warming would be most pronounced in the winter, at night, and in sub-Arctic regions like Siberia and Alaska. A warming of that magnitude, he and others insist, could not be very harmful, and would in fact confer benefits like longer growing seasons and faster plant growth.

"It should be pretty clear," he said, that the warming so far "didn't demonstrably dent health and welfare very much," and he said he saw no reason "to expect a sudden turnaround in the same over the next 50 years." After that, he said, it is impossible to predict the shape of the world's energy system and, therefore, greenhouse gas emissions.

A warming in the low end of the range predicted by the United Nations panel may well materialize, said Dr. Oppenheimer, the environmentalist. But, he said, the high end may also materialize, in which case, mainstream scientists say, there would be serious, even catastrophic, consequences for human society. "There is no compelling evidence to allow us to choose between the low end, or the high end, or the middle," Dr. Oppenheimer said.

If business continues as usual, the world is likely at some point to find out who is right.

The Great Climate Flip-flop

by WILLIAM H. CALVIN

ONE of the most shocking scientific realizations of all time has slowly been dawning on us: the earth's climate does great flip-flops every few thousand years, and with breathtaking speed. We could go back to ice-age temperatures within a decade—and judging from recent discoveries, an abrupt cooling could be triggered by our current global-warming trend. Europe's climate could become more like Siberia's. Because such a cooling would occur too quickly for us to make readjustments in agricultural productivity and supply, it would be a potentially civilization-shattering affair, likely to cause an unprecedented population crash. What paleoclimate and oceanography researchers know of the mechanisms underlying such a climate flip suggests that global warming could start one in several different ways.

For a quarter century global-warming theorists have predicted that climate creep is going to occur and that we need to prevent greenhouse gases from warming things up, thereby raising the sea level, destroying habitats, intensifying storms, and forcing agricultural rearrangements. Now we know—and from an entirely different group of scientists exploring separate lines of reasoning and data—that the most catastrophic result of global warming could be an abrupt cooling.

We are in a warm period now. Scientists have known for some time that the previous warm period started 130,000 years ago and ended 117,000 years ago, with the return of cold temperatures that led to an ice age. But the ice ages aren't what they used to be. They were formerly thought to be very gradual, with both air temperature and ice sheets changing in a slow, 100,000-year cycle tied to changes in the earth's orbit around the sun. But our current warm-up, which started about 15,000 years ago, began abruptly, with the temperature rising sharply while most of the ice was still present. We now know that there's nothing "glacially slow" about temperature change: superimposed on the gradual, long-term cycle have been dozens of abrupt warmings and coolings that lasted only centuries.

The back and forth of the ice started 2.5 million years ago, which is also when the ape-sized hominid brain began to develop into a fully human one, four times as large and reorganized for language, music, and chains of inference. Ours is now a brain able to anticipate outcomes well enough to practice ethical behavior, able to head off disasters in the making by extrapolating trends. Our civilizations began to emerge right after the continental ice sheets melted about 10,000 years ago. Civilizations accumulate knowledge, so we now know a lot about what has been going on, what has

> "Climate change" is popularly understood to mean greenhouse warming, which, it is predicted, will cause flooding, severe windstorms, and killer heat waves. But warming could lead, paradoxically, to drastic cooling—a catastrophe that could threaten the survival of civilization

---

William H. Calvin is a theoretical neurophysiologist at the University of Washington at Seattle.

10. Great Climate Flip-Flop

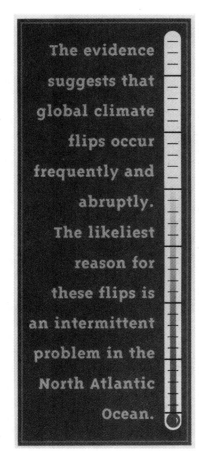

The evidence suggests that global climate flips occur frequently and abruptly. The likeliest reason for these flips is an intermittent problem in the North Atlantic Ocean.

made us what we are. We puzzle over oddities, such as the climate of Europe.

Keeping Europe Warm

EUROPE is an anomaly. The populous parts of the United States and Canada are mostly between the latitudes of 30° and 45°, whereas the populous parts of Europe are ten to fifteen degrees farther north. "Southerly" Rome lies near the same latitude, 42°N, as "northerly" Chicago—and the most northerly major city in Asia is Beijing, near 40°N. London and Paris are close to the 49°N line that, west of the Great Lakes, separates the United States from Canada. Berlin is up at about 52°, Copenhagen and Moscow at about 56°. Oslo is nearly at 60°N, as are Stockholm, Helsinki, and St. Petersburg; continue due east and you'll encounter Anchorage.

Europe's climate, obviously, is not like that of North America or Asia at the same latitudes. For Europe to be as agriculturally productive as it is (it supports more than twice the population of the United States and Canada), all those cold, dry winds that blow eastward across the North Atlantic from Canada must somehow be warmed up. The job is done by warm water flowing north from the tropics, as the eastbound Gulf Stream merges into the North Atlantic Current. This warm water then flows up the Norwegian coast, with a westward branch warming Greenland's tip, 60°N. It keeps northern Europe about nine to eighteen degrees warmer in the winter than comparable latitudes elsewhere—except when it fails. Then not only Europe but also, to everyone's surprise, the rest of the world gets chilled. Tropical swamps decrease their production of methane at the same time that Europe cools, and the Gobi Desert whips much more dust into the air. When this happens, something big, with worldwide connections, must be switching into a new mode of operation.

The North Atlantic Current is certainly something big, with the flow of about a hundred Amazon Rivers. And it sometimes changes its route dramatically, much as a bus route can be truncated into a shorter loop. Its effects are clearly global too, inasmuch as it is part of a long "salt conveyor" current that extends through the southern oceans into the Pacific.

I hope never to see a failure of the northernmost loop of the North Atlantic Current, because the result would be a population crash that would take much of civilization with it, all within a decade. Ways to postpone such a climatic shift are conceivable, however—old-fashioned dam-and-ditch construction in critical locations might even work. Although we can't do much about everyday weather, we may nonetheless be able to stabilize the climate enough to prevent an abrupt cooling.

Abrupt Temperature Jumps

THE discovery of abrupt climate changes has been spread out over the past fifteen years, and is well known to readers of major scientific journals such as *Science* and *Nature*. The abruptness data are convincing. Within the ice sheets of Greenland are annual layers that provide a record of the gases present in the atmosphere and indicate the changes in air temperature over the past 250,000 years—the period of the last two major ice ages. By 250,000 years ago *Homo erectus* had died out, after a run of almost two million years. By 125,000 years ago *Homo sapiens* had evolved from our ancestor species—so the whiplash climate changes of the last ice age affected people much like us.

In Greenland a given year's snowfall is compacted into ice during the ensuing years, trapping air bubbles, and so paleoclimate researchers have been able to glimpse ancient climates in some detail. Water falling as snow on Greenland carries an isotopic "fingerprint" of what the temperature was like en route. Counting those tree-ring-like layers in the ice cores shows that cooling came on as quickly as droughts. Indeed, were another climate flip to begin next year, we'd probably complain first about the drought, along with unusually cold winters in Europe. In the first few years the climate could cool as much as it did during the misnamed Little Ice Age (a gradual cooling that lasted from the early Renaissance until the end of the nineteenth century), with tenfold greater changes over the next decade or two.

The most recent big cooling started about 12,700 years ago, right in the midst of our last global warming. This cold period, known as the Younger Dryas, is named for the pollen of a tundra flower that turned up in a lake bed in Denmark when it shouldn't have. Things had been warming up, and half the ice sheets covering Europe and Canada had already melted. The return to ice-age temperatures lasted 1,300 years. Then, about 11,400 years ago, things suddenly warmed up again, and the earliest agricultural villages were established in the Middle East. An abrupt cooling got started 8,200 years ago, but it aborted within a century, and the temperature changes since then have been gradual in comparison. Indeed, we've had an unprecedented period of climate stability.

Coring old lake beds and examining the types of pollen trapped in sediment layers led to the discovery, early in the twentieth century, of the Younger Dryas. Pollen cores are still a primary means of seeing what regional climates were doing, even though they suffer from poorer resolution than ice cores (worms churn the sediment, obscuring records of all but the longest-lasting temperature changes). When the ice cores demonstrated the abrupt onset of the Younger Dryas, researchers wanted to know how widespread this event was. The U.S. Geological Survey took old lake-bed cores out of storage and re-examined them.

THE NORTHERN LOOP OF THE NORTH ATLANTIC CURRENT

Ancient lakes near the Pacific coast of the United States, it turned out, show a shift to cold-weather plant species at roughly the time when the Younger Dryas was changing German pine forests into scrublands like those of modern Siberia. Subarctic ocean currents were reaching the southern California coastline, and Santa Barbara must have been as cold as Juneau is now. (But the regional record is poorly understood, and I know at least one reason why. These days when one goes to hear a talk on ancient climates of North America, one is likely to learn that the speaker was forced into early retirement from the U.S. Geological Survey by budget cuts. Rather than a vigorous program of studying regional climatic change, we see the shortsighted preaching of cheaper government at any cost.)

In 1984, when I first heard about the startling news from the ice cores, the implications were unclear—there seemed to be other ways of interpreting the data from Greenland. It was initially hoped that the abrupt warmings and coolings were just an oddity of Greenland's weather—but they have now been detected on a worldwide scale, and at about the same time. Then it was hoped that the abrupt flips were somehow caused by continental ice sheets, and thus would be unlikely to recur, because we now lack huge ice sheets over Canada and Northern Europe. Though some abrupt coolings are likely to have been associated with events in the Canadian ice sheet, the abrupt cooling in the previous warm period, 122,000 years ago, which has now been detected even in the tropics, shows that flips are not restricted to icy periods; they can also interrupt warm periods like the present one.

There seems to be no way of escaping the conclusion that global climate flips occur frequently and abruptly. An abrupt cooling could happen now, and the world might not warm up again for a long time: it looks as if the last warm period, having lasted 13,000 years, came to an end with an abrupt, prolonged cooling. That's how our warm period might end too.

Sudden onset, sudden recovery—this is why I use the word "flip-flop" to describe these climate changes. They are utterly unlike the changes that one would expect from accumulating carbon dioxide or the setting adrift of ice shelves from Antarctica. Change arising from some sources, such as volcanic eruptions, can be abrupt—but the climate doesn't flip back just as quickly centuries later.

Temperature records suggest that there is some grand mechanism underlying all of this, and that it has two major states. Again, the difference between them amounts to nine to eighteen degrees—a range that may depend on how much ice there is to slow the responses. I call the colder one the "low state." In discussing the ice ages there is a tendency to think of warm as good—and therefore of warming as better. Alas, further warming might well kick us out of the "high state." It's the high state that's good, and we may need to help prevent any sudden transition to the cold low state.

Although the sun's energy output does flicker slightly, the likeliest reason for these abrupt flips is an intermittent problem in the North Atlantic Ocean, one that seems to trigger a major rearrangement of atmospheric circulation. North-south ocean currents help to redistribute equatorial heat into the temperate zones, supplementing the heat transfer by winds. When the warm currents penetrate farther than usual into the northern seas, they help to melt the sea ice that is reflecting a lot of sunlight back into space, and so the earth becomes warmer. Eventually that helps to melt ice sheets elsewhere.

The high state of climate seems to involve ocean currents that deliver an extraordinary amount of heat to the vicinity of Iceland and Norway. Like bus routes or conveyor belts, ocean currents must have a return loop. Un-

10. Great Climate Flip-Flop

like most ocean currents, the North Atlantic Current has a return loop that runs deep beneath the ocean surface. Huge amounts of seawater sink at known downwelling sites every winter, with the water heading south when it reaches the bottom. When that annual flushing fails for some years, the conveyor belt stops moving and so heat stops flowing so far north—and apparently we're popped back into the low state.

Flushing Cold Surface Water

SURFACE waters are flushed regularly, even in lakes. Twice a year they sink, carrying their load of atmospheric gases downward. That's because water density changes with temperature. Water is densest at about 39°F (a typical refrigerator setting—anything that you take out of the refrigerator, whether you place it on the kitchen counter or move it to the freezer, is going to expand a little). A lake surface cooling down in the autumn will eventually sink into the less-dense-because-warmer waters below, mixing things up. Seawater is more complicated, because salt content also helps to determine whether water floats or sinks. Water that evaporates leaves its salt behind; the resulting saltier water is heavier and thus sinks.

The fact that excess salt is flushed from surface waters has global implications, some of them recognized two centuries ago. Salt circulates, because evaporation up north causes it to sink and be carried south by deep currents. This was posited in 1797 by the Anglo-American physicist Sir Benjamin Thompson (later known, after he moved to Bavaria, as Count Rumford of the Holy Roman Empire), who also posited that, if merely to compensate, there would have to be a warmer northbound current as well. By 1961 the oceanographer Henry Stommel, of the Woods Hole Oceanographic Institution, in Massachusetts, was beginning to worry that these warming currents might stop flowing if too much fresh water was added to the surface of the northern seas. By 1987 the geochemist Wallace Broecker, of Columbia University, was piecing together the paleoclimatic flip-flops with the salt-circulation story and warning that small nudges to our climate might produce "unpleasant surprises in the greenhouse."

Oceans are not well mixed at any time. Like a half-beaten cake mix, with strands of egg still visible, the ocean has a lot of blobs and streams within it. When there has been a lot of evaporation, surface waters are saltier than usual. Sometimes they sink to considerable depths without mixing. The Mediterranean waters flowing out of the bottom of the Strait of Gibraltar into the Atlantic Ocean are about 10 percent saltier than the ocean's average, and so they sink into the depths of the Atlantic. A nice little Amazon-sized waterfall flows over the ridge that connects Spain with Morocco, 800 feet below the surface of the strait.

Another underwater ridge line stretches from Greenland to Iceland and on to the Faeroe Islands and Scotland. It, too, has a salty waterfall, which pours the hypersaline bottom waters of the Nordic Seas (the Greenland Sea and the Norwegian Sea) south into the lower levels of the North Atlantic Ocean. This salty waterfall is more like thirty Amazon Rivers combined. Why does it exist? The cold, dry winds blowing eastward off Canada evaporate the surface waters of the North Atlantic Current, and leave behind all their salt. In late winter the heavy surface waters sink en masse. These blobs, pushed down by annual repetitions of these late-winter events, flow south, down near the bottom of the Atlantic. The same thing happens in the Labrador Sea between Canada and the southern tip of Greenland.

Salt sinking on such a grand scale in the Nordic Seas causes warm water to flow much farther north than it might otherwise do. This produces a heat bonus of perhaps 30 percent beyond the heat provided by direct sunlight to these seas, accounting for the mild winters downwind, in northern Europe. It has been called the Nordic Seas heat pump.

Nothing like this happens in the Pacific Ocean, but the Pacific is nonetheless affected, because the sink in the Nordic Seas is part of a vast worldwide salt-conveyor belt. Such a conveyor is needed because the Atlantic is saltier than the Pacific (the Pacific has twice as much water with which to dilute the salt carried in from rivers). The Atlantic would be even saltier if it didn't mix with the Pacific, in long, loopy currents. These carry the North Atlantic's excess salt southward from the bottom of the Atlantic, around the tip of Africa, through the Indian Ocean, and up around the Pacific Ocean.

There used to be a tropical shortcut, an express route from Atlantic to Pacific, but continental drift connected North America to South America about three million years ago, damming up the easy route for disposing of excess salt. The dam, known as the Isthmus of Panama, may have been what caused the ice ages to begin a short time later, simply because of the forced detour. This major change in ocean circulation, along with a climate that had already been slowly cooling for millions of years, led not

> Huge amounts of seawater sink every winter in the vicinity of Iceland and Norway. When that flushing fails for some years, apparently we're popped into an abrupt cooling.

only to ice accumulation most of the time but also to climatic instability, with flips every few thousand years or so.

Failures of Flushing

FLYING above the clouds often presents an interesting picture when there are mountains below. Out of the sea of undulating white clouds mountain peaks stick up like islands.

Greenland looks like that, even on a cloudless day—but the great white mass between the occasional punctuations is an ice sheet. In places this frozen fresh water descends from the highlands in a wavy staircase.

Twenty thousand years ago a similar ice sheet lay atop the Baltic Sea and the land surrounding it. Another sat on Hudson's Bay, and reached as far west as the foothills of the Rocky Mountains—where it pushed, head to head, against ice coming down from the Rockies. These northern ice sheets were as high as Greenland's mountains, obstacles sufficient to force the jet stream to make a detour.

Now only Greenland's ice remains, but the abrupt cooling in the last warm period shows that a flip can occur in situations much like the present one. What could possibly halt the salt-conveyor belt that brings tropical heat so much farther north and limits the formation of ice sheets? Oceanographers are busy studying present-day failures of annual flushing, which give some perspective on the catastrophic failures of the past.

In the Labrador Sea, flushing failed during the 1970s, was strong again by 1990, and is now declining. In the Greenland Sea over the 1980s salt sinking declined by 80 percent. Obviously, local failures can occur without catastrophe—it's a question of how often and how widespread the failures are—but the present state of decline is not very reassuring. Large-scale flushing at both those sites is certainly a highly variable process, and perhaps a somewhat fragile one as well. And in the absence of a flushing mechanism to sink cooled surface waters and send them southward in the Atlantic, additional warm waters do not flow as far north to replenish the supply.

There are a few obvious precursors to flushing failure. One is diminished wind chill, when winds aren't as strong as usual, or as cold, or as dry—as is the case in the Labrador Sea during the North Atlantic Oscillation. This El Niño-like shift in the atmospheric-circulation pattern over the North Atlantic, from the Azores to Greenland, often lasts a decade. At the same time that the Labrador Sea gets a lessening of the strong winds that aid salt sinking, Europe gets particularly cold winters. It's happening right now: a North Atlantic Oscillation started in 1996.

Another precursor is more floating ice than usual, which reduces the amount of ocean surface exposed to the winds, in turn reducing evaporation. Retained heat eventually melts the ice, in a cycle that recurs about every five years.

Yet another precursor, as Henry Stommel suggested in 1961, would be the addition of fresh water to the ocean surface, diluting the salt-heavy surface waters before they became unstable enough to start sinking. More rain falling in the northern oceans—exactly what is predicted as a result of global warming—could stop salt flushing. So could ice carried south out of the Arctic Ocean.

There is also a great deal of unsalted water in Greenland's glaciers, just uphill from the major salt sinks. The last time an abrupt cooling occurred was in the midst of global warming. Many ice sheets had already half melted, dumping a lot of fresh water into the ocean.

> The Nordic Seas sink is part of a worldwide conveyor belt. There used to be a shortcut, but it was dammed up by the Isthmus of Panama, which may have begun the ice ages.

A brief, large flood of fresh water might nudge us toward an abrupt cooling even if the dilution were insignificant when averaged over time. The fjords of Greenland offer some dramatic examples of the possibilities for freshwater floods. Fjords are long, narrow canyons, little arms of the sea reaching many miles inland; they were carved by great glaciers when the sea level was lower. Greenland's east coast has a profusion of fjords between 70°N and 80°N, including one that is the world's biggest. If blocked by ice dams, fjords make perfect reservoirs for meltwater.

Glaciers pushing out into the ocean usually break off in chunks. Whole sections of a glacier, lifted up by the tides, may snap off at the "hinge" and become icebergs. But sometimes a glacial surge will act like an avalanche that blocks a road, as happened when Alaska's Hubbard glacier surged into the Russell fjord in May of 1986. Its snout ran into the opposite side, blocking the fjord with an ice dam. Any meltwater coming in behind the dam stayed there. A lake formed, rising higher and higher—up to the height of an eight-story building.

Eventually such ice dams break, with spectacular results. Once the dam is breached, the rushing waters erode an ever wider and deeper path. Thus the entire lake can empty quickly. Five months after the ice dam

10. Great Climate Flip-Flop

at the Russell fjord formed, it broke, dumping a cubic mile of fresh water in only twenty-four hours.

The Great Salinity Anomaly, a pool of semi-salty water derived from about 500 times as much unsalted water as that released by Russell Lake, was tracked from 1968 to 1982 as it moved south from Greenland's east coast. In 1970 it arrived in the Labrador Sea, where it prevented the usual salt sinking. By 1971–1972 the semi-salty blob was off Newfoundland. It then crossed the Atlantic and passed near the Shetland Islands around 1976. From there it was carried northward by the warm Norwegian Current, whereupon some of it swung west again to arrive off Greenland's east coast—where it had started its inch-per-second journey. So freshwater blobs drift, sometimes causing major trouble, and Greenland floods thus have the potential to stop the enormous heat transfer that keeps the North Atlantic Current going strong.

The Greenhouse Connection

OF this much we're sure: global climate flip-flops have frequently happened in the past, and they're likely to happen again. It's also clear that sufficient global warming could trigger an abrupt cooling in at least two ways—by increasing high-latitude rainfall or by melting Greenland's ice, both of which could put enough fresh water into the ocean surface to suppress flushing.

Further investigation might lead to revisions in such mechanistic explanations, but the result of adding fresh water to the ocean surface is pretty standard physics. In almost four decades of subsequent research Henry Stommel's theory has only been enhanced, not seriously challenged.

Up to this point in the story none of the broad conclusions is particularly speculative. But to address how all these nonlinear mechanisms fit together—and what we might do to stabilize the climate—will require some speculation.

Even the tropics cool down by about nine degrees during an abrupt cooling, and it is hard to imagine what in the past could have disturbed the whole earth's climate on this scale. We must look at arriving sunlight and departing light and heat, not merely regional shifts on earth, to account for changes in the temperature balance. Increasing amounts of sea ice and clouds could reflect more sunlight back into space, but the geochemist Wallace Broecker suggests that a major greenhouse gas is disturbed by the failure of the salt conveyor, and that this affects the amount of heat retained.

In Broecker's view, failures of salt flushing cause a worldwide rearrangement of ocean currents, resulting in—and this is the speculative part—less evaporation from the tropics. That, in turn, makes the air drier. Because water vapor is the most powerful greenhouse gas, this decrease in average humidity would cool things globally. Broecker has written, "If you wanted to cool the planet by 5°C [9°F] and could magically alter the water-vapor content of the atmosphere, a 30 percent decrease would do the job."

Just as an El Niño produces a hotter Equator in the Pacific Ocean and generates more atmospheric convection, so there might be a subnormal mode that decreases heat, convection, and evaporation. For example, I can imagine that ocean currents carrying more warm surface waters north or south from the equatorial regions might, in consequence, cool the Equator somewhat. That might result in less evaporation, creating lower-than-normal levels of greenhouse gases and thus a global cooling.

To see how ocean circulation might affect greenhouse gases, we must try to account quantitatively for important nonlinearities, ones in which little nudges provoke great responses. The modern world is full of objects and systems that exhibit "bistable" modes, with thresholds for flipping. Light switches abruptly change mode when nudged hard enough. Door latches suddenly give way. A gentle pull on a trigger may be ineffective, but there comes a pressure that will suddenly fire the gun. Thermostats tend to activate heating or cooling mechanisms abruptly—also an example of a system that pushes back.

We must be careful not to think of an abrupt cooling in response to global warming as just another self-regulatory device, a control system for cooling things down when it gets too hot. The scale of the response will be far beyond the bounds of regulation—more like when excess warming triggers fire extinguishers in the ceiling, ruining the contents of the room while cooling them down.

Preventing Climate Flips

THOUGH combating global warming is obviously on the agenda for preventing a cold flip, we could easily be blindsided by stability problems if we allow global warming per se to remain the main focus of our climate-change efforts. To stabilize our flip-flopping climate we'll need to identify all the important feedbacks that control climate and ocean currents—evaporation, the reflection of sunlight back into space, and so on—and then estimate their relative strengths and interactions in computer models.

Feedbacks are what determine thresholds, where one mode flips into another. Near a threshold one can sometimes observe abortive responses, rather like the act of stepping back onto a curb several times before finally running across a busy street. Abortive responses and rapid chattering between modes are common problems in nonlinear systems with not quite enough oomph—the reason that old fluorescent lights flicker. To keep a bistable system firmly in one state or the other, it should be kept away from the transition threshold.

We need to make sure that no business-as-usual climate variation, such as an El Niño or the North Atlantic Oscillation, can push our climate onto the slippery slope and into an abrupt cooling. Of particular importance are combinations of climate variations—this winter, for example, we are experiencing both an El Niño and a North Atlantic Oscillation—because such combinations can add up to much more than the sum of their parts.

We are near the end of a warm period in any event; ice ages return even without human influences on climate. The last warm period abruptly terminated 13,000 years after the abrupt warming that initiated it, and we've already gone 15,000 years from a similar starting point. But we may be able to do something to delay an abrupt cooling.

Do something? This tends to stagger the imagination, immediately conjuring up visions of terraforming on a science-fiction scale—and so we shake our heads and say, "Better to fight global warming by consuming less," and so forth.

Surprisingly, it may prove possible to prevent flip-flops in the climate—even by means of low-tech schemes. Keeping the present climate from falling back into the low state will in any case be a lot easier than trying to reverse such a change after it has occurred. Were fjord floods causing flushing to fail, because the downwelling sites were fairly close to the fjords, it is obvious that we could solve the problem. All we would need to do is open a channel through the ice dam with explosives before dangerous levels of water built up.

Timing could be everything, given the delayed effects from inch-per-second circulation patterns, but that, too, potentially has a low-tech solution: build dams across the major fjord systems and hold back the meltwater at critical times. Or divert eastern-Greenland meltwater to the less sensitive north and west coasts.

Fortunately, big parallel computers have proved useful for both global climate modeling and detailed modeling of ocean circulation. They even show the flips. Computer models might not yet be able to predict what will happen if we tamper with downwelling sites, but this problem doesn't seem insoluble. We need more well-trained people, bigger computers, more coring of the ocean floor and silted-up lakes, more ships to drag instrument packages through the depths, more instrumented buoys to study critical sites in detail, more satellites measuring regional variations in the sea surface, and perhaps some small-scale trial runs of interventions.

It would be especially nice to see another dozen major groups of scientists doing climate simulations, discovering the intervention mistakes as quickly as possible and learning from them. Medieval cathedral builders learned from their design mistakes over the centuries, and their undertakings were a far larger drain on the economic resources and people power of their day than anything yet discussed for stabilizing the climate in the twenty-first century. We may not have centuries to spare, but any economy in which two percent of the population produces all the food, as is the case in the United States today, has lots of resources and many options for reordering priorities.

Three Scenarios

FUTURISTS have learned to bracket the future with alternative scenarios, each of which captures important features that cluster together, each of which is compact enough to be seen as a narrative on a human scale. Three scenarios for the next climatic phase might be called population crash, cheap fix, and muddling through.

The population-crash scenario is surely the most appalling. Plummeting crop yields would cause some powerful countries to try to take over their neighbors or distant lands—if only because their armies, unpaid and lacking food, would go marauding, both at home and across the borders. The better-organized countries would attempt to use their armies, before they fell apart entirely, to take over countries with significant remaining resources, driving out or starving their inhabitants if not using modern weapons to accomplish the same end: eliminating competitors for the remaining food.

This would be a worldwide problem—and could lead to a Third World War—but Europe's vulnerability is particularly easy to analyze. The last abrupt cooling, the Younger Dryas, drastically altered Europe's climate as far east as Ukraine. Present-day Europe has more than 50 million people. It has excellent soils, and largely grows its own food. It could no longer do so if it lost the extra warming from the North Atlantic.

There is another part of the world with the same good soil, within the same latitudinal band, which we can use for a quick comparison. Canada lacks Europe's winter warmth and rainfall, because it has no equivalent of the North Atlantic Current to preheat its eastbound weather systems. Canada's agriculture supports about 28 million people. If Europe had weather like Canada's, it could feed only one out of twenty-three present-day Europeans.

Any abrupt switch in climate would also disrupt food supply routes. The only reason that two percent of our population can feed the other 98 percent is that we have a well-developed system of transportation and middlemen—but it is not very robust. The system allows for large urban populations in the

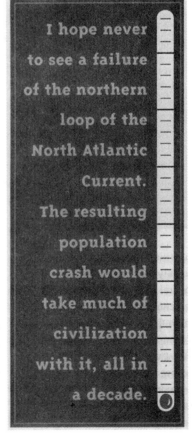

I hope never to see a failure of the northern loop of the North Atlantic Current. The resulting population crash would take much of civilization with it, all in a decade.

best of times, but not in the case of widespread disruptions.

Natural disasters such as hurricanes and earthquakes are less troubling than abrupt coolings for two reasons: they're short (the recovery period starts the next day) and they're local or regional (unaffected citizens can help the overwhelmed). There is, increasingly, international cooperation in response to catastrophe—but no country is going to be able to rely on a stored agricultural surplus for even a year, and any country will be reluctant to give away part of its surplus.

In an abrupt cooling the problem would get worse for decades, and much of the earth would be affected. A meteor strike that killed most of the population in a month would not be as serious as an abrupt cooling that eventually killed just as many. With the population crash spread out over a decade, there would be ample opportunity for civilization's institutions to be torn apart and for hatreds to build, as armies tried to grab remaining resources simply to feed the people in their own countries. The effects of an abrupt cold last for centuries. They might not be the end of *Homo sapiens*—written knowledge and elementary education might well endure—but the world after such a population crash would certainly be full of despotic governments that hated their neighbors because of recent atrocities. Recovery would be very slow.

A slightly exaggerated version of our present know-something-do-nothing state of affairs is know-nothing-do-nothing: a reduction in science as usual, further limiting our chances of discovering a way out. History is full of withdrawals from knowledge-seeking, whether for reasons of fundamentalism, fatalism, or "government lite" economics. This scenario does not require that the shortsighted be in charge, only that they have enough influence to put the relevant science agencies on starvation budgets and to send recommendations back for yet another commission report due five years hence.

A cheap-fix scenario, such as building or bombing a dam, presumes that we know enough to prevent trouble, or to nip a developing problem in the bud. But just as vaccines and antibiotics presume much knowledge about diseases, their climatic equivalents presume much knowledge about oceans, atmospheres, and past climates. Suppose we had reports that winter salt flushing was confined to certain areas, that abrupt shifts in the past were associated with localized flushing failures, *and* that one computer model after another suggested a solution that was likely to work even under a wide range of weather extremes. A quick fix, such as bombing an ice dam, might then be possible. Although I don't consider this scenario to be the most likely one, it is possible that solutions could turn out to be cheap and easy, and that another abrupt cooling isn't inevitable. Fatalism, in other words, might well be foolish.

A muddle-through scenario assumes that we would mobilize our scientific and technological resources well in advance of any abrupt cooling problem, but that the solution wouldn't be simple. Instead we would try one thing after another, creating a patchwork of solutions that might hold for another few decades, allowing the search for a better stabilizing mechanism to continue.

We might, for example, anchor bargeloads of evaporation-enhancing surfactants (used in the southwest corner of the Dead Sea to speed potash production) upwind from critical downwelling sites, letting winds spread them over the ocean surface all winter, just to ensure later flushing. We might create a rain shadow, seeding clouds so that they dropped their unsalted water well upwind of a given year's critical flushing sites—a strategy that might be particularly important in view of the increased rainfall expected from global warming. We might undertake to regulate the Mediterranean's salty outflow, which is also thought to disrupt the North Atlantic Current.

Perhaps computer simulations will tell us that the only robust solutions are those that re-create the ocean currents of three million years ago, before the Isthmus of Panama closed off the express route for excess-salt disposal. Thus we might dig a wide sea-level Panama Canal in stages, carefully managing the changeover.

Staying in the "Comfort Zone"

STABILIZING our flip-flopping climate is not a simple matter. We need heat in the right places, such as the Greenland Sea, and not in others right next door, such as Greenland itself. Man-made global warming is likely to achieve exactly the opposite—warming Greenland and cooling the Greenland Sea.

A remarkable amount of specious reasoning is often encountered when we contemplate reducing carbon-dioxide emissions. That increased quantities of greenhouse gases will lead to global warming is as solid a scientific prediction as can be found, but other things influence climate too, and some people try to escape confronting the consequences of our pumping more and more greenhouse gases into the atmosphere by supposing that something will come along miraculously to counteract them. Volcanos spew sulfates, as do our own smokestacks, and these reflect some sunlight back into space, particularly over the North Atlantic and Europe. But we can't assume that anything like this will counteract our longer-term flurry of carbon-dioxide emissions. Only the most naive gamblers bet against physics, and only the most irresponsible bet with their grandchildren's resources.

To the long list of predicted consequences of global warming—stronger storms, methane release, habitat changes, ice-sheet melting, rising seas, stronger El Niños, killer heat waves—we must now add an abrupt, catastrophic cooling. Whereas the familiar consequences of global warming will force expensive but gradual adjustments, the abrupt cooling promoted by man-made warming looks like a particularly efficient means of committing mass suicide.

We cannot avoid trouble by merely cutting down on our present warming trend, though that's an excellent place to start. Paleoclimatic records reveal that any notion we may once have had that the climate will remain the same unless pollution changes it is wishful thinking. Judging from the duration of the last warm period, we are probably near the end of the current one. Our goal must be to stabilize the climate in its favorable mode and ensure that enough equatorial heat continues to flow into the waters around Greenland and Norway. A stabilized climate must have a wide "comfort zone," and be able to survive the El Niños of the short term. We can design for that in computer models of climate, just as architects design earthquake-resistant skyscrapers. Implementing it might cost no more, in relative terms, than building a medieval cathedral. But we may not have centuries for acquiring wisdom, and it would be wise to compress our learning into the years immediately ahead. We have to discover what has made the climate of the past 8,000 years relatively stable, and then figure out how to prop it up.

> Those who will not reason
> Perish in the act:
> Those who will not act
> Perish for that reason.
>
> —W. H. Auden

Breaking the Global-Warming GRIDLOCK

by
DANIEL SAREWITZ
and ROGER PIELKE JR.

Both sides on the issue of greenhouse gases frame their arguments in terms of science, but each new scientific finding only raises new questions—dooming the debate to be a pointless spiral. It's time, the authors argue, for a radically new approach: if we took practical steps to reduce our vulnerability to today's weather, we would go a long way toward solving the problem of tomorrow's climate

IN THE LAST WEEK OF OCTOBER, 1998, HURRICANE Mitch stalled over Central America, dumping between three and six feet of rain within forty-eight hours, killing more than 10,000 people in landslides and floods, triggering a cholera epidemic, and virtually wiping out the economies of Honduras and Nicaragua. Several days later some 1,500 delegates, accompanied by thousands of advocates and media representatives, met in Buenos Aires at the fourth Conference of the Parties to the United Nations Framework Convention on Climate Change. Many at the conference pointed to Hurricane Mitch as a harbinger of the catastrophes that await us if we do not act immediately to reduce emissions of carbon dioxide and other so-called greenhouse gases. The delegates passed a resolution of "solidarity with Central America" in which they expressed concern "that global warming may be contributing to the worsening of weather" and urged "governments, ... and society in general, to continue their efforts to find permanent solutions to the factors which cause or may cause climate events." Children wandering bereft in the streets of Tegucigalpa became unwitting symbols of global warming.

But if Hurricane Mitch was a public-relations gift to environmentalists, it was also a stark demonstration of the failure of our current approach to protecting the environment. Disasters like Mitch are a present and historical reality, and they will become more common and more deadly regardless of global warming. Underlying the havoc in Central America were poverty, poor land-use practices, a degraded local environment, and inadequate emergency preparedness—conditions that will not be alleviated by reducing greenhouse-gas emissions.

At the heart of this dispiriting state of affairs is a vitriolic debate between those who advocate action to reduce global warming and those who oppose it. The controversy is informed by strong scientific evidence that the earth's surface has warmed over the past century. But the controversy, and the science, focus on the wrong issues, and distract attention from what needs to be done. The enormous scientific, political, and financial resources now aimed at the problem of global warming create the perfect conditions for international and domestic political gridlock, but they can have little effect on the root causes of global environmental degradation, or on the human suffering that so often accompanies it. Our goal is to move beyond the gridlock and stake out some common ground for political dialogue and effective action.

Framing the Issue

In politics everything depends on how an issue is framed: the terms of debate, the allocation of power and resources, the potential courses of action. The issue of global warming has been framed by a single question: Does the carbon dioxide emitted by industrialized societies threaten the earth's climate? On one side are the doomsayers, who foretell environmental disaster unless carbon-dioxide emissions are immediately reduced. On the other side are the cornucopians, who blindly insist that society can continue to pump billions of tons of greenhouse gases into the atmosphere with no ill effect, and that any effort to reduce emissions will stall the engines of industrialism that protect us from a Hobbesian wilderness. From our perspective, each group is operating within a frame that has little to do with the practical problem of how to protect the global environment in a world of six billion people (and counting). To understand why global-warming policy is a comprehensive and dangerous failure, therefore, we must begin with a look at how the issue came to be framed in this way. Two converging trends are implicated: the evolution of scientific research on the earth's climate, and the maturation of the modern environmental movement.

Since the beginning of the Industrial Revolution the combustion of fossil fuels—coal, oil, natural gas—has powered economic growth and also emitted great quantities of carbon dioxide and other greenhouse gases. More than a century ago the Swedish chemist Svante Arrhenius and the American geologist T. C. Chamberlin independently recognized that industrialization could lead to rising levels of carbon dioxide in the atmosphere, which might in turn raise the atmosphere's temperature by trapping solar radiation that would otherwise be reflected back into space—a "greenhouse effect" gone out of control. In the late 1950s the geophysicist Roger Revelle, arguing that the world was making itself the subject of a giant "geophysical experiment," worked to establish permanent stations for monitoring carbon-dioxide levels in the atmosphere. Monitoring documented what theory had predicted: atmospheric carbon dioxide was increasing.

In the United States the first high-level government mention of global warming was buried deep within a 1965 White House report on the nation's environmental problems. Throughout the 1960s and 1970s global warming—at that time typically referred to as "inadvertent modification of the atmosphere," and today embraced by the term "climate change"—remained an intriguing hypothesis that caught the attention of a few scientists but generated little concern among the public or environmentalists. Indeed, some climate researchers saw evidence for global cooling and a future ice age. In any case, the threat of nuclear war was sufficiently urgent, plausible, and horrific to crowd global warming off the catastrophe agenda.

Continued research, however, fortified the theory that fossil-fuel combustion could contribute to global warming. In 1977 the nonpartisan National Academy of Sciences issued a study called *Energy and Climate*, which carefully suggested that the possibility of global warming "should lead neither to panic nor to complacency." Rather, the study continued, it should "engender a lively sense of urgency in getting on with the work of illuminating the issues that have been identified and resolving the scientific uncertainties that remain." As is typical with National Academy studies, the primary recommendation was for more research.

In the early 1980s the carbon-dioxide problem received its first sustained attention in Congress, in the form of hearings organized by Representative Al Gore, who had become concerned about global warming when he took a college course with Roger Revelle, twelve years earlier. In 1983 the Environmental Protection Agency released a report detailing some of the possible threats posed by the anthropogenic, or human-caused, emission of carbon dioxide, but the Reagan Administration decisively downplayed the document. Two years later a prestigious international scientific conference in Villach, Austria, concluded that climate change deserved the attention of policymakers worldwide. The following year, at a Senate fact-finding hearing stimulated by the conference, Robert Watson, a climate scientist at NASA, testified, "Global warming is inevitable. It is only a question of the magnitude and the timing."

At that point global warming was only beginning to insinuate itself into the public consciousness. The defining event came in June of 1988, when another NASA climate scientist, James Hansen, told Congress with "ninety-nine percent confidence" that "the greenhouse effect has been detected, and it is changing our climate now." Hansen's proclamation made the front pages of major newspapers, ignited a firestorm of public debate, and elevated the carbon-dioxide problem to pre-eminence on the environmental agenda, where it remains to this day. Nothing had so galvanized the environmental community since the original Earth Day, eighteen years before.

Historically, the conservation and environmental movements have been rooted in values that celebrate the intrinsic worth of unspoiled landscape and propagate the idea that the human spirit is sustained through communion with nature. More than fifty years ago Aldo Leopold, perhaps the most important environmental voice of the twentieth century, wrote, "We face the question whether a still higher 'standard of living' is worth its cost in things natural, wild, and free. For us of the minority, ... the chance to find a pasque-flower is a right as inalienable as free speech." But when global warming appeared, environmentalists thought they had found a justification better than inalienable rights—they had found facts and rationality, and they fell head over heels in love with science.

Of course, modern environmentalists were already in the habit of calling on science to help advance their agenda. In 1967, for example, the Environmental Defense Fund was founded with the aim of using science to support environmental protection through litigation. But global warming was, and is, different. It exists as an environmental issue only because of science. People can't directly sense global warming, the way they can see a clear-cut forest or feel the sting of urban smog in their throats. It is not a discrete event, like an oil spill or a nuclear accident. Global warming is so abstract that scientists argue over how they would know if they actually observed it. Scientists go to great lengths to measure and derive something called the "global average temperature" at the earth's surface, and the total rise in this temperature over the past century—an increase of about six tenths of a degree Celsius as of 1998—does suggest warming. But people and ecosystems experience local and regional temperatures, not the global average. Furthermore, most of the possible effects of global warming are not apparent in the present; rather, scientists predict that they will occur decades or even centuries hence. Nor is it likely that scientists will ever be able to attribute any isolated event—a hurricane, a heat wave—to global warming.

A central tenet of environmentalism is that less human interference in nature is better than more. The imagination of the environmental community was ignited not by the observation that greenhouse-gas concentrations were increasing but by the scientific conclusion that the increase was caused by human beings. The Environmental Defense Fund, perhaps because of its explicitly scientific bent, was one of the first advocacy groups to make this connection. As early as 1984 its senior scientist, Michael Oppenheimer, wrote on the op-ed page of *The New York Times*,

> With unusual unanimity, scientists testified at a recent Senate hearing that using the atmosphere as a garbage dump is about to catch up with us on a global scale.... Carbon dioxide emissions from fossil fuel combustion and other "greenhouse" gases are throwing a blanket over the Earth. ... The sea level will rise as land ice melts and the ocean expands. Beaches will erode while wetlands will largely disappear.... Imagine life in a sweltering, smoggy New York without Long Island's beaches and you have glimpsed the world left to future generations.

Preserving tropical jungles and wetlands, protecting air and water quality, slowing global population growth—goals that had all been justified for independent reasons, often by independent organizations—could now be linked to a single fact, anthropogenic carbon-dioxide emissions, and advanced along a single political front, the effort to reduce those emissions. Protecting forests, for example, could help fight global warming because forests act as "sinks" that absorb carbon dioxide. Air pollution could be addressed in part by promoting the same clean-energy sources that would reduce carbon-dioxide emissions. Population growth needed to be controlled in order to reduce demand for fossil-fuel combustion. And the environmental community could reinvigorate its energy-conservation agenda, which had flagged since the early 1980s, when the effects of the second Arab oil shock wore off. Senator Timothy Wirth, of Colorado, spelled out the strategy in 1988: "What we've got to do in energy conservation is try to ride the global warming issue. Even if the theory of global warming is wrong, to have approached global warming as if it is real means energy conservation, so we will be doing the right thing anyway in terms of economic policy and environmental policy." A broad array of environmental groups and think tanks, including the Environmental Defense Fund, the Sierra Club, Greenpeace, the World Resources Institute, and the Union of Concerned Scientists, made reductions in carbon-dioxide emissions central to their agendas.

The moral problem seemed clear: human beings were causing the increase of carbon dioxide in the atmosphere. But the moral problem existed only because of a scientific fact—a fact that not only provided justification for doing many of the things that environmentalists wanted to do anyway but also dictated the overriding course of action: reduce carbon-dioxide emissions. Thus science was used to rationalize the moral imperative, unify the environmental agenda, and determine the political solution.

Research as Policy

THE summer of 1988 was stultifyingly hot even by Washington, D.C., standards, and the Mississippi River basin was suffering a catastrophic drought. Hansen's proclamation that the greenhouse effect was "changing our climate now" generated a level of public concern sufficient to catch the attention of many politicians. George Bush, who promised to be "the environmental President" and to counter "the greenhouse effect with the White House effect," was elected that November. Despite his campaign rhetoric, the new President was unprepared to offer policies that would curtail fossil-fuel production and consumption or impose economic costs for uncertain political gains. Bush's advisers recognized that support for scientific research offered the best solution politically, because it would give the appearance of action with minimal political risk.

With little debate the Republican Administration and the Democratic Congress in 1990 created the U.S. Global Change Research Program. The program's annual budget reached $1 billion in 1991 and $1.8 billion in 1995, making it one of the largest science initiatives ever undertaken by the U.S. government. Its goal, according to Bush Administration documents, was "to establish the scientific basis for national and international policymaking related to natural and human-induced changes in the global Earth system." A central scientific objective was to "support national and international policy-making by developing the ability to predict the nature and consequences of changes in the Earth system, particularly climate change." A decade and more than $16 billion later, scientific research remains the principal U.S. policy response to climate change.

Meanwhile, the marriage of environmentalism and science gave forth issue: diplomatic efforts to craft a global strategy to reduce carbon-dioxide emissions. Scientists, environmentalists, and government officials, in an attempt to replicate the apparently successful international response to stratospheric-ozone depletion that was mounted in the mid-1980s, created an institutional structure aimed at formalizing the connection between science and political action. The Intergovernmental Panel on Climate Change was established through the United Nations, to provide snapshots of the evolving state of scientific understanding. The IPCC issued major assessments in 1990 and 1996; a third is due early next year. These assessments provide the basis for action under a complementary mechanism, the United Nations Framework Convention on Climate Change. Signed by 154 nations at the 1992 "Earth Summit" in Rio de Janeiro, the convention calls for voluntary reductions in carbon-dioxide emissions. It came into force as an international treaty in March of 1994, and has been ratified by 181 nations. Signatories continue to meet in periodic Conferences of the Parties, of which the most significant to date occurred in Kyoto in 1997, when binding emissions reductions for industrialized countries were proposed under an agreement called the Kyoto Protocol.

The IPCC defines climate change as any sort of change in the earth's climate, no matter what the cause. But the Framework Convention restricts its definition to changes that result from the anthropogenic emission of greenhouse gases. This restriction has profound implications for the framing of the issue. It makes all action under the convention hostage to the ability of scientists not just to document global warming but to attribute it to human causes. An apparently simple question, Are we causing global warming or aren't we?, has become the obsessional focus of science—and of policy.

Finally, if the reduction of carbon-dioxide emissions is an organizing principle for environmentalists, scientists, and environmental-policy makers, it is also an organizing principle for all those whose interests might be threatened by such a reduction. It's easy to be glib about who they might be—greedy oil and coal companies, the rapacious logging industry, recalcitrant automobile manufacturers, corrupt foreign dictatorships—and easy as well to document the excesses and absurdities propagated by some representatives of these groups. Consider, for example, the Greening Earth Society, which "promotes the optimistic scientific view that CO_2 is beneficial to humankind and all of nature," and happens to be funded by a coalition of coal-burning utility companies. One of the society's 1999 press releases reported that "there will only be sufficient food for the world's projected population in 2050 if atmospheric concentrations of carbon dioxide are permitted to increase, unchecked." Of course, neither side of the debate has a lock on excess or distortion. The point is simply that the climate-change problem has been framed in a way that catalyzes a determined and powerful opposition.

The Problem With Predictions

WHEN anthropogenic carbon-dioxide emissions became the defining fact for global environmentalism, scientific uncertainty about the causes and consequences of global warming emerged as the apparent central obstacle to action. As we have seen, the Bush Administration justified its huge climate-research initiative explicitly in terms of the need to reduce uncertainty before taking action. Al Gore, by then a senator, agreed, explaining that "more research and better research and better targeted research is absolutely essential if we are going to eliminate the remaining areas of uncertainty and build the broader and stronger political consensus necessary for the unprecedented actions required to address this problem." Thus did a Republican Administration and a Democratic Congress—one side looking for reasons to do nothing, the other seeking justification for action—converge on the need for more research.

How certain do we need to be before we take action? The answer depends, of course, on where our interests lie. Environmentalists can tolerate a good deal more uncertainty on this issue than can, say, the executives of utility or automobile companies. Science is unlikely to overcome such a divergence in interests. After all, science is not a fact or even a set of facts; rather, it is a process of inquiry that generates more questions than answers. The rise in anthropogenic greenhouse-gas emissions, once it was scientifically established, simply pointed to other questions. How rapidly might carbon-dioxide levels rise in the future? How might climate respond to this rise? What might be the effects of that response? Such questions are inestimably complex, their answers infinitely contestable and always uncertain, their implications for human action highly dependent on values and interests.

Having wedded themselves to science, environmentalists must now cleave to it through thick and thin. When research results do not support their cause, or are simply uncertain, they cannot resort to values-based arguments, because their political opponents can portray such arguments as an opportunistic abandonment of rationality. Environmentalists have tried to get out of this bind by invoking the "precautionary principle"— a dandified version of "better safe than sorry"—to advance the idea that action in the presence of uncertainty is justified if potential harm is great. Thus uncertainty itself becomes an argument for action. But nothing is gained by this tactic either, because just as attitudes toward uncertainty are rooted in individual values and interests, so are attitudes toward potential harm.

Charged by the Framework Convention to search for proof of harm, scientists have turned to computer models of the atmosphere and the oceans, called general circulation models, or GCMs. Carbon-dioxide levels and atmospheric temperatures are measures of the physical state of the atmosphere. GCMs, in contrast, are mathematical representations that scientists use to try to understand past climate conditions and predict future ones. With GCMs scientists seek to explore how climate might respond under different influences—for example, different rates of carbon-dioxide increase. GCMs have calculated global average temperatures for the past century that closely match actual surface-temperature records; this gives climate modelers some confidence that they understand how climate behaves.

Computer models are a bit like Aladdin's lamp—what comes out is very seductive, but few are privy to what goes on inside. Even the most complex models, however, have one crucial quality that non-experts can easily understand: their accuracy can be fully evaluated only after seeing what happens in the real world over time. In other words, predictions of how climate will behave in the future cannot be proved accurate today. There are other fundamental problems with relying on GCMs. The ability of many models to reproduce temperature records may in part reflect the fact that the scientists who designed them already "knew the answer." As John Firor, a former director of the National Center for Atmospheric Research, has observed, climate models "are made by humans who tend to shape or use their models in ways that mirror their own notion of what a desirable outcome would be." Although various models can reproduce past temperature records, and yield similar predictions of future temperatures, they are unable to replicate other observed aspects of climate, such as cloud behavior and atmospheric temperature, and they diverge widely in predicting specific regional climate phenomena, such as precipitation and the frequency of extreme weather events. Moreover, it is simply not possible to know far in advance if the models agree on future temperature because they are similarly right or similarly wrong.

In spite of such pitfalls, a fundamental assumption of both U.S. climate policy and the UN Framework Convention is that increasingly sophisticated models, run on faster computers and supported by more data, will yield predictions that can resolve political disputes and guide action. The promise of better predictions is irresistible to champions of carbon-dioxide reduction, who, after all, must base their advocacy on the claim that anthropogenic greenhouse-gas emissions will be harmful in the future. But regardless of the sophistication of such predictions, new findings will almost inevitably be accompanied by new uncertainties— that's the nature of science—and may therefore act to fuel, rather than to quench, political debate. Our own prediction is that increasingly complex mathematical models that delve ever more deeply into the intricacies and the uncertainties of climate will only hinder political action.

An example of how more scientific research fuels political debate came in 1998, when a group of prominent researchers released the results of a model analyzing carbon-dioxide absorption in North America. Their controversial findings, published in the prestigious journal *Science,* suggested that the amount of carbon dioxide absorbed by U.S. forests might be greater than the amount emitted by the nation's fossil-fuel combustion. This conclusion has two astonishing implications. First, the United States—the world's most profligate energy consumer—may not be directly contributing to rising atmospheric levels of carbon dioxide. Second, the atmosphere seems to be benefiting from young forests in the eastern United States that are particularly efficient at absorbing carbon dioxide. But these young forests exist only because old-growth forests were clear-cut in the eighteenth and nineteenth centuries to make way for farms that were later abandoned in favor of larger, more efficient midwestern farms. In other words, the possibility that the United States is a net carbon-dioxide sink does not reflect efforts to protect the environment; on the contrary, it reflects a history of deforestation and development.

Needless to say, these results quickly made their way into the political arena. At a hearing of the House Resources Committee, Representative John E. Peterson, of Pennsylvania, a Republican, asserted, "There are recent studies that show that in the Northeast, where we have continued to cut timber, and have a regenerating, younger forest, that the greenhouse gases are less when they leave the forest.... So a young, growing, vibrant forest is a whole lot better for clean air than an old dying forest." George Frampton, the director of the White House Council on Environmental Quality, countered, "The science on this needs a lot of work.... We need more money for scientific research to undergird that point of view." How quickly the tables can turn: here was a conservative politician wielding (albeit with limited coherence) the latest scientific results to justify logging old-growth forests in the name of battling global warming, while a Clinton Administration official backpedaled in the manner more typically adopted by opponents of action on climate change—invoking the need for more research.

That's a problem with science—it can turn around and bite you. An even more surprising result has recently emerged from the study of Antarctic glaciers. A strong argument in favor of carbon-dioxide reduction has been the possibility that if temperatures rise owing to greenhouse-gas emissions, glaciers will melt, the sea level will rise, and populous coastal zones all over the world will be inundated. The West Antarctic Ice Sheet has been a subject of particular concern, both because of evidence that it is now retreating and because of geologic studies showing that it underwent catastrophic collapse at least once in the past million years or so. "Behind the reasoned scientific estimates," Greenpeace warns, "lies the possibility of . . . the potential catastrophe of a six metre rise in sea level." But recent research from Antarctica shows that this ice sheet has been melting for thousands of years. Sealevel rise is a problem, but anthropogenic global warming is not the only culprit, and reducing emissions cannot be the only solution.

To make matters more difficult, some phenomena, especially those involving human behavior, are intrinsically unpredictable. Any calculation of future anthropogenic global warming must include an estimate of rates of fossil-fuel combustion in the coming decades. This means that scientists must be able to predict not only the amounts of coal, oil, and natural gas that will be consumed but also changes in the mixture of fossil fuels and other energy sources, such as nuclear, hydro-electric, and solar. These predictions rest on interdependent factors that include energy policies and prices, rates of economic growth, patterns of industrialization and technological innovation, changes in population, and even wars and other geopolitical events. Scientists have no history of being able to predict any of these things. For example, their inability to issue accurate population projections is "one of the best-kept secrets of demography," according to Joel Cohen, the director of the Laboratory of Populations at Rockefeller University. "Most professional demographers no longer believe they can predict precisely the future growth rate, size, composition and spatial distribution of populations," Cohen has observed.

Predicting the human influence on climate also requires an understanding of how climate behaved "normally," before there was any such influence. But what are normal climate patterns? In the absence of human influence, how stationary is climate? To answer such questions, researchers must document and explain the behavior of the pre-industrial climate, and they must also determine how the climate would have behaved over the past two centuries had human beings not been changing the composition of the atmosphere. However, despite the billions spent so far on climate research, Kevin Trenberth, a senior scientist at the National Center for Atmospheric Research, told the *Chicago Tribune* last year, "This may be a shock to many people who assume that we do know adequately what's going on with the climate, but we don't." The National Academy of Sciences reported last year that "deficiencies in the accuracy, quality, and continuity of the [climate] records . . . place serious limitations on the confidence" of research results.

If the normal climate is non-stationary, then the task of identifying the human fingerprint in global climate change becomes immeasurably more difficult. And the idea of a naturally stationary climate may well be chimerical. Climate has changed often and dramatically in the recent past. In the 1940s and 1950s, for example, the East Coast was hammered by a spate of powerful hurricanes, whereas in the 1970s and 1980s hurricanes were much less common. What may appear to be "abnormal" hurricane activity in recent years is abnormal only in relation to this previous quiet period. As far as the ancient climate goes, paleoclimatologists have found evidence of rapid change, even over periods as short as several years. Numerous influences could account for these changes. Ash spewed high into the atmosphere by large volcanoes can reflect solar radiation back into space and result in short-term cooling, as occurred after the 1991 eruption of Mount Pinatubo. Variations in the energy emitted by the sun also affect climate, in ways that are not yet fully understood. Global ocean currents, which move huge volumes of warm and cold water around the world and have a profound influence on climate, can speed up, slow down, and maybe even die out over very short periods of time—perhaps less than a decade. Were the Gulf Stream to shut down, the climate of Great Britain could come to resemble that of Labrador.

Finally, human beings have been changing the surface of the earth for millennia. Scientists increasingly realize that deforestation, agriculture, irrigation, urbanization, and other human activities can lead to major changes in climate on a regional or perhaps even a global scale. Thomas Stohlgren, of the U.S. Geological Survey, has written, "The effects of land use practices on regional climate may overshadow larger-scale temperature changes commonly associated with observed increases in carbon dioxide." The idea that climate may constantly be changing for a variety of reasons does not itself undercut the possibility that anthropogenic carbon dioxide could seriously affect the global climate, but it does confound scientific efforts to predict the consequences of carbon-dioxide emissions.

The Other 80 Percent

IF predicting how climate will change is difficult and uncertain, predicting how society will be affected by a changing climate—especially at the local, regional, and national levels, where decision-making takes place—is immeasurably more so. And predicting the impact on climate of reducing carbon-dioxide emissions is so uncertain as to be meaningless. What we do know about climate change suggests that there will be winners and losers, with some areas and nations potentially benefiting from, say, longer growing seasons or more rain, and others suffering from more flooding or drought. But politicians have no way to accurately calibrate the effects—human and economic—of global warming, or the benefits of reducing carbon-dioxide emissions.

Imagine yourself a leading policymaker in a poor, overpopulated, undernourished nation with severe environmental problems. What would it take to get you worried about global warming? You would need to know not just that global warming would make the conditions in your country worse but also that any of the scarce resources you applied to reducing carbon-dioxide emissions would lead to more benefits than if they were applied in another area, such as industrial development or housing construction. Such knowledge is simply unavailable. But you do know that investing in industrial development or better housing would lead to concrete political, economic, and social benefits.

More specifically, suppose that many people in your country live in shacks on a river's floodplain. Floodplains are created and sustained by repeated flooding, so floods are certain to occur in the future, regardless of global warming. Given a choice between building new houses away from the floodplain and converting power plants from cheap local coal to costlier imported fuels, what would you do? New houses would ensure that lives and homes would be saved; a new power plant would reduce carbon-dioxide emissions but leave people vulnerable to floods. In the developing world the carbon-dioxide problem pales alongside immediate environmental and developmental problems. The *China Daily* reported during the 1997 Kyoto Conference:

> The United States . . . and other nations made the irresponsible demand . . . that the developing countries should make commitments to limiting greenhouse gas emissions. . . . As a developing country, China has 60 million poverty-stricken people and China's per capita gas emissions are only one-seventh of the average amount of more developed countries. Ending poverty and developing the economy must still top the agenda of [the] Chinese government.

For the most part, the perspectives of those in the developing world—about 80 percent of the planet's population—have been left outside the frame of the climate-change discussion. This is hardly surprising, considering that the frame was defined mainly by environmentalists and scientists in affluent nations. Developing nations, meanwhile, have quite reasonably refused to agree to the targets for carbon-dioxide reduction set under the Kyoto Protocol. The result may feel like a moral victory to some environmentalists, who reason that industrialized countries, which caused the problem to begin with, should shoulder the primary responsibility for solving it. But the victory is hollow, because most future emissions increases will come from the developing world. In affluent nations almost everyone already owns a full complement of energy-consuming devices. Beyond a certain point increases in income do not result in proportional increases in energy consumption; people simply trade in the old model for a new and perhaps more efficient one. If present trends continue, emissions from the developing world are likely to exceed those from the industrialized nations within the next decade or so.

Twelve years after carbon dioxide became the central obsession of global environmental science and politics, we face the following two realities:

First, atmospheric carbon-dioxide levels will continue to increase. The Kyoto Protocol, which represents the world's best attempt to confront the issue, calls for industrialized nations to reduce their emissions below 1990 levels by the end of this decade. Political and technical realities suggest that not even this modest goal will be achieved. To date, although eighty-four nations have signed the Kyoto Protocol, only twenty-two nations—half of them islands, and none of them major carbon-dioxide emitters—have ratified it. The United States Senate, by a vote of 95-0 in July of 1997, indicated that it would not ratify any climate treaty that lacked provisions requiring developing nations to reduce their emissions. The only nations likely to achieve the emissions commitments set under Kyoto are those, like Russia and Ukraine, whose economies are in ruins. And even successful implementation of the treaty would not halt the progressive increase in global carbon-dioxide emissions.

Second, even if greenhouse-gas emissions could somehow be rolled back to pre-industrial levels, the impacts of climate on society and the environment would continue to increase. Climate affects the world not just through phenomena such as hurricanes and droughts but also because of societal and environmental vulnerability to such phenomena. The horrific toll of Hurricane Mitch reflected not an unprecedented climatic event but a level of exposure typical in developing countries where dense and rapidly increasing populations live in environmentally degraded conditions. Similar conditions underlay more-recent disasters in Venezuela and Mozambique.

If these observations are correct, and we believe they are essentially indisputable, then framing the problem of global warming in terms of carbon-dioxide reduction is a political, environmental, and social dead end. We are not suggesting that humanity can with impunity emit billions of tons of carbon dioxide into the atmosphere each year, or that reducing those emissions is not a good idea. Nor are we making the nihilistic point that since climate undergoes changes for a variety of reasons, there is no need to worry about additional changes imposed by human beings. Rather, we are arguing that environmentalists and scientists, in focusing their own, increasingly congruent interests on carbon-dioxide emissions, have framed the problem of global environmental protection in a way that can offer no realistic prospect of a solution.

Redrawing the Frame

LOCAL weather is the day-to-day manifestation of global climate. Weather is what we experience, and lately there has been plenty to experience. In recent decades human, economic, and environmental losses from disasters related to weather have increased dramatically. Insurance-industry data show that insured losses from weather have been rising steadily. A 1999 study by the German firm Munich Reinsurance Company compared the 1960s with the 1990s and concluded that "the number of great natural catastrophes increased by a factor of three, with economic losses—taking into account the effects of inflation—increasing by a factor of more than eight and insured losses by a factor of no less than sixteen." And yet scientists have been unable to observe a global increase in the number or the severity of extreme weather events. In 1996 the IPCC concluded, "There is no evidence that extreme weather events, or climate variability, has increased, in a global sense, through the 20th century, although data and analyses are poor and not comprehensive."

What has unequivocally increased is society's vulnerability to weather. At the beginning of the twentieth century the earth's population was about 1.6 billion people; today it is about six billion people. Almost four times as many people are exposed to weather today as were a century ago. And this increase has, of course, been accompanied by enormous increases in economic activity, development, infrastructure, and interdependence. In the past fifty years, for example, Florida's population rose fivefold; 80 percent of this burgeoning population lives within twenty miles of the coast. The great Miami hurricane of 1926 made landfall over a small, relatively poor community and caused about $76 million worth of damage (in inflation-adjusted dollars). Today a storm of similar magnitude would strike a sprawling, affluent metropolitan area of two million people, and could cause more than $80 billion worth of damage. The increase in vulnerability is far more dramatic in the developing world, where in an average year tens of thousands of people die in weather-related disasters. According to the *World Disasters Report 1999,* 80 million people were made homeless by weather-related disasters from 1988 to 1997. As the population and vulnerability of the developing world continue to rise, such numbers will continue to rise as well, with or without global warming.

Environmental vulnerability is also on the rise. The connections between weather impacts and environmental quality are immediate and obvious—much more so than the connections between global warming and environmental quality. Deforestation, the destruction of wetlands, and the development of fragile coastlines can greatly magnify flooding; floods, in turn, can mobilize toxic chemicals in soil and storage facilities and cause devastating pollution of water sources and harm to wildlife. Poor agricultural, forest-management, and grazing practices can exacerbate the effects of drought, amplify soil erosion, and promote the spread of wildfires. Damage to the environment due to deforestation directly contributed to the devastation wrought by Hurricane Mitch, as denuded hillsides washed away in catastrophic landslides, and excessive development along unmanaged floodplains put large numbers of people in harm's way.

Our view of climate and the environment draws on people's direct experience and speaks to widely shared values. It therefore has an emotional and moral impact that can translate into action. This view is framed by four precepts. First, the impacts of weather and climate are a serious threat to human welfare in the present and are likely to get worse in the future. Second, the only way to reduce these impacts is to reduce societal vulnerability to them. Third, reducing vulnerability can be achieved most effectively by encouraging democracy, raising standards of living, and improving environmental quality in the developing world. Fourth, such changes offer the best prospects not only for adapting to a capricious climate but also for reducing carbon-dioxide emissions.

The implicit moral imperative is not to prevent human disruption of the environment but to ameliorate the social and political conditions that lead people to behave in environmentally disruptive ways. This is a critical distinction—and one that environmentalists and scientists embroiled in the global-warming debate have so far failed to make.

To begin with, any global effort to reduce vulnerability to weather and climate must address the environmental conditions in developing nations. Poor land-use and natural-resource-management practices are, of course, a reflection of poverty, but they are also caused by government policies, particularly those that encourage unsustainable environmental activities. William Ascher, a political scientist at Duke University, has observed that such policies typically do not arise out of ignorance or lack of options but reflect conscious tradeoffs made by government officials faced with many competing priorities and political pressures. Nations, even poor ones, have choices. It was not inevitable, for example, that Indonesia would promote the disastrous exploitation of its forests by granting subsidized logging concessions to military and business leaders. This was the policy of an autocratic government seeking to manipulate powerful sectors of society. In the absence of open, democratically responsive institutions, Indonesian leaders were not accountable for the costs that the public might bear, such as increased vulnerability to floods, landslides, soil erosion, drought, and fire. Promoting democratic institutions in developing nations could be the most important item on an agenda aimed at protecting the global environment and reducing vulnerability to climate. Environmental groups concerned about the consequences of climate change ought to consider reorienting their priorities accordingly.

Such long-term efforts must be accompanied by activities with a shorter-term payoff. An obvious first step would be to correct some of the imbalances created by the obsession with carbon dioxide. For example, the U.S. Agency for International Development has allocated $1 billion over five years to help developing nations quantify, monitor, and reduce greenhouse-gas emissions, but is spending less than a tenth of that amount on programs to prepare for and prevent disasters. These priorities should be rearranged. Similarly, the United Nations' International Strategy for Disaster Reduction is a relatively low-level effort that should be elevated to a status comparable to that of the Framework Convention on Climate Change.

Intellectual and financial resources are also poorly allocated in the realm of science, with research focused disproportionately on understanding and predicting basic climatic processes. Such research has yielded much interesting information about the global climate system. But little priority is given to generating and disseminating knowledge that people and communities can use to reduce their vulnerability to climate and extreme weather events. For example, researchers have made impressive strides in anticipating the impacts of some relatively short-term climatic phenomena, notably El Niño and La Niña. If these advances were accompanied by progress in monitoring weather, identifying vulnerable regions and populations, and communicating useful information, we would begin to reduce the toll exacted by weather and climate all over the world.

A powerful international mechanism for moving forward already exists in the Framework Convention on Climate Change. The language of the treaty offers sufficient flexibility for new priorities. The text states that signatory nations have an obligation to "cooperate in preparing for adaptation to the impacts of climate change [and to] develop and elaborate appropriate and integrated plans for coastal zone management, water resources and agriculture, and for the protection and rehabilitation of areas . . . affected by drought and desertification, as well as floods."

The idea of improving our adaptation to weather and climate has been taboo in many circles, including the realms of international negotiation and political debate. "Do we have so much faith in our own adaptability that we will risk destroying the integrity of the entire global ecological system?" Vice President Gore asked in his book *Earth in the Balance* (1992). "Believing that we can adapt to just about anything is ultimately a kind of laziness, an arrogant faith in our ability to react in time to save our skin." For environmentalists, adaptation represents a capitulation to the momentum of human interference in nature. For their opponents, putting adaptation on the table would mean acknowledging the reality of global warming. And for scientists, focusing on adaptation would call into question the billions of tax dollars devoted to research and technology centered on climate processes, models, and predictions.

Yet there is a huge potential constituency for efforts focused on adaptation: everyone who is in any way subject to the effects of weather. Reframing the climate problem could mobilize this constituency and revitalize the Framework Convention. The revitalization could concentrate on coordinating disaster relief, debt relief, and development assistance, and on generating and providing information on climate that participating countries could use in order to reduce their vulnerability.

An opportunity to advance the cause of adaptation is on the horizon. The U.S. Global Change Research Program is now finishing its report on the National Assessment of the Potential Consequences of Climate Variability and Change. The draft includes examples from around the United States of why a greater focus on adaptation to climate makes sense. But it remains to be seen if the report will redefine the terms of the climate debate, or if it will simply become fodder in the battle over carbon-dioxide emissions.

Finally, efforts to reduce carbon-dioxide emissions need not be abandoned. The Framework Convention and its offshoots also offer a promising mechanism for promoting the diffusion of energy-efficient technologies that would reduce emissions. Both the convention and the Kyoto Protocol call on industrialized nations to share new energy technologies with the developing world. But because these provisions are coupled to carbon-dioxide-reduction mandates, they are trapped in the political gridlock. They should be liberated, promoted independently on the basis of their intrinsic environmental and economic benefits, and advanced through innovative funding mechanisms. For example, as the United Nations Development Programme has suggested, research into renewable-energy technologies for poor countries could be supported in part by a modest levy on patents registered under the World Intellectual Property Organization. Such ideas should be far less divisive than energy policies advanced on the back of the global-warming agenda.

As an organizing principle for political action, vulnerability to weather and climate offers everything that global warming does not: a clear, uncontroversial story rooted in concrete human experience, observable in the present, and definable in terms of unambiguous and widely shared human values, such as the fundamental rights to a secure shelter, a safe community, and a sustainable environment. In this light, efforts to blame global warming for extreme weather events seem maddeningly perverse—as if to say that those who died in Hurricane Mitch were symbols of the profligacy of industrialized society, rather than victims of poverty and the vulnerability it creates.

Such perversity shows just how morally and politically dangerous it can be to elevate science above human values. In the global-warming debate the logic behind public discourse and political action has been precisely backwards. Environmental prospects for the coming century depend far less on our strategies for reducing carbon-dioxide emissions than on our determination and ability to reduce human vulnerability to weather and climate.

Daniel Sarewitz is a research scholar at Columbia University's Center for Science, Policy and Outcomes. **Roger Pielke Jr.** is a scientist with the Environmental and Societal Impacts Group at the National Center for Atmospheric Research. They are the editors, with Radford Byerly Jr., of *Prediction: Science, Decision Making, and the Future of Nature* (2000).

Beyond the Valley of
THE DAMMED

A strange alliance of fish lovers, tree huggers, and bureaucrats say what went up must come down

BY BRUCE BARCOTT

By God but we built some dams! We backed up the Kennebec in Maine and the Neuse in North Carolina and a hundred creeks and streams that once ran free. We stopped the Colorado with the Hoover, high as 35 houses, and because it pleased us we kept damming and diverting the river until it no longer reached the sea. We dammed our way out of the Great Depression with the Columbia's Grand Coulee; a dam so immense you had to borrow another fellow's mind because yours alone wasn't big enough to wrap around it. Then we cleaved the Missouri with a bigger one still, the Fort Peck Dam, a jaw dropper so outsized they put it on the cover of the first issue of *Life*. We turned the Tennessee, the Columbia, and the Snake from continental arteries into still bathtubs. We dammed the Clearwater, the Boise, the Santiam, the Deschutes, the Skagit, the Willamette, and the McKenzie. We dammed Crystal River and Muddy Creek, the Little River and the Rio-Grande. We dammed the Minnewawa and the Minnesota and we dammed the Kalamazoo. We dammed the Swift and we dammed the Dead.

One day we looked up and saw 75,000 dams impounding more than half a million miles of river. We looked down and saw rivers scrubbed free of salmon and sturgeon and shad. Cold rivers ran warm, warm rivers ran cold, and fertile muddy banks turned barren.

And that's when we stopped talking about dams as instruments of holy progress and started talking about blowing them out of the water.

Surrounded by a small crowd, Secretary of the Interior Bruce Babbitt stood atop McPherrin Dam, on Butte Creek, not far from Chico, California, in the hundred-degree heat of the Sacramento Valley. The constituencies represented—farmers, wildlife conservationists, state fish and game officials, irrigation managers—had been wrangling over every drop of water in this naturally arid basin for most of a century. On this day, however, amity reigned.

With CNN cameras rolling, Babbitt hoisted a sledgehammer above his head and—with "evident glee," as one reporter later noted—brought this tool of destruction down upon the dam. Golf claps all around.

The secretary's hammer strike in July 1998 marked the beginning of the end for that ugly concrete plug and three other Butte Creek irrigation dams. All were coming out to encourage the return of spring-run chinook salmon, blocked from their natural spawning grounds for more than 75 years. Babbitt then flew to Medford, Oregon, and took a swing at 30-year-old Jackson Street Dam on Bear Creek. Last year alone, Babbitt cracked the concrete at four dams on Wisconsin's Menominee River and two dams on Elwha River in Washington state; at Quaker Neck Dam on North Carolina's Neuse River; and at 160-year-old Edwards Dam on the Kennebec in Maine.

By any reckoning, this was a weird inversion of that natural order. Interior secretaries are supposed to christen dams, not smash them. Sixty years ago, President Franklin D. Roosevelt and his interior secretary, Harold Ickes, toured the West to dedicate four of the largest dams in the history of civilization. Since 1994, Babbitt, who knows his history, has been following in their footsteps, but this secretary is preaching the gospel of

12. Beyond the Valley of the Dammed

What we know now that we didn't know then is that **a river isn't a water pipe.**

dam-going-away. "America overshot the mark in our dam-building frenzy," he told the Ecological Society of America. "The public is now learning that we have paid a steadily accumulating price for these projects. . . . We did not build them for religious purposes and they do not consecrate our values. Dams do, in fact, outlive their function. When they do, some should go."

Many dams continue, of course, to be invaluable pollution-free power plants. Hydroelectric dams provide 10 percent of the nation's electricity (and half of our renewable energy). In the Northwest, dams account for 75 percent of the region's power and bestow the lowest electrical rates in the nation. In the past the public was encouraged to believe that hydropower was almost free; but as Babbitt has been pointing out, the real costs can be enormous.

What we know now that we didn't know in 1938 is that a river isn't a water pipe. Dam a river and it will drop most of the sediment it carries into a still reservoir, trapping ecologically valuable debris such as branches, wood particles, and gravel. The sediment may be mixed with more and more pollutants—toxic chemicals leaching from abandoned mines, for example, or naturally occurring but dangerous heavy metals. Once the water passes through the dam it continues to scour, but it can't replace what it removes with material from upstream. A dammed river is sometimes called a "hungry" river, one that eats its bed and banks. Riverbeds and banks may turn into cobblestone streets, large stones cemented in by the ultrafine silt that passes through the dams. Biologists call this "armoring."

Naturally cold rivers may run warm after the sun heats water trapped in the reservoir; naturally warm rivers may run cold if their downstream flow is drawn from the bottom of deep reservoirs. Fish adapted to cold water won't survive in warm water, and vice versa.

As the toll on wild rivers became more glaringly evident in recent decades, opposition to dams started to go mainstream. By the 1990s, conservation groups, fishing organizations, and other river lovers began to call for actions that had once been supported only by environmental extremists and radical groups like Earth First! Driven by changing economics, environmental law, and most of all the specter of vanishing fish, government policy makers began echoing the conservationists. And then Bruce Babbitt, perhaps sensing the inevitable tide of history, began to support decommissioning as well.

So far, only small dams have been removed. Babbitt may chip away at all the little dams he wants, but when it comes to ripping major federal hydropower projects out of Western rivers, that's when the politics get national and nasty. Twenty-two years ago, when President Jimmy Carter suggested pulling the plug on several grand dam projects, Western senators and representatives politically crucified him. Although dam opponents have much stronger scientific and economic arguments on their side in 1999, the coming dam battles are apt to be just as nasty.

Consider the Snake River, where a major confrontation looms over four federal hydropower dams near the Washington-Idaho state line. When I asked Babbitt about the Snake last fall, he almost seemed to be itching for his hammer. "The escalating debate over dams is going to focus in the coming months on the Snake River," he declared. "We're now face to face with this question: Do the people of this country place more value on Snake River salmon or on those four dams? The scientific studies are making it clear that you can't have both."

Brave talk—but only a couple of weeks later, after a bruising budget skirmish with congressional dam proponents who accused him of planning to tear down dams across the Northwest, Babbitt sounded like a man who had just learned a sobering lesson in the treacherous politics of dams. The chastened interior secretary assured the public that "I have never advocated, and do not advocate, the removal of dams on the main stem of the Columbia-Snake river system."

Showdown on the Snake

Lewiston, Idaho, sits at the confluence of the Snake and Clearwater Rivers. It's a quiet place of 33,000 solid citizens, laid out like a lot of towns these days: One main road leads into the dying downtown core, the other to a thriving strip of Wal-Marts, gas stations, and fast-food greaseries. When Lewis (hence the name) and Clark floated through here in 1805, they complained about the river rapids—"Several of them verry bad," the spelling-challenged Clark scrawled in his journal. Further downriver, where the Snake meets the Columbia, the explorers were amazed to see the local Indians catching and drying incredible numbers of coho salmon headed upriver to spawn.

The river still flows, though it's been dammed into a lake for nearly 150 miles. Between 1962 and 1975, four federal hydroelectric projects were built on the river by the Army Corps of Engineers: Ice Harbor Dam, Lower Monumental Dam, Little Goose Dam, and Lower Granite Dam. The dams added to the regional power supply, but more crucially, they turned the Snake from a whitewater roller-coaster into a navigable waterway. The surrounding wheat farmers could now ship their grain on barges to Portland, Oregon, at half the cost of overland transport, and other industries also grew to depend on this cheap highway to the sea.

Like all dams, however, they were hell on the river and its fish—the chinook, coho, sockeye, and steelhead. True, some salmon species still run up the river to spawn, but by the early 1990s the fish count had dwindled from 5 million to less than

20,000. The Snake River coho have completely disappeared, and the sockeye are nearing extinction.

In and around Lewiston, the two conflicting interests—livelihoods that depend on the dams on the one side, the fate of the fish on the other—mean that just about everyone is either a friend of the dams or a breacher. The Snake is the dam-breaching movement's first major test case, but it is also the place where dam defenders plan to make their stand. Most important, depending in part on the results of a study due later this year, the lower Snake could become the place where the government orders the first decommissioning of several big dams.

In the forefront of those who hope this happens is Charlie Ray, an oxymoron of a good ol' boy environmentalist whose booming Tennessee-bred baritone and sandy hair lend him the aspect of Nashville Network host. Ray makes his living as head of salmon and steelhead programs for Idaho Rivers United, a conservationist group that has been raising a fuss about free-flowing rivers since 1991. At heart he's not a tree hugger, but a steelhead junkie: "You hook a steelhead, man, you got 10,000 years of survival instinct on the end of that line."

Despite Ray's bluff good cheer, it's not easy being a breacher in Lewiston. Wheat farming still drives a big part of the local economy, and the pro-dam forces predict that breaching would lead to financial ruin. Lining up behind the dam defenders are Lewiston's twin pillars of industry: the Potlatch Corporation and the Port of Lewiston. Potlatch, one of the country's largest paper producers, operates its flagship pulp and paper mill in Lewiston, employing 2,300 people. Potlatch executives will tell you the company wants the dams mainly to protect the town's economy, but local environmentalists say the mill would find it more difficult to discharge warm effluent into a free-flowing, shallow river.

Potlatch provides Charlie Ray with a worthy foil in company spokesman Frank Carroll, who was hired after spending 17 years working the media for the U.S. Forest Service. Frankie and Charlie have been known to scrap. At an antibreaching rally in Lewiston last September, Carroll stood off-camera watching Ray being interviewed by a local TV reporter. Fed up with hearing Ray's spin, Carroll started shouting "Bullshit, Charlie, that's bullshit!" while the video rolled. Ray's nothing more than a "paid operative," Carroll says. Ray's reaction: "Yeah, like Frankie's not."

"A lot of people are trying to trivialize the social and economic issues," Carroll says, "trying to tell us the lives people have here don't count, that we'll open up a big bait shop and put everyone to work hooking worms. We resent that. Right now, there's a blanket of prosperity that lies across this whole region, and that prosperity is due to the river in its current state—to its transportation."

Ever since the dams started going up along the Snake River, biologists and engineers have been trying to revive the rapidly declining salmon runs. Their schemes include fish ladders, hatcheries, and a bizarre program in which young smolts are captured and shipped downriver to the sea in barges. By the late 1980s, it was clear that nothing was working; the fish runs continued to plummet. In 1990, the Shoshone-Bannock Indians, who traditionally fished the Snake's sockeye run, successfully petitioned the National Marine Fisheries Service to list the fish as endangered. Every salmon species in the Snake River is now officially threatened or endangered, which means the agencies that control the river must deal with all kinds of costly regulations.

In 1995, under pressure from the federal courts, the National Marine Fisheries Service and the Army Corps of Engineers (which continues to operate the dams) agreed to launch a four-year study of the four lower Snake River dams. In tandem with the Fisheries Service, the Corps made a bombshell announcement. The study would consider three options: maintain the status quo, turbocharge the fish-barging operation, or initiate a "permanent natural river drawdown"—breaching. The study's final report is due in December, but whatever its conclusions, that initial statement marked a dramatic shift. Suddenly, an action that had always seemed unthinkable was an officially sanctioned possibility.

Two separate scientific studies concluded that breaching presented the best hope for saving the river. In 1997 the *Idaho Statesman,* the state's largest newspaper, published a three-part series arguing that breaching the four dams would net local taxpayers and the region's economy $183 million a year. The dams, the paper concluded, "are holding Idaho's economy hostage."

"That series was seismic," says Reed Burkholder, a Boise-based breaching advocate. Charlie Ray agrees. "We've won the scientific argument," he says. "And we've won the economic argument. We're spending more to drive the fish to extinction than it would cost to revive them."

In fact, the economic argument is far from won. The *Statesman*'s numbers are not unimpeachable. The key to their prediction, a projected $248 million annual boost in recreation and fishing, assumes that the salmon runs will return to pre-1960s levels. Fisheries experts say that could take up to 24 years, if it happens at all. The $34 million lost at the Port of Lewiston each year, however, would be certain and immediate.

The Northwest can do without the power of the four lower Snake River dams: They account for only about 4 percent of the region's electricity supply. The dams aren't built for flood control, and contrary to a widely held belief, they provide only a small amount of irrigation water to the region's farmers. What the issue comes down to, then, is the Port of Lewiston. You take the dams out, says port manager Dave Doeringsfeld, "and transportation costs go up 200 to 300 percent."

To breach or to blow?

The pro-dam lobbyists know they possess a powerful, not-so-secret weapon: Senator Slade Gorton, the Washington Republican who holds the commanding post of chairman of the Subcommittee on Interior Appropriations. Gorton has built his political base by advertising himself as the foe of liberal Seattle environmentalists, and with his hands on Interior's purse strings, he can back up the role with real clout. As determined as Bruce Babbitt is to bring down a big dam, Slade Gorton may be more determined to stop him.

12. Beyond the Valley of the Dammed

During last October's federal budget negotiations, Gorton offered to allocate $22 million for removing two modest dams in the Elwha River on the Olympic Peninsula, a salmon-restoration project dear to the hearts of dam-breaching advocates. But Gorton agreed to fund the Elwha breaching if—and only if—the budget included language forbidding federal officials from unilaterally ordering the dismantling of any dam, including those in the Columbia River Basin. Babbitt and others balked at Gorton's proposal. As a result, the 1999 budget includes zero dollars for removal of the Elwah dams.

Gorton's Elwha maneuver may have been hardball politics for its own sake, but it was also a clear warning: If the Army Corps and the National Marine Fisheries Service recommend breaching on the Snake in their study later this year, there will be hell to pay.

Meanwhile, here's a hypothetical question: If you're going to breach, how do you actually do it? How do you take those behemoths out? It depends on the dam, of course, but the answer on the Snake is shockingly simple.

"You leave the dam there," Charlie Ray says. We're standing downstream from Lower Granite Dam, 35 million pounds of steel encased in concrete. Lower Granite isn't a classic ghastly curtain like Hoover Dam; it resembles nothing so much as an enormous half-sunk harmonica. Ray points to a berm of granite boulders butting up against the concrete structure's northern end. "Take out the earthen portion and let the river flow around the dam. This is not high-tech stuff. This is front-end loaders and dump trucks."

It turns out that Charlie is only a few adjectives short of the truth. All you do need are loaders and dump trucks—really, really big ones, says Steve Tatro of the Army Corps of Engineers. Tatro has the touchy job of devising the best way to breach his agency's own dams. First, he says, you'd draw down the reservoir, using the spillways and the lower turbine passages as drains. Then you'd bypass the concrete and steel entirely and excavate the dam's earthen portion. Depending on the dam, that could mean excavating as much as 8 million cubic yards of material.

Tatro's just-the-facts manner can't disguise the reality that there is something deeply cathartic about the act he's describing. Most environmental restoration happens at the speed of nature. Which is to say, damnably slow. Breaching a dam—or better yet, blowing a dam—offers a rare moment of immediate gratification.

The Glen Canyon story

From the Mesopotamian canals to Hoover Dam, it took the human mind about 10,000 years to figure out how to stop a river. It has taken only 60 years to accomplish the all-too-obvious environmental destruction.

Until the 1930s, most dam projects were matters of trial and (often) error, but beginning with Hoover Dam in 1931, dam builders began erecting titanic riverstoppers that approached an absolute degree of reliability and safety. In *Cadillac Desert,* a 1986 book on Western water issues, author Marc Reisner notes that from 1928 to 1956, "the most fateful transformation that has ever been visited on any landscape, anywhere, was wrought." Thanks to the U.S. Bureau of Reclamation, the Tennessee Valley Authority, and the Army Corps, dams lit a million houses, turned deserts into wheat fields, and later powered the factories that built the planes and ships that beat Hitler and the Japanese. Dams became monuments to democracy and enlightenment during times of bad luck and hunger and war.

Thirty years later, author Edward Abbey became the first dissenting voice to be widely heard. In *Desert Solitaire* and *The Monkey Wrench Gang,* Abbey envisioned a counterforce of wilderness freaks wiring bombs to the Colorado River's Glen Canyon Dam, which he saw as the ultimate symbol of humanity's destruction of the American West. Kaboom! Wildness returns to the Colorado.

Among environmentalists, the Glen Canyon Dam has become an almost mythic symbol of riparian destruction. All the symptoms of dam kill are there. The natural heavy metals that the Colorado River used to disperse into the Gulf of California now collect behind the dam in Lake Powell. And the lake is filling up: Sediment has reduced the volume of the lake from its original 27 million acre-feet to 23 million. One million acre-feet of water are lost to evaporation every year—enough, environmentalists note, to revive the dying upper reaches of the Gulf of California. The natural river ran warm and muddy, and flushed its channel with floods; the dammed version runs cool, clear, and even. Trout thrive in the Colorado. This is like giraffes thriving on tundra.

Another reason for the dam's symbolic power can be traced to its history. For decades ago, David Brower, then executive director of the Sierra Club, agreed to a compromise that haunts him to this day: Conservationists would not oppose Glen Canyon and 11 other projects if plans for the proposed Echo Park and Split Mountain dams, in Utah and Colorado, were abandoned. In 1963, the place Wallace Stegner once called "the most serenely beautiful of all the canyons of the Colorado" began disappearing beneath Lake Powell. Brower led the successful fight to block other dams in the Grand Canyon area, but he remained bitter about the compromise. "Glen Canyon died in 1963," he later wrote, "and I was partly responsible for its needless death."

In 1981 Earth First! inaugurated its prankster career by unfurling an enormous black plastic "crack" down the face of Glen Canyon Dam. In 1996 the Sierra Club rekindled the issue by calling for the draining of Lake Powell. With the support of Earth Island Institute (which Brower now chairs) and other environmental groups, the proposal got a hearing before a subcommittee of the House Committee on Resources in September 1997. Congress has taken no further action, but a growing number of responsible voices now echo the monkey-wrenchers' arguments. Even longtime Bureau of Reclamation supporter Barry Goldwater admitted, before his death last year, that he considered Glen Canyon Dam a mistake.

Defenders of the dam ask what we would really gain from a breach. The dam-based ecosystem has attracted peregrine falcons, bald eagles, carp, and catfish. Lake Powell brings in $400 million

> In 1963, the most beautiful of all the canyons of the Colorado **began disappearing beneath Lake Powell.**

a year from tourists enjoying houseboats, powerboats, and personal watercraft—a local economy that couldn't be replaced by the thinner wallets of rafters and hikers.

"It would be completely foolhardy and ridiculous to deactivate that dam," says Floyd Dominy during a phone conversation from his home in Boyce, Virginia. Dominy, now 89 years old and retired since 1969, was the legendary Bureau of Reclamation commissioner who oversaw construction of the dam in the early 1960s. "You want to lose all that pollution-free energy? You want to destroy a world-renowned tourist attraction—Lake Powell—that draws more than 3 million people a year?"

It goes against the American grain: the notion that knocking something down and returning it to nature might be progress just as surely as replacing wildness with asphalt and steel. But 30 years of environmental law, punctuated by the crash of the salmon industry, has shifted power from the dam builders to the conservationists.

The most fateful change may be a little-noticed 1986 revision in a federal law. Since the 1930s, the Federal Energy Regulatory Commission has issued 30- to 50-year operating licenses to the nation's 2,600 or so privately owned hydroelectric dams. According to the revised law, however, FERC must consider not only power generation, but also fish and wildlife, energy conservation, and recreational uses before issuing license renewals. In November 1997, for the first time in its history, FERC refused a license against the will of a dam owner, ordering the Edwards Manufacturing Company to rip the 160-year-old Edwards Dam out of Maine's Kennebec River. More than 220 FERC hydropower licenses will expire over the next 10 years.

If there is one moment that captures the turning momentum in the dam wars, it might be the dinner Richard Ingebretsen shared with the builder of Glen Canyon Dam, Floyd Dominy himself. During the last go-go dam years, from 1959 to 1969, this dam-building bureaucrat was more powerful than any Western senator or governor. Ingebretsen is a Salt Lake City physician, a Mormon Republican, and a self-described radical environmentalist. Four years ago, he founded the Glen Canyon Institute to lobby for the restoration of Glen Canyon. Ingebretsen first met Dominy when the former commissioner came to Salt Lake City in 1995 to debate David Brower over the issue of breaching Glen Canyon Dam. To his surprise, Ingebretsen found that he liked the man. "I really respect him for his views," he says.

Their dinner took place in Washington, D.C., in early 1997. At one point Dominy asked Ingebretsen how serious the movement to drain Lake Powell really was. Very serious, Ingebretsen replied. "Of course I'm opposed to putting the dam in mothballs," Dominy said. "But I heard what Brower wants to do." (Brower had suggested that Glen Canyon could be breached by coring out some old water bypass tunnels that had been filled in years ago.). "Look," Dominy continued, "those tunnels are jammed with 300 feet of reinforced concrete. You'll never drill that out."

With that, Dominy pulled out a napkin and started sketching a breach. "You want to drain Lake Powell?" he asked. "What you need to do is drill new bypass tunnels. Go through the soft sandstone around and beneath the dam and line the tunnels with waterproof plates. It would be an expensive, difficult engineering feat. Nothing like this has ever been done before, but I've done a lot of thinking about it, and it will work. You can drain it."

The astonished Ingebretsen asked Dominy to sign and date the napkin. "Nobody will believe this," he said. Dominy signed.

Of course, it will take more than a souvenir napkin to return the nation's great rivers to their full wildness and health. Too much of our economic infrastructure depends on those 75,000 dams for anyone to believe that large numbers of river blockers, no matter how obsolete, will succumb to the blow of Bruce Babbitt's hammer anytime soon. For one thing, Babbitt himself is hardly in a position to be the savior of the rivers. Swept up in the troubles of a lame-duck administration and his own nagging legal problems (last spring Attorney General Janet Reno appointed an independent counsel to look into his role in an alleged Indian casino-campaign finance imbroglio), this interior secretary is not likely to fulfill his dream of bringing down a really big dam. But a like-minded successor just might. It will take a president committed and powerful enough to sway both Congress and the public, but it could come to pass.

Maybe Glen Canyon Dam and the four Snake River dams won't come out in my lifetime, but others will. And as more rivers return to life, we'll take a new census of emancipated streams: We freed the Neuse, the Kennebec, the Allier, the Rogue, the Elwah, and even the Tuolumne. We freed the White Salmon and the Souradabscook, the Ocklawaha and the Genesee. They will be untidy and unpredictable, they will flood and recede, they will do what they were meant to do: run wild to the sea.

Bruce Barcott is the author of The Measure of a Mountain: Beauty and Terror on Mount Rainier (*Sasquatch, 1997*).

ENVIRONMENT

A River Runs Through It

75,000 dams were built. Now a few are coming down.

BY ANDREW MURR

As CHURCH BELLS PEALED AND thousands cheered, the backhoe scooped out a dollop of dirt and gravel that had been packed against Maine's Edwards Dam. And suddenly the Kennebec River did something it hasn't done since Andrew Jackson sat in the White House: first in a trickle and then in a torrent, it flowed freely to the Atlantic, through a 60-foot hole that workers had earlier punched underneath the dirt-and-gravel bandage. The 917-foot dam is the first that federal regulators have ordered razed against the owner's wishes—because the environmental damage it did by preventing salmon, striped bass, shad and six other species from reaching spawning grounds outweighed the benefits of the .1 percent of Maine's power it provided. Interior Secretary Bruce Babbitt was on hand for last week's milestone. "If someone's got a dam that's going down," he tells friends, "I'll be there."

Babbitt can expect to rack up the frequent-flier miles. About 75,000 big dams block American rivers, testaments to the conviction that any river flowing to sea unimpeded is a waste of water and power. But that attitude is under attack. Many of the aging dams kill millions of valuable salmon migrating to sea. As a result, the Federal Energy Regulatory Commission (FERC) is refusing to relicense dams where the environmental costs outweigh the value of the hydropower, or demanding that a dam be retrofitted with fish ladders. That's often so expensive that the owner opts to tear down the dam instead. Portland (Ore.) General Electric will raze dams on the Sandy and Little Sandy rivers, for instance, rather than make repairs. Demolition will free 22 miles of salmon and steelhead spawning grounds.

This year dams will fall from California to Connecticut (map). It shouldn't take long to see results. The Quaker Neck Dam on the Neuse River in North Carolina came down in 1997–98; bass and striped shad are already running again. On Butte Creek in northern California, removing three dams beginning in 1997 allowed the salmon run to jump from zero to 20,000. Those numbers have environmentalists and fishermen eying dams in the Olympic Peninsula, where dozens of populations of salmon are endangered or extinct. The Elwha and Glines Canyon dams, for example, cut the annual runs of salmon and steelhead on the Elwha River from 380,000 early this century to zero. The biggest targets are four hydroelectrics on the Snake River. But many locals oppose razing these giants, which supply 5 percent of the region's electricity, because doing so could raise rates the same amount. Call it salmon vs. watts.

With SHARON BEGLEY

Tearing Down the Water Walls

This year, dams will come down from Maine to California. In August, tourists will even be able to watch the demolition of the Cascade Dam in Yosemite Valley.

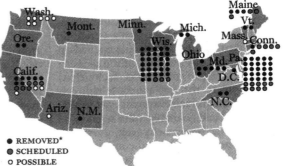

● REMOVED*
● SCHEDULED
○ POSSIBLE

*SINCE 1990. SOURCE FOR REMOVED DAMS: AMERICAN RIVERS

THERE GOES the NEIGHBORHOOD!

ATLANTA IS THE HOTTEST CITY *in the New South*, *with its population doubling* IN THE PAST 30 YEARS. *But to make room for new suburbs*, 50 ACRES OF FOREST *are felled every day. Can anything stop the sprawl?*

BY FEN MONTAIGNE

NESTLED IN THE HILLS OF NORTH GEORGIA IS THE 20-acre farm of Herb and Kathy Epperson, a swath of pasture and forest that has been in his family for 165 years. Downtown Atlanta is just 40 miles to the south, but when you visit the Epperson spread it feels a world away. A cedar home sits in a grove of tall pines, and out back, behind a meadow, are two 400-foot-long poultry houses in which the family raises about 250,000 chickens a year. Woods and fields and chicken coops surround the property.

The Eppersons knew that Atlanta was heading their way. They just didn't realize how fast it was coming. Living in central Cherokee County, they had watched as the fields, farms, and forests in the southern part of the county became subdivisions, strip shopping centers, and superstores. Then, in the fall of 1998, Atlanta came knocking on their front door.

A neighbor gave them the news: A developer had secured an option on a 300-acre property across the road and was rushing to rezone it for 1,200 houses and apartments. The development, the Eppersons realized, would turn the quiet community of Buffington on its head. Torrey Homes' several thousand new residents would more than double the town's population. The narrow country roads and the tiny Buffington Elementary School would be swamped.

"The heart of our community would be tore out, and for what?" says Herb Epperson, a rangy 50-year-old with a neatly trimmed beard who teaches drafting at a nearby high school. "We think we're very fortunate to be able to live here, and we don't want such drastic change."

"We were at a meeting, and the developers were trying to sell us on how wonderful this would be and how we'd be able to ride our bikes on their sidewalks," recalls Kathy Epperson in her fast-moving Georgia twang. "That was ridiculous. I said to their lawyer, 'This is going to ruin our way of life.' And he said, 'Well, you can sell out and move to Florida.' That made me sick. I was fuming for a few days. Then we started making calls."

Buffington's residents had only a few weeks before county officials would consider the rezoning request for Torrey Homes. But a shift in the public mood had taken place in Cherokee County, and it would work in the community's favor. Just a few months earlier, fed up with the unrestrained growth that had clogged the roads and schools of southern Cherokee and erased much of its rural character, voters had turned against the pro-growth chairman of the county commission and elected in his place Emily M. Lemcke, a relative newcomer who had campaigned on a platform of slow growth.

A series of bureaucratic delays allowed the Buffington residents to postpone a vote on the development until January 1999, after the new five-member county commission had taken office. In one of its first acts, the commission, led by Lemcke, voted against the Torrey Homes project.

But what once seemed to be a clear-cut victory has become a murky conflict that illustrates the relentless development pressure in Atlanta's exurbs. After the vote, D. R. Horton-Torrey took the county to court, and a judge ordered the commission to take another look at the rezoning request. Last December, over Lemcke's objections,

14. There Goes the Neighborhood

> *While the metro* POPULATION HAS GROWN *at about 4 percent a year, the region has* BEEN DEVOURING LAND *at four times that rate.*

the county commission voted 3–2 to allow Torrey to build one house per acre on 250 acres and to develop the remaining 50 acres commercially. That's a far less dense development than Torrey had sought, but it nonetheless means that yet another Cherokee County pasture could vanish.

D. R. Horton-Torrey is now deciding whether it's economical to proceed with the scaled-down project. For his part, Herb Epperson says he could live with one house per acre. But the experience has left him uneasy. "The people don't want this county to be overrun by development," he says. "Growth can be good, but let's control it. I just feel like we're growing too fast."

THE EPPERSONS ARE ON THE FRONT LINES OF A BATTLE over sprawl that has taken center stage in the Atlanta area and in much of the country. And as recent events in Cherokee County show, Americans are at last beginning to question what a half-century flood of suburban expansion has wrought. Distressed about traffic jams, deteriorating air quality, and the loss of forests and rural lands to cookie-cutter suburbs, residents in places like Cherokee County are saying enough is enough—and electing politicians who will heed their call. America is starting to witness "the beginning of the end of sprawl," says Christopher B. Leinberger, a nationally known developer and real estate consultant. "The American dream is once again changing," says Leinberger. "The aging baby boomers and the Gen X'ers want a different story," one that includes a return to city life and the construction of villagelike, pedestrian-oriented developments with plenty of open land nearby.

If such a change is indeed on the way, it's happening none too soon for places like Atlanta. The metropolitan area has become the poster child for a helter-skelter pattern of growth that has done grievous harm to the environment and eroded a once-vaunted quality of life. Unfettered by geographic boundaries and fueled by a superheated Sun Belt economy, Atlanta has become one of the fastest spreading metropolitan areas in history. With its population having doubled in the past 30 years, to 3.6 million, and with as many as 95,000 newcomers arriving each year, metro Atlanta has marched outward at such a rapid rate that its north-south diameter has grown from 65 miles to 110 miles in a decade. While the metro population has grown at about 4 percent a year, the region has been devouring land at four times that rate.

Since 1973 the Atlanta area has lost nearly 25 percent of its tree cover, or roughly 350,000 acres, satellite photographs show. Every day 50 acres of trees fall to development. Tens of thousands of acres of pasture have been transformed into homes and stores. Such habitat loss and fragmentation have taken a heavy toll on the region's bird populations, as many nesting and breeding grounds have either been destroyed or whittled down so drastically that predators can easily kill avian species. Deep-woodland nesters and grassland species have been particularly hard-hit. In north Georgia and the western Carolinas—the Piedmont—development has hurt species such as the bobwhite quail, Bachman's sparrow, loggerhead shrike, eastern meadowlark, and prairie warbler, all of which have shown steady declines over the past two decades.

"The entire Piedmont is heading toward one big area of sprawl, and when you lose habitat outright you lose birds. Period," says Chuck Hunter, the U.S. Fish and Wildlife Service's Southeast coordinator for non-game-bird species. "You did not start out with the healthiest environment here, because of cotton farming and logging earlier in the century. The difference is that with farming or logging you can always restore it. But once you've got houses and Wal-Marts, there's no return. It's the final nail in the coffin."

AS TREES ARE FELLED AND HILLSIDES FLATTENED, rivers such as Cherokee County's Etowah are becoming choked with silt. Chemicals and pesticides flow into the water from new parking lots and lawns, harming rare species of darters and mussels. The region's air quality has deteriorated so much that Washington has cut off federal highway funds to the 13-county metropolitan area.

But if Atlanta has become a symbol of out-of-control growth, its leaders are now moving to turn their region into an example of how sprawl can be contained. In 1998 Georgia's newly elected governor, Roy E. Barnes, persuaded the legislature to approve the Georgia Regional Transportation Authority (GRTA), an antisprawl superagency. The GRTA can issue up to $2 billion in bonds for mass transit and land preservation, block road projects that encourage sprawl, and control important land-use decisions throughout the region. Barnes is now pushing for a requirement that counties preserve 20 percent of their land as open space or risk a cutoff in state funds.

Local developers, sensing that the public is weary of endlessly metastasizing tract homes, are at last beginning to embrace in-town development and "conservation subdivisions," which preserve large areas of common green space. The population of Atlanta proper, after falling from 495,000 in 1970 to 384,000 in 1990, has leveled off and is now climbing back toward 400,000. Last year 1,752 new housing units were built within the city limits.

"Georgia has always sold itself on cheap land, cheap labor, cheap gasoline," says Rand Wentworth, the director of the Atlanta office of the Trust for Public Land, which is working to create a 180-mile "green" corridor along the Chattahoochee River. "We're now realizing that we don't want to lead the race to the bottom. In the next century the important race is the race to the top, and the top is defined by quality of life."

DRIVING PAST ROLLING MEADOWS ON HIGHWAY 20, Emily Lemcke—tall, 48, with blond hair that falls to her shoulders—surveys the scene and says, "Isn't this land pretty? I love it. And it was rezoned commercial under the last commission. Doesn't it make you sick to think it's going to be developed?"

Hardly Chamber of Commerce patter. But if her predecessor saw the landscape and thought economic development, Lemcke looks at Cherokee's open spaces and wonders how she can keep them that way.

"Cherokee County is not destroyed," says Lemcke. "We still have a lot of beautiful land. I know there's going to be more growth. To deny that is foolish. But what we have to have now is quality growth . . . that is considerate of the topography and vegetation and capitalizes on the existing infrastructure."

A native of Maryland, Lemcke moved to Atlanta 20 years ago to work as a banker. In 1982, seeking a more tranquil life, she moved to Cherokee County, a rural community of 464 square miles whose lifeblood was farming, poultry, and textiles. For decades the county's population had held steady at fewer than 20,000, but by 1982 it had climbed to about 52,000.

In the mid-1990s, southern Cherokee County was becoming as heavily developed as its densely populated neighbor to the south, Cobb County. And the county commission, led by one Hollis Lathem, was rezoning land and approving new subdivisions and shopping centers at a dizzying rate. Residents who had moved to Cherokee for its bucolic lifestyle found that the rat race had caught up with them, and they were angry.

"People felt like they were helpless, that the developers were more important to the county commissioners than the quality of life of Cherokee's citizens," says Ron James, a 42-year-old employee of Lockheed-Martin Aerospace, who grew up in Cobb County and moved to Cherokee County in 1985. "Folks were getting tired of moving and moving. So we wanted to draw a line in the sand."

Sharing similar concerns, Lemcke—who had no previous political experience—decided to oppose Lathem.

Pledging smart growth and an end to the dominance of Cherokee's business oligarchy, she struck a chord with the county's 135,000 citizens, most of whom were relative newcomers. Although she spent only one-fifth as much money as Lathem did, Lemcke won, with 57 percent of the vote.

One of the first acts of the Lemcke commission was to impose a moratorium on the rezoning of land for residential use. That ban will be lifted after the county devises an impact fee on home construction to help cover the cost of providing county services. The commission has also tightened zoning laws.

But it has inherited a government that is straining to keep up with growth. The public school system, for example, is filled to bursting, with more than a third of the district's 24,446 students spending at least part of the day in portable classrooms. School officials say the county must immediately build four elementary schools and one middle school, at a total cost of about $60 million.

Lathem, for his part, defends his record, pointing out that the majority of residential development during his four years as chairman was approved by previous commissions. He says he aggressively sought new commercial development in order to create jobs and give county residents a place to shop near home. "There's no question that Cherokee County is a far better place to live now," says Lathem. "It's much better off economically." He points out that the most vocal opponents of development are the very people who recently moved to the county: "The basic mentality of the newcomers is, 'Let's build a fence around the county.' "

Cherokee County suffers from no lack of finger-pointing, and even the current "slow growth" commission is split over the best way to handle future development. Some commissioners oppose dense residential projects; others share Lemcke's goal of mixed-use clusters and large areas of open space.

But there's no question that rapid growth has taken a toll on Cherokee County's environment. Lake Allatoona, a reservoir on the Etowah that provides drinking water for 300,000 residents in Cherokee and neighboring counties, is in critical condition. A study by the A. L. Burruss Institute of Public Service at Kennesaw State University, in Marietta, Georgia, predicted that if present rates of siltation and pollution continue, Lake Allatoona will be eutrophic—essentially dead—within 10 years.

"If you took 125 million people and stood them on the shore of Lake Allatoona and they each dumped a 15-pound bag of soil and phosphorus into the lake, that's how much sedimentation goes into Allatoona every year," says Harry McGinnis, the institute's director.

A visit to the interchange of I-575 and Highway 92 in southwestern Cherokee County shows why the area's waterways are in such bad shape. There, in a flurry of development, three enormous superstores—a "Big K" K-Mart, a Wal-Mart, and a Home Depot—have been built.

14. There Goes the Neighborhood

Land for a fourth project, a Target store, is being cleared. I visited the construction site with Keith Parsons, president of the Georgia River Network, a conservation organization. Part of the site, which sits on a hilltop, was still laced with clear streams that ran through ravines shaded by groves of gums, oaks, loblolly pines, and tulip poplars. But the hill was in the process of being flattened and the streams encased in culverts that would be covered by the store and the parking lot. The future Target was a spreading scar of red Georgia clay. Soon the creeks—which during my visit were still home to frogs and salamanders—will be little more than biologically sterile ditches that carry sediment and gasoline-spiked runoff into Noonday Creek, which flows into Lake Allatoona.

"There are probably 100 developers on 100 tracts of land in the Atlanta area doing the same thing," says Parsons, a biologist who wears his graying hair in a ponytail. "This place is a classic example of the resources we're losing. Most people drive by these strip malls and have no idea what's been lost. These little creeks and streams are the last pristine, free-flowing systems in these counties. If you destroy them, there's not much hope of restoring the larger river systems."

The environmental degradation caused by recent development, coupled with decades of damage from dams, mining, and pollution, have left the Etowah one of the South's most beleaguered rivers. Fifteen of its 100 fish species have been extirpated, and 17 other species, such as the Etowah and Cherokee darters, are in danger of disappearing from the river. Scientists estimate that at least two-thirds of the river's approximately 60 mussel species have been wiped out.

The Etowah and the larger river system to which it belongs, the Coosa, "may hold the dubious distinction of having more recent extinctions of aquatic organisms than any other equally sized river system in the United States," a team of scientists reported recently.

The state of Georgia has made some effort to protect rivers like the Etowah, requiring that no trees be cleared within 25 feet of streams and waterways, but a recent study by the Georgia River Network showed that developers and state regulators have virtually ignored the law. Over a three-year period, developers applied for 795 waivers, and regulators approved 763.

THE REAL ESTATE BOOM THREATENING CHEROKEE County's environment is being fed, of course, by land sales, and many of the sellers are lifelong residents of the county. For much of the 20th century, Cherokee County was a relatively poor place where farm families eked out a living. Many areas didn't have electricity until the late 1940s. So now, with land selling for $20,000 an acre, it's easy to understand why many families have decided to finally make a profit off their farms.

One of those families is Bernese and Albert Cagle, who, with three of their sons, operate Cagle's Dairy, the last dairy in Georgia that still produces and processes its own milk. In 1996, for $1.1 million, the Cagles sold 56 acres of land that has now been turned into Carrington Farms, a development of moderately priced, boxy houses. SWIM AND TENNIS: LOW 100's, the sign reads.

The Cagles—he is 64, she is 60—still own 130 acres and lease an additional 500 acres for their 400 head of cattle. Their sons plan to continue to run the dairy, and they say the land sale gave their parents the nest egg they had never been able to accumulate. "We're using that money to let Mama and Daddy slow down and do the things they want to after all these years," says Scott Cagle, 40.

The Cagles do not believe that the responsibility of preserving open space in Cherokee County should fall on the shoulders of farmers and large landholders. Rather, they favor a program of transferable development rights, under which developers would essentially pay landowners to keep their land open in return for the right to build higher-density projects in designated portions of the county. The current county commission is working on such a program, but in the meantime it is seeking other ways to preserve open space, including a one-cent sales-tax increase that would fund the purchase of $5 million in land along the Etowah River.

Equally important, says Lemcke, is attracting developers who will embrace ideas such as conservation subdivisions. One such project, currently under construction, is Governor's Preserve, a development along 3.6 miles of the Etowah that will retain 505 of its 862 acres in forests and meadows connected by nature trails. Another is Orange Shoals, a 400-acre subdivision with 100 acres in a common, forested "greenbelt" accessible to every home. Developer Chaunkee A. Venable has placed birdhouses and feeders throughout Orange Shoals, and the developers of Governor's Preserve will build more than 200 birdhouses and nesting boxes. The Atlanta Audubon Society has certified Orange Shoals as a wildlife sanctuary and is likely to do the same with Governor's Preserve.

GOVERNOR'S PRESERVE AND ORANGE SHOALS REPRE-sent an effort to bring together what have long seemed to be irreconcilable goals: the demands of the market, concern for the environment, and the desire of people to commune with more than just a quarter-acre patch of Kentucky bluegrass. Whether conservation subdivisions are the wave of the future or merely oases in an ever-widening sea of sprawl remains to be seen. But for those alarmed at the continuing loss of America's open spaces, events in Cherokee County offer some reason for hope. The mere fact that the nation's 50-year model of suburban growth is being widely challenged is a sign that change is in the air.

Ursula Cox, whose family came to Cherokee County in the 1830s, is a prophet of such change. She spends

What You Can Do

STOP THE BULLDOZERS!

Push your town or county to plan its growth more carefully. Urban and suburban sprawl eats up more than 3 million acres of farmland and wildland each year. And new developments often cost more in services (schools, sewer lines, fire protection) than they add in tax revenues. **When you move,** choose a house or an apartment in a city or a "traditional neighborhood development," where houses are close enough to each other and to stores and parks that you can walk, bike, or take mass transportation to most of your destinations. Many zoning codes actually work against compact development, separating stores from residential areas and requiring houses to be set on large lots—which eat up wildlife habitat. **If you own** a farm or other undeveloped property, donate or sell a conservation easement on it. You'll give up rights to develop the land in perpetuity, often in return for a lower tax rate and/or tax reduction. For more information, call the Land Trust Alliance (below) or visit its web site. **Vote for open-space** protection, since buying land is often the only way to guarantee that it won't become a subdivision. And press your congressmembers to fully fund the federal Land and Water Conservation Fund, which was set up to use revenues from oil and gas leases to purchase important land.

SPRAWL BUSTERS

Environmental Protection Agency
312-353-2000; www.epa.gov/region5/sprawl

Land Trust Alliance
202-638-4725; www.lta.org

National Audubon Society 518-766-0167;
http://ny.audubon.org/smart.html

Planning Commissioners Journal
888-475-3328; www.plannersweb.com

Preservation Institute
510-848-7827; www.preservenet.com

Sierra Club, Challenge to Sprawl Campaign
415-977-5500; www.sierra-club.org/sprawl

Smart Growth Network
202-962-3591; www.smartgrowth.org

Sprawl Watch Clearinghouse
202-974-5133; www.sprawlwatch.org

Sustainable Communities Network
628-681-1955 or 202-328-8160;
www.sustainable.org

Trust for Public Land
800-714-5263; www.tpl.org

---

much of her time trying to convince her fellow citizens that land has a value other than that fixed by real estate appraisers. Cox, 46 and an accomplished equestrienne, shares her 80-acre farm with a menagerie of horses, geese, dogs, cats, chickens, and a Vietnamese potbellied pig. Over the past several years, she has gone to battle against a number of projects that she says epitomize the thoughtless development threatening the county. Among them is the "northern arc," a perimeter highway that would leap over Atlanta's current beltway and open up central and northern Cherokee County to further sprawl.

Sitting at an old picnic table in front of a 19th-century farmhouse splotchy with peeling white paint, Cox says that as more fields and forests are consumed by development, people are realizing that unfettered growth leaves a bleak landscape in its wake.

Gesturing toward the old cedars and magnolias that tower above her house, Cox says, "This place is the same as it's been for decades. To me, it's a place of shelter and inspiration. But now land has become a pure commodity. Real estate has become big business, and that shouldn't be a county's mainstay. People tell me I should sell my place and make a fortune. But I look at land as something you love. People say I just want to stop growth; that's not so. I just want to see it managed. I just want to make this a nice place to live."

Veteran journalist Fen Montaigne moved to Atlanta in 1998. He and his family live in a 30-year-old house—within the city limits.

the RACE to SAVE OPEN SPACE

BY WESTON KOSOVA

Christine Todd Whitman, New Jersey's governor, has heard all the jokes. WELCOME TO THE GARDEN STATE, goes one bumper sticker. "How long can you hold your breath?" With 8 million people on 5 million acres, the country's most densely populated state is infamous for its sclerotic turnpike and bleak cityscapes. But these days Whitman is more interested in trumpeting the state's lesser-known attributes. "We have a Revolutionary War trail that runs from the northern part of the state on down. We've got natural watersheds and miles of bike trails."

All of which Governor Whitman is determined to keep out of the hands of developers. Last year she introduced the Garden State Preservation Trust, a 10-year, $1 billion proposal to set aside a million acres of land—roughly a fifth of the state—for parks, farmland, and wildlife habitat. New Jersey voters enthusiastically approved the plan 2–1. "This is about quality of life," Whitman says. "It's not just for people in rural areas or about protecting land in distant places for people in the cities to visit. It's land we can all use and enjoy." The trust plans to get started by spending $140,000 to create a wildlife refuge on a reservoir in Clark and Scotch Plains, and $240,000 for 22 acres of Pine Barrens habitat in Burlington County as part of a plan for a preserve that will expand to 4,000 acres.

New Jersey's ambitious experiment is just one example of a growing movement across the country to preserve open spaces and prevent developers from pushing into the diminishing farmland and forestland on the outskirts of many cities. In prosperous but sprawl-fatigued America, limiting pavement is becoming a hot political issue. In the past year voters in nearly all 50 states approved hundreds of plans to set aside green space—despite the tax increases required to pay for many of them. President Clinton has made wilderness preservation a top priority in his home-stretch year; his Land Legacy program would set aside $1 billion in federal money to protect open lands. He has also said he'll fight to fully fund the Land and Water Conservation Fund (LWCF) with profits from offshore oil drilling. Meanwhile, Vice President Al Gore's Livability Agenda is a centerpiece of his own campaign for the White House.

Many Republicans in Congress who spent the past decade fighting efforts to protect wilderness and habitat were among the biggest supporters of the Conservation and Reinvestment Act, which would preserve land for, among other things, hunting and fishing. "Open space land protection isn't Democrats versus Republicans," says Daniel P. Beard, vice president for public policy of the National Audubon Society. "It's a nonpartisan issue." In November Democrats and Republicans on the House Resources Committee approved a 15-year, $45 billion land-preservation plan using funds from the LWCF. "Baby boomers like me are interested in being able to see songbirds, get to a creekside, and enjoy the quiet of the outdoors, without having to drive two hours," says Rep. Mark Udall (D-CO). "A lot of my colleagues—many Republicans included—are starting to think the same way. Maybe the land won't be of pristine quality, but it can be enough to experience the smells and sights and sounds of the natural world."

The economic and development boom of the 1990s gobbled up farms and forests at an astonishing rate. A recent Agriculture Department study highlighted not only the trend (see "Who's Losing the Race?") but another troubling byproduct of sprawl: Urban areas across the country are losing trees. In 1973 trees covered about 37 percent of the Washington, D.C., area. By 1997 that had dwindled to a mere 13 percent. Many other cities nationwide have suffered comparable losses.

In recent years, however, a movement has sprung up to conserve the patches of land the bulldozers have missed. In 1998 there were 240 initiatives on state ballots nationwide intended to preserve green space or curb development, according to Phyllis Myers, whose group, State Resource Strategies, tracks environmental politics. Seventy-two percent of the measures passed, giving states and local governments the green light to spend more than $7 billion on land protection. In Oregon, voters approved a citizen-led initiative to set aside $45 million a year for parks, watersheds, river corridors, and salmon habitat. In Minnesota, voters overwhelmingly agreed to extend the state's Environmental and Natural Resources Trust Fund to 2025, with the aim of setting aside more land. Even Arizona voters, a predominantly conservative lot, approved a plan to spend $220 million over 11 years to purchase environmentally sensitive areas.

The open-space movement shows no signs of slowing down. Although 1999 wasn't an election year in most states, Myers says activists in 22 states managed to bring 139 habitat, parkland, and open-space measures to the ballot; 77 percent passed, authorizing more than $1 billion in new spending. The largest allocations of funds were in Long Island, New York; Glendale, Arizona; and Larimer County, Colorado. Ballot watchers expect to see dozens more initiatives in this November's presidential election.

Whether this political activity is the start of a green revolution is unclear. "I don't think everyone who votes for these measures is worried about improving the environment as a whole," Myers says. "It's more that they want to improve their community and their own way of life." Frank Luntz, a Republican pollster who surveyed Americans about preserving green space last year, says people overwhelmingly wanted more of it—but not necessarily because they feared for the planet or hated developers. "You want to know what this is about?" Luntz says. "It's about Americans screaming, 'I want it all! I want the big house, I want the TV, I want the jobs, I want the economy, and I want green and open spaces.' People's wallets are filled with cash, and they want to have a place they can go to get back to nature. Not that they will—but they want to know they can."

No matter what the motive, the result is that green space is being preserved. "Hey, if people are only interested in saving the woods so they can hike and play in them, that's fine with me," says Melanie Griffin of the Sierra Club, which released a widely read report on open space and has worked extensively to protect it. "After all, the woods still get saved."

WHO'S LOSING THE RACE?

Land that's been built upon has only the slimmest chance of ever being "undeveloped." Between 1992 and 1997, the U.S. Department of Agriculture reports, the rate of development of privately held land in this country more than doubled. The state-by-state breakdown below shows the average number of acres developed each year in the time frame indicated. Data for Alaska were not available at press time.

U.S. States	1982–1992	**1992–1997**
Texas	139,250	243,900
Pennsylvania	43,110	224,640
Georgia	76,630	210,640
Florida	116,310	189,060
North Carolina	93,580	156,300
California	80,020	138,960
Tennessee	44,110	122,320
Michigan	46,230	110,160
South Carolina	40,010	107,940
Ohio	46,860	104,240
New York	22,510	98,480
Virginia	45,360	93,440
Alabama	32,090	89,060
Kentucky	36,280	70,820
Washington	28,830	70,000
New Mexico	16,630	69,700
Mississippi	14,430	62,520
Minnesota	23,560	62,260
Missouri	20,450	62,100
Illinois	24,600	58,440
New Jersey	29,860	56,640
Wisconsin	24,760	56,560
Massachusetts	23,310	56,300
West Virginia	11,410	55,120
Indiana	22,830	54,880
Arkansas	9,580	47,440
Oklahoma	15,680	44,900
Maryland	14,690	44,460
Arizona	37,460	39,880
Kansas	11,700	38,500
Louisiana	26,320	34,420
Maine	9,250	33,560
Puerto Rico	12,460	30,620
Oregon	16,450	30,080
Montana	7,960	24,540
Idaho	8,590	24,160
Colorado	30,740	24,060
New Hampshire	14,920	21,460
Utah	10,690	21,020
Iowa	5,230	20,580
Nebraska	3,920	16,240
South Dakota	6,060	15,340
Connecticut	8,420	12,680
Wyoming	3,370	10,540
North Dakota	8,600	9,940
Nevada	8,270	8,300
Delaware	3,530	7,020
Vermont	6,490	5,200
Rhode Island	2,650	2,040
Hawaii	2,360	1,740
TOTAL ACRES DEVELOPED	**1,388,410**	**3,193,200**

Past and Present Land Use and Land Cover in the USA

WILLIAM B. MEYER

Dr. William B. Meyer is a geographer currently employed on the research faculty of the George Perkins Marsh Institute at Clark University in Worcester, Massachusetts. His principal interests lie in the areas of global environmental change with particular emphasis on land use and land cover change, in land use conflict, and in American environmental history.

"Land of many uses," runs a motto used to describe the National Forests, and it describes the United States as a whole just as well. "Land of many covers" would be an equally apt, but distinct, description. *Land use* is the way in which, and the purposes for which, human beings employ the land and its resources: for example, farming, mining, or lumbering. *Land cover* describes the physical state of the land surface: as in cropland, mountains, or forests. The term land cover originally referred to the kind and state of vegetation (such as forest or grass cover), but it has broadened in subsequent usage to include human structures such as buildings or pavement and other aspects of the natural environment, such as soil type, biodiversity, and surface and groundwater. A vast array of physical characteristics—climate, physiography, soil, biota—and the varieties of past and present human utilization combine to make every parcel of land on the nation's surface unique in the cover it possesses and the opportunities for use that it offers. For most practical purposes, land units must be aggregated into quite broad categories, but the frequent use of such simplified classes should not be allowed to dull one's sense of the variation that is contained in any one of them.

Land cover is affected by natural events, including climate variation, flooding, vegetation succession, and fire, all of which can sometimes be affected in character and magnitude by human activities. Both globally and in the United States, though, land cover today is altered principally by direct human use: by agriculture and livestock raising, forest harvesting and management, and construction. There are also incidental impacts from other human activities such as forests damaged by acid rain from fossil fuel combustion and crops near cities damaged by tropospheric ozone resulting from automobile exhaust.

Changes in land cover by land use do not necessarily imply a degradation of the land. Indeed, it might be presumed that any change produced by human use is an improvement, until demonstrated otherwise, because someone has gone to the trouble of making it. And indeed, this has been the dominant attitude around the world through time. There are, of course, many reasons why it might be otherwise. Damage may be done with the best of intentions when the harm inflicted is too subtle to be perceived by the land user. It may also be done when losses produced by a change in land use spill over the boundaries of the parcel involved, while the gains accrue largely to the land user. Economists refer to harmful effects of this sort as *negative externalities,* to mean secondary or unexpected consequences that may reduce the net value of production of an activity and displace some of its costs upon other parties. Land use changes can be undertaken because they return a net profit to the land user, while the impacts of negative externalities such as air and water pollution, biodiversity loss, and increased flooding are borne by others. Conversely, activities that result in secondary benefits (or *positive externalities*) may not be undertaken by landowners if direct benefits to them would not reward the costs.

Over the years, concerns regarding land degradation have taken several overlapping (and occasionally conflicting) forms. *Conservationism* emphasized the need for careful and efficient management to guarantee a sustained supply of productive land resources for future generations. *Preservationism* has sought to protect scenery and ecosystems in a state as little human-altered as possible. Modern *environmentalism* subsumes many of these goals and adds new concerns that cover the varied secondary effects of land use both on land cover and on other related aspects of the global environment. By and large, American attitudes in the past century have shifted from a tendency to interpret human use as improving the condition of the land towards a tendency to see human impact as primarily destructive. The term "land reclamation" long denoted the conversion of land from its natural cover; today it is more often used to describe the restoration and repair of land damaged by human use. It would be easy, though, to exaggerate the shift in attitudes. In truth, calculating the balance of costs and benefits from many land use and land cover changes is enormously difficult. The full extent and consequences of proposed changes are often less than certain, as is their possible irreversibility and thus their lasting significance for future generations.

WHERE ARE WE?

The United States, exclusive of Alaska and Hawaii, assumed its present size and shape around the middle of the 19th century. Hawaii is relatively small, ecologically distinctive, and profoundly affected by a long and distinctive history of human use; Alaska is huge and little affected to date by direct land use. In this review assessment we therefore survey land use and land cover change, focusing on the past century and a half, only in the conterminous or lower 48 states. Those states

"The adjustments that are made in land use and land cover in coming years will in some way alter the life of nearly every living thing on Earth."

cover an area of almost 1900 million acres, or about 3 million square miles.

How land is *used,* and thus how *land cover* is altered, depends on who owns or controls the land and on the pressures and incentives shaping the behavior of the owner. Some 400 million acres in the conterminous 48 states—about 21% of the total—are federally owned. The two largest chunks are the 170 million acres of western rangeland controlled by the Bureau of Land Management and the approximately equal area of the National Forest System. Federal land represents 45% of the area of the twelve western states, but is not a large share of any other regional total. There are also significant land holdings by state governments throughout the country.

Most of the land in the United States is privately owned, but under federal, state, and local restrictions on its use that have increased over time. The difference between public and private land is important in explaining and forecasting land use and land coverage change, but the division is not absolute, and each sector is influenced by the other. Private land use is heavily influenced by public policies, not only by regulation of certain uses but through incentives that encourage others. Public lands are used for many private activities; grazing on federal rangelands and timber extraction from the national forests by private operators are the most important and have become the most controversial. The large government role in land use on both government and private land means that policy, as well as economic forces, must be considered in explaining and projecting changes in the land. Economic forces are of course significant determinants of policy—perhaps the most significant—but policy remains to some degree an independent variable.

There is no standard, universally accepted set of categories for classifying land by either use or cover, and the most commonly used, moreover, are hybrids of land cover and land use. Those employed here, which are by and large those of the U.S. National Resources Inventory conducted every five years by relevant federal agencies, are cropland, forest, grassland (pasture and rangeland), wetlands, and developed land.

- *Cropland* is land in farms that is devoted to crop production; it is not to be confused with total farmland, a broad land use or land ownership category that can incorporate many forms of land cover.
- *Forest land* is characterized by a predominance of tree cover and is further divided by the U.S. Census into timberland and non-timberland. By definition, the former must be capable of producing 20 cubic feet of industrial wood per acre per year and remain legally open to timber production.
- *Grassland* as a category of land cover embraces two contrasting Census categories of use: pasture (enclosed and what is called improved grassland, often closely tied to cropland and used for intensive livestock raising), and range (often unenclosed or unimproved grazing land with sparser grass cover and utilized for more extensive production).
- *Wetlands* are not a separate Census or National Resources Inventory category and are included within other categories: swamp, for example, is wetland forest. They are defined by federal agencies as lands covered all or part of the year with water, but not so deeply or permanently as to be classified as water surface *per se*.
- The U.S. government classifies as *developed* land urban and built-up parcels that exceed certain size thresholds. "Developed" or "urban" land is clearly a use rather than a cover category. Cities and suburbs as they are politically defined have rarely more than half of their area, and often much less, taken up by distinctively "urban" land cover such as buildings and pavement. Trees and grass cover

Land Use and Cover in the Conterminous U.S.

Land Class	Area in Million Acres	Fraction of Total Area
Privately Owned (shown in diagram below)		
Cropland	422	22.4%
Rangeland	401	21.3
Forest	391	20.8
Pasture	129	6.9
Developed	77	4.1
Other Catagories[1]	60	3.2
Federally Owned[2]	404	21.4
TOTAL[3]	1884	100%

[1] Other minor covers and surface water
[2] Federal land is approximately half forest and half rangeland
[3] Included in various catagories is about 100 million acres of wetland, covering about 5% of the national area

Table 1 Source: U.S. 1987 National Resources Inventory, published in 1989. U.S. Government Printing Office.

substantial areas of the metropolitan United States; indeed, tree cover is greater in some settlements than in the rural areas surrounding them.

By the 1987 U.S. National Resources Inventory, non-federal lands were divided by major land use and land cover classes as follows: cropland, about 420 million acres (22% of the entire area of the 48 states); rangeland about 400 million (21%); forest, 390 million (21%); pasture, 130 million (7%); and developed land, 80 million (4%). Minor covers and uses, including surface water, make up another 60 million acres (Table 1). The 401 million acres of federal land are about half forest and half range. Wetlands, which fall within these other Census classes, represent approximately 100 million acres or about five percent of the national area; 95 percent of them are freshwater and five percent are coastal.

These figures, for even a single period, represent not a static but a dynamic total, with constant exchanges among uses. Changes in the area and the location of cropland, for example, are the result of the *addition* of new cropland from conversion of grassland, forest, and wetland and its *subtraction* either by abandonment of cropping and reversion to one of these less intensive use/cover forms or by conversion to developed land. The main causes of forest *loss* are clearing for agriculture, logging, and clearing for development; the main cause of forest *gain* is abandonment of cropland followed by either passive or active reforestation. Grassland is converted by the creation of pasture from forest, the interchange of pasture and cropland, and the conversion of rangeland to cropland, often through irrigation.

Change in wetland is predominantly loss through drainage for agriculture and construction. It also includes natural gain and loss, and the growing possibilities for wetland creation and restoration are implicit in the Environmental Protection Agency's "no *net* loss" policy (emphasis added). Change in developed land runs in only one direction: it expands and is not, to any significant extent, converted to any other category.

Comparison of the American figures with those for some other countries sets them in useful perspective. The United States has a greater relative share of forest and a smaller relative share of cropland than does Europe as a whole and the United Kingdom in particular.

Though Japan is comparable in population density and level of development to Western Europe, fully two-thirds of its area is classified as forest and woodland, as opposed to ten percent in the United Kingdom; it preserves its largely mountainous forest area by maintaining a vast surplus of timber imports over exports, largely from the Americas and Southeast Asia.

Regional patterns within the U.S. (using the four standard government regions of Northeast, Midwest, South, and West) display further variety. The Northeast, though the most densely populated region, is the most heavily wooded, with three-fifths of its area in forest cover. It is also the only region of the four in which "developed" land, by the Census definition, amounts to more than a minuscule share of the total; it covers about eight percent of the Northeast and more than a quarter of the state of New Jersey. Cropland, not surprisingly, is by far the dominant use/cover in the Midwest, accounting for just under half of its expanse. The South as a whole presents the most balanced mix of land types: about 40 percent forest, 20 percent each of cropland and rangeland, and a little more than ten percent pasture. Western land is predominantly rangeland, with forest following and cropland a distant third. Wetlands are concentrated along the Atlantic seaboard, in the Southeast, and in the upper Midwest. Within each region, of course, there is further variety at and below the state level.

WHERE HAVE WE BEEN?

The public domain, which in 1850 included almost two-thirds of the area of the present conterminous states, has gone through two overlapping phases of management goals. During the first, dominant in 1850 and long thereafter, the principal goal of management was to transfer public land into private hands, both to raise revenue and to encourage settlement and land improvements. The government often attached conditions (which were sometimes complied with) to fulfill other national goals, such as swamp drainage, timber planting, and railroad construction in support of economic development.

The second phase, that of federal retention and management of land, began with the creation of the world's first national park, Yellowstone, shortly after the Civil War. It did not begin to be a significant force, however, until the 1890s, when 40 million acres in the West were designated as federal forest reserves, the beginning of a system that subsequently expanded into other regions of the country as well. Several statutory vestiges of the first, disposal era remain (as in mining laws, for example), but the federal domain is unlikely to shrink noticeably in coming decades, in spite of repeated challenges to the government retention of public land and its regulation of private land. In recent years, such challenges have included the "Sagebrush Rebellion" in the rangelands of the West in the 1970s and 1980s calling for the withdrawal of federal control, and legal efforts to have many land use regulations classified as "takings," or as exercises of the power of eminent domain. This classification, where it is

16. Past and Present Land Use and Land Cover in the USA

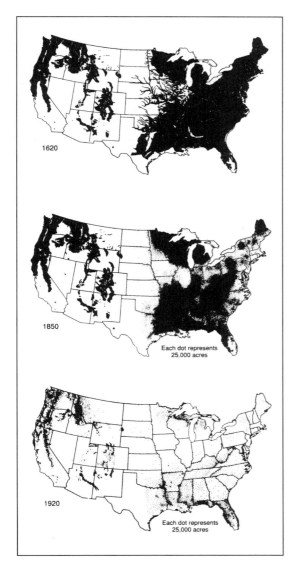

Figure 1 Area of virgin forest: top to bottom 1620, 1850, and 1920 as published by William B. Greeley, "The Relation of Geography to Timber Supply," Economic Geography, vol. 1, pp. 1–11 (1925). The depiction of U.S. forests in the later maps may be misleading in that they show only old-growth forest and not total tree cover.

granted, requires the government to compensate owners for the value of development rights lost as a result of the regulation.

Cropland

Total cropland rose steadily at the expense of other land covers throughout most of American history. It reached a peak during the 1940s and has subsequently fluctuated in the neighborhood of 400 million acres, though the precise figure depends on the definition of cropland used. Long-term regional patterns have displayed more variety. Cropland abandonment in some areas of New England began to be significant in some areas by the middle of the nineteenth century. Although total farmland peaked in the region as late as 1880 (at 50%) and did not decline sharply until the turn of the century, a steady decline in the subcategory of cropland and an increase in other farmland covers such as woodland and unimproved pasture was already strongly apparent. The Middle Atlantic followed a similar trajectory, as, more recently, has the South. Competition from other, more fertile sections of the country in agricultural production and within the East from other demands on land and labor have been factors; a long-term rise in agricultural productivity caused by technological advances has also exerted a steady downward pressure on total crop acreage even though population, income, and demand have all risen.

Irrigated cropland on a significant scale in the United States extends back only to the 1890s and the early activities in the West of the Bureau of Reclamation. Growing rapidly through about 1920, the amount of irrigated land remained relatively constant between the wars, but rose again rapidly after 1945 with institutional and technological developments such as the use of center-pivot irrigation drawing on the Ogallala Aquifer on the High Plains. It reached 25 million acres by 1950 and doubled to include about an eighth of all cropland by about 1980. Since then the amount of irrigated land has experienced a modest decline, in part through the decline of aquifers such as the Ogallala and through competition from cities for water in dry areas.

Forests

At the time of European settlement, forest covered about half of the present 48 states. The greater part lay in the eastern part of the country, and most of it had already been significantly altered by Native American land use practices that left a mosaic of different covers, including substantial areas of open land.

Forest area began a continuous decline with the onset of European settlement that would not be halted until the early twentieth century. Clearance for farmland and harvesting for fuel, timber, and other wood products represented the principal sources of pressure. From an estimated 900 million acres in 1850, the wooded area of the entire U.S. reached a low point of 600 million acres around 1920 (Fig. 1).

It then rose slowly through the postwar decades, largely through abandonment of cropland and regrowth on cutover areas, but around 1960 began again a modest decline, the result of settlement expansion and of higher rates of timber extraction through mechanization. The agricultural censuses recorded a drop of 17 million acres in U.S. forest cover between 1970 and 1987 (though data uncertainties and the small size of the changes relative to the total forest area make a precise dating of the reversals difficult). At the same time, if the U.S. forests have been shrinking in area they have been growing in density and volume. The trend in forest biomass has been consistently upward; timber stock measured in the agricultural censuses from 1952 to 1987 grew by about 30%.

National totals of forested area again represent the aggregation of varied regional experiences. Farm abandonment in much of the East has translated directly into forest recovery, beginning in the mid- to late-nineteenth century (Fig. 2). Historically, lumbering followed a regular pattern of harvesting one region's resources and moving on to the next; the once extensive old-growth forest of the Great Lakes, the South, and the Pacific Northwest represented successive and overlapping frontiers. After about 1930, frontier-type exploitation gave way to a greater emphasis on permanence and management of stands by timber companies. Wood itself has declined in importance as a natural resource, but forests have been increasingly valued and protected for a range of other services, including wildlife habitat, recreation, and streamflow regulation.

Grassland
The most significant changes in grassland have involved impacts of grazing on the western range. Though data for many periods are scanty or suspect, it is clear that rangelands have often been seriously overgrazed, with deleterious consequences including soil erosion and compaction, increased streamflow variability, and floral and faunal biodiversity loss as well as reduced value for production. The net value of grazing use on the western range is nationally small, though significant locally, and pressures for tighter management have increasingly been guided by ecological and preservationist as well as production concerns.

Wetland
According to the most recent estimates, 53% of American wetlands were lost between the 1780s and the 1980s, principally to drainage for agriculture. Most of the conversion presumably took place during the twentieth century; between the 1950s and the 1970s alone, about 11 million acres were lost. Unassisted private action was long thought to drain too little; since mid-century, it has become apparent that the opposite is true, that unfettered private action tends to drain too much, i.e., at the expense of now-valued wetland. The positive externalities once expected from drainage—improved public health and beautification of an unappealing natural landscape—carry less weight today than the negative ones that it produces. These include the decline of wildlife, greater extremes of streamflow, and loss of a natural landscape that is now seen as more attractive than a human-modified one. The rate of wetland loss has now been cut significantly by regulation and by the removal of incentives for drainage once offered by many government programs.

Developed land
As the American population has grown and become more urbanized, the land devoted to settlement has increased in at least the same degree. Like the rest of the developed world, the United States now has an overwhelmingly non-farm population residing in cities, suburbs, and towns and villages. Surrounding urban areas is a classical frontier of rapid and sometimes chaotic land use and land cover change. Urban impacts go beyond the mere subtraction of land from other land uses and land covers for settlement and infrastructure; they also involve the mining of building materials, the disposal of wastes, the creation of parks and water supply reservoirs, and the introduction of pollutants in air, water, and soil. Long-term data on urban use and cover trends are unfortunately not available. But the trend in American cities has undeniably been one of residential dispersal and lessened settlement densities as transportation technologies have improved; settlement has thus required higher amounts of land per person over time.

WHERE ARE WE GOING?

The most credible projections of changes in land use and land cover in the United States over the next fifty years have come from recent assessments produced under the federal laws that now mandate regular national inventories of resource stocks and prospects. The most recent inquiry into land resources, completed by the Department of Agriculture in 1989 (and cited at the end of this article), sought to project their likely extent and condition a half-century into the future, to the year 2040. The results indicated that only slow

16. Past and Present Land Use and Land Cover in the USA

changes were expected nationally in the major categories of land use and land cover: a loss in forest area of some 5% (a slower rate of loss than was experienced in the same period before); a similarly modest decline in cropland; and an increase in rangeland of about 5% through 2040. Projections are not certainties, however: they may either incorrectly identify the consequences of the factors they consider or fail to consider important factors that could alter the picture. Because of the significant impacts of policy, its role—notoriously difficult to forecast and assess— demands increased attention, in both its deliberate and its inadvertent effects.

Trends in the United States stand in some contrast to those in other parts of the developed world. While America's forest area continues to decline somewhat, that of many comparable countries has increased in modest degree, while the developing world has seen significant clearance in the postwar era. There has been substantial stability, with slow but fluctuating decline, in cropland area in the United States. In contrast, cropland and pasture have declined modestly in the past several decades in Western Europe and are likely to decline sharply there in the future as longstanding national and European Community agricultural policies subsidizing production are revised; as a result, the European countryside faces the prospect of radical change in land use and cover and considerable dislocation of rural life.

WHY DOES IT MATTER?

Land use and land cover changes, besides affecting the current and future supply of land resources, are important sources of many other forms of environmental change. They are also linked to them through synergistic connections that can amplify their overall effect.

Loss of plant and animal biodiversity is principally traceable to land transformation, primarily through the fragmentation of natural habitat. Worldwide trends in land use and land cover change are an important source of the so-called greenhouse gases, whose accumulation in the atmosphere may bring about global climate change. As much as 35% of the increase in atmospheric CO_2 in the last 100 years can be attributed to land use change, principally through deforestation. The major known sources of increased methane—rice paddies, landfills, biomass burning, and cattle—are all related to land use. Much of the increase in nitrous oxide is now thought due to a collec-

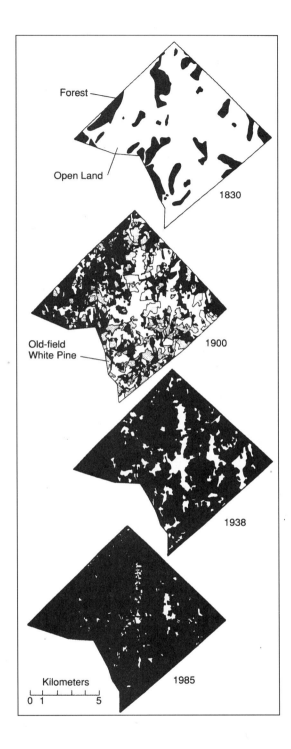

Figure 2 Modern spread of forest (shown in black) in the township of Petersham, Massachusetts, 1830 through 1985. White area is that considered suitable for agriculture; shaded portions in 1900 map indicate agricultural land abandoned between 1870 and 1900 that had developed forest of white pine in this period. From "Land-use History and Forest Transformations," by David R. Foster, in *Humans as Components of Ecosystems,* edited by M. J. McDonnell and S.T.A. Pickett, Springer-Verlag, New York, pp. 91–110, 1993.

tion of sources that also depend upon the use of the land, including biomass burning, livestock raising, fertilizer application and contaminated aquifers.

Land use practices at the local and regional levels can dramatically affect soil condition as well as water quality and water supply. And finally, vulnerability or sensitivity to existing climate hazards and possible climate change is very much affected by changes in land use and cover. Several of these connections are illustrated below by examples.

Carbon emissions

In most of the world, both fossil fuel combustion and land transformation result in a net release of carbon dioxide to the atmosphere. In the United States, by contrast, present land use and land cover changes are thought to absorb rather than release CO_2 through such processes as the rapid growth of relatively youthful forests. In balance, however, these land-use-related changes reduce U.S. contributions from fossil fuel combustion by only about 10%. The use of carbon-absorbing tree plantations to help diminish global climate forcing has been widely discussed, although many studies have cast doubt on the feasibility of the scheme. Not only is it a temporary fix (the trees sequester carbon only until the wood is consumed, decays, or ceases to accumulate) and requires vast areas to make much of a difference, but strategies for using the land and its products to offset some of the costs of the project might have large and damaging economic impacts on other land use sectors of the economy.

Effects on arable land

The loss of cropland to development aroused considerable concern during the 1970s and early 1980s in connection with the 1981 National Agricultural Lands Study, which estimated high and sharply rising rates of conversion. Lower figures published in the 1982 National Resource Inventory, and a number of associated studies, have led most experts to regard the conversion of cropland to other land use categories as representing something short of a genuine crisis, likely moreover to continue at slower rather than accelerating rates into the future. The land taken from food and fiber production and converted to developed land has been readily made up for by conversion of land from grassland and forest. The new lands are not necessarily of the same quality as those lost, however, and some measures for the protection of prime farmland are widely considered justified on grounds of economics as well as sociology and amenities preservation.

Vulnerability to climate change

Finally, patterns and trends in land use and land cover significantly affect the degree to which countries and regions are vulnerable to climate change—or to some degree, can profit from it. The sectors of the economy to which land use and land cover are most critical—agriculture, livestock, and forest products—are, along with fisheries, among those most sensitive to climate variation and change. How vulnerable countries and regions are to climate impacts is thus in part a function of the importance of these activities in their economies, although differences in ability to cope and adapt must also be taken into account.

These three climate-sensitive activities have steadily declined in importance in recent times in the U.S. economy. In the decade following the Civil War, agriculture still accounted for more than a third of the U.S. gross domestic product, or GDP. In 1929, the agriculture-forest-fisheries sector represented just under ten percent of national income. By 1950, it had fallen to seven percent of GDP, and it currently represents only about two percent. Wood in 1850 accounted for 90 percent of America's total energy consumption; today it represents but a few percent. These trends suggest a lessened macroeconomic vulnerability in the U.S. to climate change, though they may also represent a lessened ability to profit from it to the extent that change proves beneficial. They say nothing, however, about primary or secondary impacts of climate change on other sectors, about ecological, health, and amenity losses, or about vulnerability in absolute rather than relative terms, and particularly the potentially serious national and global consequences of a decline in U.S. food production.

The same trend of lessening vulnerability to climate changes is apparent even in regions projected to be the most exposed to the more harmful of them, such as reduced rainfall. A recent study examined agro-economic impacts on the Missouri-Iowa-Nebraska-Kansas area of the Great Plains, were the "Dust Bowl" drought and heat of the 1930s to recur today

Shifting patterns of land use in the U.S. and throughout the world are a proximate cause of many of today's environmental concerns.

or under projected conditions of the year 2030. It found that although agricultural production would be substantially reduced, the consequences would not be severe for the regional economy overall: partly because of technological and institutional adaptation and partly because of the declining importance of the affected sectors, as noted above. The 1930s drought itself had less severe and dramatic effects on the population and economy of the Plains than did earlier droughts in the 1890s and 1910s because of land use, technological, and institutional changes that had taken place in the intervening period.

Shifting patterns in human settlement are another form of land use and land cover change that can alter a region's vulnerability to changing climate. As is the case in most other countries of the world, a disproportionate number of Americans live within a few miles of the sea. In the postwar period, the coastal states and counties have consistently grown faster than the country as a whole in population and in property development. The consequence is an increased exposure to hazards of hurricanes and other coastal storms, which are expected by some to increase in number and severity with global warming, and to the probable sea-level rise that would also accompany an increase in global surface temperature. It is unclear to what extent the increased exposure to such hazards might be balanced by improvements in the ability to cope, through better forecasts, better construction, and insurance and relief programs. Hurricane fatalities have tended to decline, but property losses per hurricane have steadily increased in the U.S., and the consensus of experts is that they will continue to do so for the foreseeable future.

CONCLUSIONS

How much need we be concerned about changes in land use and land cover in their own right? How much in the context of other anticipated environmental changes?

As noted above, shifting patterns of land use in the U.S. and throughout the world are a proximate cause of many of today's environmental concerns. How land is used is also among the human activities most likely to feel the effects of possible climate change. Thus if we are to understand and respond to the challenges of global environmental change we need to understand the dynamics of land transformation. Yet those dynamics are notoriously difficult to predict, shaped as they are by patterns of individual decisions and collective human behavior, by history and geography, and by tangled economic and political considerations. We should have a more exact science of how these forces operate and how to balance them for the greatest good, and a more detailed and coherent picture of how land in the U.S. and the rest of the world is used.

The adjustments that are made in land use and land cover in coming years, driven by worldwide changes in population, income, and technology, will in some way alter the life of nearly every living thing on Earth. We need to understand them and to do all that we can to ensure that policy decisions that affect the use of land are made in the light of a much clearer picture of their ultimate effects.

FOR FURTHER READING

Americans and Their Forests: A Historical Geography, by Michael Williams. Cambridge University Press, 599 pp, 1989.

An Analysis of the Land Situation in the United States: 1989–2040. USDA Forest Service General Technical Report RM-181. U.S. Government Printing Office, Washington, D.C., 1989.

Changes in Land Use and Land Cover: A Global Perspective. W. B. Meyer and B. L. Turner II, editors. Cambridge University Press, 537 pp, 1994.

"Forests in the Long Sweep of American History," by Marion Clawson. *Science*, vol. 204, pp 1168–1174, 1979.

INFRASTRUCTURE

Operation Desert Sprawl

The biggest issue in booming Las Vegas isn't growth. It's finding somebody to pay the staggering costs of growth.

BY WILLIAM FULTON AND PAUL SHIGLEY

On a late spring afternoon, the counters at the Las Vegas Development Services Center are only slightly less crowded than those at the nearby McDonald's. Here, in a nondescript office building some eight blocks from City Hall, a small army of planners occupies counters and cubicles, standing ready to process the daily avalanche of building projects. Several times a minute, people with blueprints tucked under their arms hurry in or out the door.

Upstairs, Tim Chow, the city's planning and development director, shakes his head and smiles. He says he has "the toughest planning job in the country," and he may be right. Two hundred new residents arrive in Las Vegas every day; a house is built every 15 minutes. Last year alone, the city issued 7,700 residential building permits, plus permits for $200 million worth of commercial construction—enough to build a good-sized Midwestern county seat from scratch.

Before he came to Nevada in April, Chow held a similar position in a smaller county in California. There, he points out, the planning process grinds slowly. State law—and local politics—require extensive environmental studies and public hearings before planners approve subdivisions and retail centers.

But this isn't California. It's Las Vegas, the nation's fastest-growing community, and the policy is to build first and ask questions later. "We just don't have time to do the kinds of rigorous analysis done in other places with regard to compatibility, impacts, infrastructure, coordination," Chow says. "Sometimes you make mistakes, and the impacts of those mistakes are felt many years later."

Tim Chow's planning counter isn't the only one in this area that looks like a fast-food restaurant. Fifteen miles to the southeast, his counterpart in Henderson, Mary Kay Peck, is presiding over the rapid creation of the second-largest city in Nevada. Henderson officials claim to have "the high-

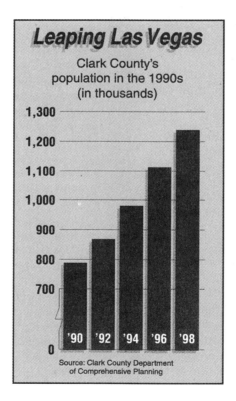

est development standards in the Las Vegas Valley," but that hasn't stopped the city from growing seven-fold in the past two decades, to 170,000 people. The city recently annexed 2,500 acres to accommodate a new Del Webb project, and is lining up a federal land exchange that will allow the addition of 8,000 more acres near McCarran International Airport.

Henderson is in a running argument with well-established Reno over which city is second most populous in the state after Las Vegas. But that argument won't last much longer. "We have room," Peck says, "to grow and grow and grow." Henderson is expected to add another 100,000 people in the next decade—pushing it far past Reno. By 2010, Henderson will be as big as Las Vegas was in 1990.

The new residential subdivisions built all over the Las Vegas Valley in the past few years have created an enormous unsatisfied demand for parks, transportation, and water delivery systems. "Traffic is probably 100 times worse than it was 10 years ago," says Bobby Shelton, spokesman for the Clark County Public Works Department.

Local governments are trying to cope with the onslaught. A rail system connecting downtown, the casino-lined Strip and the airport is on the drawing board. Voters recently approved two tax increases—one for transportation projects, one for water projects—that together will produce some $100 million a year. But even with these projects moving forward Vegas-style—ready, fire, aim—the problem is getting worse, not better. Even the local building industry acknowledges that growth has gotten far ahead of the infrastructure that is needed to support it.

"Someone back in the '70s and '80s should have said, 'Hey, we are going to need parks and schools and roads,'" says Joanne Jensen, of the Southern Nevada Home Builders, who moved to Las Vegas from Chicago in 1960. "We're playing catch-up." After two months in town, Planning Director Chow uses the same words to describe the situation.

Take the infrastructure needs faced by any fast-growing American community, multiply by a factor of about 20, and you get a rough idea of what is going on in Las Vegas. It is similar to other places in that it has come to realize that residential development does not pay for itself. The difference is in the magnitude of its problem. No other city of comparable size is taking on people at anything remotely close to the Las Vegas

17. Operation Desert Sprawl

rate. No other city is being challenged to build so much so quickly.

The infrastructure will be built. That is not really the issue. Las Vegas and Henderson will continue to grow. The question is who will pay the bill.

As in other American cities, property owners have paid for community infrastructure during most of Las Vegas' history. But they have grown weary of the expense, and the one obvious way to grant them relief is to make developers cover more of the cost. Local homebuilders estimate that "impact fees"—fees paid by developers to cover the cost of community infrastructure—already account for a quarter of the cost of a new house in Las Vegas.

It was no surprise this year that a budding politician decided to make a name for himself by proposing that Las Vegas solve its growth problems by soaking its developers. The surprise was who that politician turned out to be, and how potent his message proved.

Nine months ago, Oscar Goodman was known around town as the classic Vegas mob lawyer—a veteran criminal defense attorney, given to wearing dark, double-breasted suits and a short-cropped silver beard. Today, on the strength of his soak-the-developer campaign rhetoric, Oscar Goodman is mayor of Las Vegas.

Ever since the mid 1980s, civic leaders here have worked hard to bury the Bugsy Siegel image—downplaying the gangster past, building suburban-style homes and office parks at a furious pace, and generally trying to re-position Vegas as an affordable, high-energy Sun Belt city attractive to everyone from retirees to young families. A few years ago, the *New York Times Magazine* heralded this new era in Las Vegas by reporting the city's transformation "from vice to nice."

It's true that the gambling industry remains the local economy's bedrock foundation. Las Vegas boasts 110,000 hotel rooms—one for every 11 residents of the region—and attracts 30 million visitors a year. But a fresh-scrubbed image was necessary to catapult Vegas past its sleazy resort-town reputation and support a new wave of mainstream urban growth. And so even as it tried to remain affordable, Las Vegas began to go upscale as well.

Perhaps the most highly publicized success story is Summerlin, the Howard Hughes Corp.'s 35-square-mile development on the valley's west side. On a tract of land that was nothing but empty desert when Hughes bought it, 35,000 people now live in the earth-tone houses and apartments lining cul-de-sacs. The company predicts the population could reach 180,000 by buildout. And lower-end Summerlin knock-offs line the roads leading from the big development toward downtown Las Vegas and the Strip.

Summerlin's sales literature boasts of a school and recreation system that starts with T-ball fields and continues on up to college scholarships for the local youth. "Summerlin," one of the brochures says, "offers a surprising number of public schools and private academies, offering a full range of close-to-home choices for preschool/kindergarten through high school education." It is clever promotion like this that has enabled Las Vegas to become the nation's fastest-growing Sun Belt city without sacrificing its economic base of tourism and gambling. But the cost of all of that growth has been high nevertheless. There was no way to finesse the need to build public infrastructure at an exponential rate. And for most of the past decade, the local governments swallowed hard and arranged for that infrastructure in the old-fashioned way: by raising taxes.

Three years ago, for example, traffic congestion had become so bad that it was decided to speed up completion of the 53-mile Las Vegas Beltway—by 17 years—so it could open in 2003. In Henderson, giant belly scrapers building the road rumble back and forth just beyond the walled-in backyards of brand-new houses.

Amazingly, the $1.5 billion beltway is being finished without federal funds. A 1 percent motor vehicle privilege tax and a "new home fee" of about $500 per house are generating $50 million a year, which goes toward bonds issued to raise capital for the construction.

Seven years ago, Las Vegas had no bus system at all. Today, the Citizen Area Transit system carries 128,000 passengers a day—the same volume as the busiest part of the beltway. Passengers pay 50 percent of the system's costs at the farebox—a higher percentage than in almost any other American city. The Regional Transportation System hopes to begin construction of a 5.2-mile rail line around downtown in 2001—mostly with federal funds and sales-tax revenues—and tie it to a privately funded monorail along the Strip.

Meanwhile, the Southern Nevada Water Authority, which serves as a wholesaler to cities and water districts in the Las Vegas area, is building a $2 billion water delivery project. The project involves a "second straw" from Lake Mead, which Hoover Dam creates about 30 miles southeast of Las Vegas, and 87 miles of large water mains in the Las Vegas Valley. The water agency's goal is to provide enough water for an additional 2 million people.

There is one catch so far as water supply is concerned: Las Vegas still lacks the legal right to draw additional water from Lake Mead. Changing this arrangement will require a massive political deal to overturn the 75-year-old agreement among seven Western states along the Colorado River. But that didn't stop Clark County voters from approving a quarter-cent sales-tax increase last year—expected to generate close to $50 million a year—to pay for the new straw and the other improvements.

So it can't be said that Las Vegas area residents have been unwilling to open their wallets and spend money for growth—they have spent heavily for it. But they realize all too well how much of the bill remains to be paid. This is the issue that is driving Las Vegas politics, and producing its unexpected results.

Even before this year's mayoral campaign, critics of the growth machine began speaking more loudly in the Las Vegas Valley, and their demand that developers pay impact fees had begun to resonate with the voters.

The ringleaders of this new movement have been Jan Laverty Jones, who was Goodman's predecessor as mayor, and Dina Titus, a political science professor at the University of Nevada–Las Vegas, who is Democratic leader in the Nevada Senate. Because Las Vegas has a city manager form of government, the mayor's job was never viewed as important. But Jones, a former Vegas businesswoman who used to appear as Little Bo Peep in television car ads, adopted a high profile—especially on growth issues—at exactly the moment when growth was creating massive frustration.

Jones pushed to require developers to bear more of the cost of parks, roads and other community improvements. Meanwhile, Titus took an aggressive approach in the legislature. In 1997, she floated a "Ring Around the Valley" urban growth boundary concept. That bill failed, but last year she managed to push through a bill requiring local governments to conduct an impact analysis for large commercial and residential projects. Although the bill doesn't require mitigation of impacts, Titus hopes it will encourage local officials in the Las Vegas area to impose mitigation requirements more often and more rationally. "Right now, it's very capricious," she says. "Somebody has to put in street lights, another developer doesn't have to put in anything. It all depends on the political situation at hand."

Into this combustible situation stepped two mayoral candidates with exactly the wrong credentials to deal with it. Arnie Adamsen, a three-term city council member, was a title company executive; Mark Fine was one of the creators of Summerlin. Their ties to the real estate industry could not have been more conspicuous. Both Fine and Adamsen acknowledged that growth was a problem, but refused to recommend that developers pay more of its cost. That opened the door for Oscar Goodman, an unlikely political leader even in a town as unusual as Las Vegas.

At age 59, Goodman had lived in Las Vegas far longer than most of the voters—more than 30 years. A native of Philadelphia, he once clerked for Arlen Specter, then a prosecutor, now a U.S. senator. In the early '60s, however, Goodman headed for Las Vegas, looking for a place where he could make a reputation on his own.

Before long, he gained a reputation as a colorful—and effective—defender of accused mobsters. He once persuaded a judge to drop mob financier Meyer Lansky as a defendant in a case, apparently because of Lansky's failing health. In his most celebrated victory, he kept Tony "The Ant" Spilotro out of jail in the face of multiple murder and racketeering charges. Spilotro became the model for the character played by Joe Pesci in the movie "Casino"; Goodman played himself in the same movie.

Of course, Goodman always insisted that he had a wide-ranging criminal law practice, and only 5 percent of his clients were alleged mobsters. Furthermore, Goodman and his wife were regarded around town as good citizens who often contributed to, or even spearheaded, civic causes. Even so, when Goodman first announced his candidacy for mayor, it was surprising that he even wanted the job.

Given the structure of Las Vegas' government, most mayors have been part-timers who served as glorified presiding officers for the city council. On the other hand, the most recent occupant of the seat had shown that it could be more than a ceremonial position. "Jones made it a more important job," says Eugene Moehring, author of *Resort City in the Sunbelt,* a history of Las Vegas. "She showed that you could use it as a platform. I think Oscar saw that and decided to go for it."

Goodman announced that, if elected mayor, he would give up his law practice. Even so, he wasn't taken seriously at first by the Las Vegas political establishment. Then he started talking about sprawl, real estate developers and impact fees. "Growth has to pay for growth," Goodman said in his radio and television commercials, using his fast-talking West Philadelphia accent to get his message across. "Either it has to come from taxpayers or developers.... It's time to make developers pay impact fees when they build new homes."

Goodman argued that even a modest impact fee would bring in more than $15 million a year to help pay for sprawl costs and revitalize Las Vegas' downtown. He recommended a fee of roughly $2,000 for each new home, and said he would use it to revitalize neglected downtown neighborhoods. Such a scheme would require a change in state law—and a big fight with the real estate lobby—but Goodman promised to take on both the legislature and the builders. Polls showed that 80 percent of Las Vegans supported higher development fees.

> It wasn't ties to organized crime that concerned voters in the Las Vegas mayor's race; it was ties to real estate.

Goodman was widely perceived in Las Vegas as a "Jesse Ventura candidate"—a celebrity from another field with a populist message and enough personal wealth to assure that he would not be bought by special interests. His opponents essentially played into Goodman's hands. Adamsen, his opponent in the June runoff, insisted that the fees would merely be passed on to homebuyers. "Development fees are very popular with the public because they want someone else to pay for it," he said. Adamsen focused on Goodman's criminal connections, saying they would harm the city's new, family-oriented image. If anybody in town knew about extracting money from a community through nefarious means, he charged, it was Goodman, not the developers.

In the end, though, Goodman's contacts with organized crime were a non-issue. In the Las Vegas of 1999, impact fees are a bigger issue than crime, organized or otherwise. What mattered to most voters was that the candidate didn't have any close ties to the real estate business. Goodman won 64 percent of the vote in his runoff with Adamsen. The Jesse Ventura approach worked so well that even many leading members of the Las Vegas political establishment embraced Goodman during the runoff.

On the last Monday in June, Oscar Goodman took the oath of office before an overflow crowd in the Las Vegas City Council chambers, then stepped outside for his first press conference as mayor. At 10:30 in the morning, it was 95 degrees, with a hot, dry desert wind blowing across the City Hall courtyard. Wearing a trademark double-breasted suit, Goodman was unfazed by what was, for Las Vegas, a typical June morning.

And he also seemed unfazed by his quick introduction to the real world of municipal politics in America's most transient city. In the three short weeks between the election and the inauguration, one of Goodman's campaign aides was accused of attempting to charge $150,000 to arrange an appointment to the City Council. The new mayor was faced with the question of what to do about renewing the contract of the city's waste hauler, who had contributed heavily to his campaign. And, of course, during the Goodman interregnum, 4,000 new residents had arrived in Las Vegas.

Goodman insists he will take his plan for a $2,000-a-house impact fee to the legislature. But the next regular session does not begin until January of 2001, and even with 18 months to build support, his task will not be easy. Current state law prevents any such scheme from being implemented—as it does many other development fees—and the real estate lobby has vowed to oppose it.

Meanwhile, says Goodman with his typical glib charm, "I've sat down with the developers, and I've told them we're going to run an efficient City Hall operation that will meet their day-to-day needs. Those guys are going to *want* impact fees to pay us back for all the good things we're going to do for them in the next two years."

Even Goodman sympathizers agree that Goodman—like that other populist, Jesse Ventura—faces a difficult challenge in learning how to govern on the job. "Growth is a tough issue, and he's going to have to hit the ground running on it," says former Mayor Jan Jones. "It's going to be a big learning curve." In particular, Goodman will have to cultivate regional agencies and suburban governments in Las Vegas—with which Jones often clashed—in order to curb growth on the metropolitan fringe and encourage renewal of older areas in Las Vegas itself.

Increasingly, however, there is agreement among both politicians and developers that the future of Las Vegas—like the future of most American communities—involves developers paying for the impact of growth somehow: There simply is no other way to finance the infrastructure costs. Whatever Goodman does, the Southern Nevada Water Authority is planning to levy a hookup fee of several thousand dollars a unit to help pay for the second straw from Lake Mead. That will only increase the average home price in the metro area, which is already up to $142,000.

Oscar Goodman may or may not succeed as mayor, but it's clear that the political sentiment that he tapped into is here to stay. In wide-open Las Vegas—as in so many of the more conventional cities across the country—the crucial votes no longer lie with the upwardly mobile families desperate for a place to live. They lie with the established middle-class residents who are tired of paying the bill.

Turning the Tide

A coastal refuge holds its ground as sprawl spills into southern Maine.

By Richard M. Stapleton

"The shore has a dual nature," Rachel Carson wrote, "changing with the swing of the tides, belonging now to the land, now to the sea." At Wells, Maine, a daysail south of the harbor where Carson wrote *The Edge of the Sea*, the Merriland River enters an expanse of tidal wetlands, slows, and doubles back and forth, oxbow upon oxbow.

This is not Maine's picture-postcard coastline of craggy granite standing firm against the pounding sea; that's farther east. The state's southern coast is softer, built of sand and gravel bulldozed here by Ice Age glaciers, and with each reversal, the Merriland River gnaws at its banks, collecting silicate souvenirs to be carried out to sea.

The wetlands that fill the Merriland River estuary, along with those of the Mousam, Kennebunk, Ogunquit, and other rivers that drain the mountains to the north and west, are partially protected as part of the Rachel Carson National Wildlife Refuge. The refuge is divided into ten units, strung like jewels from Kittery on the Maine–New Hampshire border through Cape Elizabeth, just south of Portland. Currently the refuge encompasses nearly 5,000 acres of protected coastal wildlife habitat, including significant wetlands.

Their adjacent uplands, however, remain largely at risk. When the refuge was first established, in 1966, the need to include uplands in the refuge boundary went largely unrecognized. So today, standing at an overlook at the Rachel Carson Refuge, we see not only the Merriland salt marsh, and perhaps a belted kingfisher or a great blue heron, but also a stark new house jutting from the bank across the way. The home, built on spec, is both a symbol and product of another sea change—sprawl—that threatens to overrun not only the Rachel Carson Refuge but all of south coastal Maine.

A Tenuous Refuge

"The first building sets the precedent, and then others follow," says Debra Kimbrell-Anderson, the refuge's deputy manager. "If that house were not there"—she points to the white house across the way—"there would be no septic tank, no lawn fertilizer, no car drippings on the driveway, no roof runoff, no dogs running free...." Kimbrell-Anderson tallies the problems caused by development of the immediate uplands. "We are losing the critical buffer at the wetland's edge."

Left unsaid is that the house itself is at risk. Just behind us, part of the refuge's wheelchair-accessible trail had to be relocated after a huge chunk of pine-forested upland ledge, undercut by an oxbow, simply sloughed off and fell into the river, taking the trail with it. "There is tremendous development pressure here to allow creative structures—houses on pillars—with the tidal flow beneath them, in essence flushing their basements into the watershed," Kimbrell-Anderson adds. And indeed, the Trust for Public Land is presently working with the refuge to protect 72 acres at Goose Rocks Corner, where a subdivision is planned on a tract crisscrossed by wetlands.

The refuge supports an incredible diversity of wildlife. There are 47 known mammal species, from mink and ermine to moose and black bear, and 35 reptile and amphibian species,

---

The influx of "people from away," as the locals refer to out-of-staters, is just part of the pressure that is driving sprawl up Maine's southern coast.

Fisherman find they no longer can afford to live by the sea... farmers find their fields sprouting houses.

including half a dozen endangered or threatened turtles. And it is a birder's paradise. Songbird residents are joined by shorebird migrations in spring and summer, waterfowl concentrations in winter and early spring, and raptor migrations in early fall. In all, some 250 species, including the endangered piping plover, peregrine falcon, and least and roseate terns, are refuge regulars.

But their habitat is tenuous. With its ten separate units, the refuge has a circuitous boundary and a lot more neighbors than most preserves. All this development has countless consequences. "People want a better view, so they will cut brush and trees in the refuge," Kimbrell-Anderson says. "We find our boundary signs missing, and then we find lawns and gardens extending into the refuge."

Not all impact is so obvious. Development redirects the flow of water. The salinity of the marsh is reduced, causing invasive plants such as phragmites and purple loosestrife to displace the native cattail. While native plants such as cattail are richly supportive of wildlife, a field of phragmites reeds is as barren as the driest desert.

The carving-up of the countryside has communities scrambling to protect spaces at imminent risk, while working longer term to inventory open space and secure funding for land preservation. The Trust for Public Land, long active on all three fronts, recently opened a Maine field office to increase its ability to help. "The irony," says Project Manager Jennifer Melville, "is that for all its open space, Maine ranks nearly last in the percentage of publicly owned land."

The Price Paid for Sprawl

The issue of sprawl at first seems oxymoronic here. Maine's land mass is equal to that of the five other New England states combined, while its population is less than a quarter of Boston's. But increasingly when the people of Boston, barely an hour's drive south of the Maine border, seek a "getaway," they get away to Maine's coast and lakes. So do a lot of others; a recent *New York Times* story, "Second-Home Buyers Go Longer Distances," featured a home belonging to the chancellor of New York City schools. The chancellor's retreat is located in Greenwood, Maine, some 400 miles from Manhattan.

The influx of "people from away," as the locals refer to out-of-staters, is just part of the pressure that is driving sprawl up Maine's southern coast. People are simply spreading out. While the state's population, 1.2 million, has grown less than 1 percent a year over the past three decades, residential land use increased nearly 150 percent during the same period, resulting in what a report by the Maine State Planning Office called "unanticipated and unintended consequences."

The consequences start with economic costs. Maine taxpayers have had to spend $300 million building new classrooms in rural and suburban areas, even as the state's total school population has gone down. Meanwhile, the shrinking urban populations—often elderly and poor—are hard-pressed to maintain existing facilities. Portland's historic St. Dominic's Church, its neighborhood population cut by half, is being closed even as the Roman Catholic Diocese is building new churches in the surrounding suburban ring.

Then there are livelihood issues. With shoreline proximity at a premium, fishermen find they no longer can afford to live by the sea. And with cleared land easiest to develop, farmers find their fields sprouting houses—two-thirds of new development in Maine's southernmost counties occurred on prime agricultural land.

Finally, in this state which trades heavily on its pristine environment, there are environmental consequences. Sprawl brings fragmentation—reducing the countryside to islands

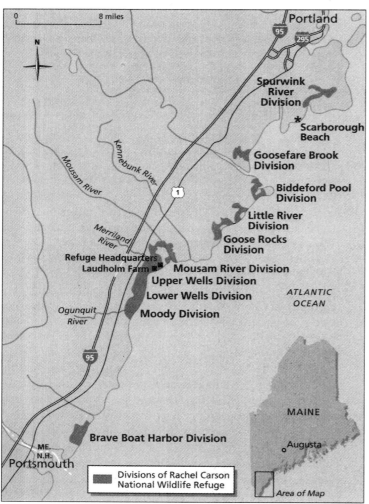

MAGELLAN GEOGRAPHIX/MAPS.COM

The Rachel Carson Refuge is fragmented into ten separate units, making it vulnerable to the effects of development.

of open space too small to support wildlife's foraging—and it brings pollution. "In part because of the added vehicle miles being driven every day, southern Maine's air exceeds federal standards for the hydrocarbons and nitrogen oxides that cook in the summertime to become smog," says Evan Richert, director of the Maine State Planning Office.

Calling in his inaugural address last year for a "great public discussion" of the issue of sprawl, Governor Angus King called it a price paid in terms of quality of life. "How much do any of us gain," he asked, "if we're richer in money and poorer in spirit, poorer in the things that make Maine the special place that it has been for so many years?"

Funds for Maine's Future

Maine's voters seem to agree. In 1987 they approved a referendum establishing a $35 million Land for Maine's Future program to acquire land for open space and recreation. In the following decade, more than 65,000 acres were protected. When the fund was depleted last year, a consortium of conservation and outdoors groups, supported by businesses dependent on outdoor recreation, rallied to campaign for a $50-million bond referendum to replenish it. TPL joined The Nature Conservancy and Maine Coast Heritage Trust in helping to staff and fund the campaign. Outdoor retailer L. L. Bean jump-started the campaign funding, mounted a major display at its Freeport headquarters, and played a high-profile leadership role throughout. The November vote was not even close; the referendum passed in every county, and overall the approval margin was better than two to one.

Even as the bond campaign unfolded, TPL was working to protect land for which time was running out. In one of its most visible projects, TPL purchased and held the privately owned Scarborough Beach State Park. Scarborough Beach is 1,800 gently curving feet of glistening white sand backed by barrier dunes, a 15-acre salt pond, and teeming marshlands. The beach, just a short drive from Portland, attracts more than 60,000 people a year. And for some 80 years, it had been the private property of one family, with Maine leasing it as a state park. "My great-grandfather had an interest in conserving the beach and keeping it in its pristine state," says Seth Sprague, "and essentially the family has been involved in doing just that ever since." The Spragues didn't want the beach developed, but they did want to sell it. They offered the land to the state, but Maine, its land acquisition coffers empty, was powerless to act. In stepped TPL, purchasing 62 acres and holding the land until the state could arrange financing.

"Saving Scarborough Beach clearly illustrated the benefits of the Land for Maine's Future bond to the voters," Governor King says. "It played a key role in gaining passage of the $50-million bond, and it would not have happened without the leadership of TPL."

Mapping the Watershed

Even with the bond act approved, resources are limited; one of the problems is knowing just which property to protect. At Laudholm Farm, next door to the Rachel Carson Refuge headquarters, the Wells National Estuarine Research Reserve is undertaking what it calls the Coastal Mosaic Project, a program designed to develop region-wide conservation plans and strategies. The name comes from the map created when land use is plotted; open space in southern Maine shows up like random green tiles scattered in a large mosaic. Eighteen local land trusts, conservation commissions, and environmental organizations are working with the reserve to connect existing open space and protect water quality.

"We are taking a watershed approach to conservation," says Reserve Director Kent Kirkpatrick, "to help communities develop proactive conservation strategies across political boundaries. The quality of the salt marsh at the bottom of the watershed depends on what happens upstream." Adds Reserve Volunteer Coordinator Nancy Bayse, "We should not be operating in a vacuum in making land preservation decisions. Sometimes a great-looking piece of land may not be the most important in the biological or ecological sense." The Coastal Mosaic Project will help conservationists better evaluate properties that come on the market as well as identify critical tracts in need of protection.

The reserve knows firsthand the threat posed by sprawl. Its home is a saltwater farm dating back to the 1640s. Listed in the National Register of Historic Places, Laudholm Farm has been carefully restored—its buildings now house the reserve's research laboratory and offices as well as classrooms, exhibit space, and a public auditorium. The reserve operates on 1,600 acres of federal, state, town, and private land and includes saltwater marshes, estuaries, woodlands, meadows, and ocean beach. Yet the meadow flanking the roadway to its farmstead has sprouted two new houses, and the reserve risks losing its view of the sea should two privately held properties within its boundaries be lost to development.

The trails at both the Wells Reserve and the Rachel Carson Refuge attract a steady flow of birdwatchers and hikers, but perhaps their greatest impact is on children. At the refuge, overlooking the Merriland River, we run into a group of chattering kindergarteners. "We've brought 160 kids here in the past two days," says teacher Ellen Barter. "We combined a unit on habitat with one on the five senses. The children are recording all the things we saw, heard, touched, smelled, and tasted." ("Tasted," she quickly adds, is covered by snacks!)

"Education is important to every wildlife refuge, but it's especially appropriate here," says Refuge Manager Ward Feurt. Rachel Carson's courageous book *Silent Spring*, which publicized the deadly effects of DDT and other pesticides on birds and other wildlife, is credited with inspiring the environmental movement in America. "Carson's ecological message was that all life is interdependent," Feurt says. "We need to keep teaching that message here; we also need to act on it."

Richard M. Stapleton is a sailor, gardener, and freelance journalist living in New Jersey.

A greener, or browner, Mexico?

CIUDAD JUAREZ

NAFTA purports to be the world's first environmentally friendly trade treaty, but its critics claim it has made Mexico dirtier. There is evidence on both sides

GIVEN the industrial invasion that the North American Free-Trade Agreement has brought to the cities that line Mexico's border with the United States, one might expect the skies of Ciudad Juarez to be brown with pollution, and its watercourses solid with toxic sludge. But no. The centre of Ciudad Juarez looks like a poorer version of El Paso, Texas, its cross-border neighbour: flat, dull and full of shopping malls. Since NAFTA took effect on January 1st 1994, Mexico has passed environmental laws similar to those of the United States and Canada, its NAFTA partners; has set up a fully-fledged environment ministry; and has started to benefit from several two- and three-country schemes designed to fulfil "side accords" on the environment—the first such provisions in a trade agreement.

But the debate over NAFTA's impact on Mexico's environment remains polarised. Certainly NAFTA has had the hoped-for result of encouraging industries to move to Mexico. Since 1994, the number of *maquiladoras* (factories using imported raw materials to make goods for re-export, usually to the United States) has doubled, and half of the new factories are in the 100km-wide (60 mile) northern border region.

But that has put pressure on water supplies. The influx of migrant workers is slowly drying out cities like Juarez, which shares its only water source, an underground aquifer, with El Paso (which has other water supplies). Growth is aggravating old deficiencies: 18% of Mexican border towns have no drinking water, 30% no sewage treatment, and 43% inadequate rubbish disposal, according to Franco Barreno, head of the Border Environment Co-operation Commission (BECC), a bilateral agency set up under NAFTA. Putting that right, which is the BECC's job, will cost $2 billion–3 billion.

The problem of toxic waste is more serious still, and is being ignored. Mexico has just one landfill site for hazardous muck, which can take only 12% of the estimated waste from non-*maquila* industry; the rest is dumped illegally. Mexican law says *maquiladoras* must send their toxic waste back to the country the raw materials came from but, with enforcement weak and repatriation expensive, many are probably ignoring the law, says Victoriano Garza, of the Autonomous University of Ciudad Juarez.

Mexico has long been a dumping-ground for unwanted rubbish from the United States. But in fact less than 1% of the country's toxic gunge is generated in the border region. After 1994, despite the growth in *maquiladoras,* Mexico's exports of toxic waste to the United States dropped sharply—suggesting that tougher rules were making companies adopt greener manufacturing policies. But then, in 1997, waste exports mysteriously shot up again (see chart).

Even greens are split about NAFTA. "In 15 years as an environmentalist, I haven't noticed any change in policy," laments Homero Aridjis, a writer. Quite the contrary, says Tlahoga Ruge, of the North American Centre for Environmental Information and Communication: "NAFTA has brought the environment into the mainstream in Mexico, as something to be taken seriously."

Such is the complexity of the issue, only in June did the Commission for Environmental Co-operation (CEC), a Montreal-based body created in 1994 to implement the environmental side-agreement, publish its methodology for assessing NAFTA's impact. An accompanying case study, on maize farming in Mexico, illustrates the difficulties. Lower protective tariffs are forcing Mexico's subsistence maize farmers to modernise, change their crops or look for other work; each choice has potential impacts, such as soil deterioration, depletion of the maize gene pool, or migration to cities. Social and environmental impacts go hand-in-hand.

Critics say that, as well as being slow, the CEC has too few powers. It can investigate citizens' complaints about breaches of each NAFTA coun-

19. Greener, or Browner, Mexico?

Bubbling Mexico's *maquiladora* plants: employment; transfer of toxic waste* to the United States
Sources: Mexican trade and industry ministry; United States Environmental Protection Agency

try's national environmental laws, but governments are free to ignore its recommendations. Moreover, the side-agreement's gentle urgings are out-gunned by the main accord's firm admonition, in its Chapter 11, to protect foreign investors from uncertainty. A recent report by the International Institute for Sustainable Development, a group based in Winnipeg, Canada, warns that this provision is increasingly being used by business to the detriment of environmental protection. And unlike the CEC, the arbitration bodies that settle Chapter 11 disputes make binding decisions.

Companies have challenged environmental laws in all three NAFTA countries, but Mexico is particularly vulnerable. Its laws are new, its institutions untested and its environmental culture undeveloped. To these handicaps was added the collapse of Mexico's peso in 1995, from which it is still trying to recover. Beefing up environmental institutions has not been its top priority. In Juarez, for example, there are only 15 federal environmental inspectors (and they cover the whole of Chihuahua state). Mexico also lags in data-collection: next week the CEC is due to publish a report on industrial emissions and air pollution, but in the United States and Canada only, because Mexican factories do not have to report their emissions.

These things need to change. NAFTA is supposed to lead to completely free trade after 15 years; its full impact has yet to be felt. In 2001, the *maquiladoras* will lose some of their tax breaks. Increasingly, American firms may opt to set up ordinary factories in Mexico, which would not have to send their waste back home. Mexico badly needs somewhere to put this stuff, but so far attempts to find sites for new toxic-waste landfills have been scuppered by not-in-my-backyard opposition (and, ironically, by the strict stipulations of the new environmental laws).

But not all is murk. Besides better laws and institutions, Mexico has seen a rapid rise in non-governmental organisations, themselves partly an import from the north, which work on the education and awareness-raising that they say the government is neglecting. Ciudad Juarez may be running out of water to drink, but Mexico City's aquifer is so low that the whole place is slowly sinking. And, unlike the capital, Juarez benefits from all kinds of two-country efforts to fix its problem, thanks mainly to NAFTA.

Unit 3

Unit Selections

20. **The Rise of the Region State,** Kenichi Ohmae
21. **Continental Divide,** Torsten Wohlert
22. **Russia's Fragile Union,** Matthew Evangelista
23. **Beyond the Kremlin's Walls,** The Economist
24. **The Delicate Balkan Balance,** The Economist
25. **Sea Change in the Arctic,** Richard Monastersky
26. **Greenville: From Back Country to Forefront,** Eugene A. Kennedy
27. **Flood Hazards and Planning in the Arid West,** John Cobourn
28. **The Rio Grande: Beloved River Faces Rough Waters Ahead,** Steve Larese
29. **Does It Matter Where You Are?** The Economist

Key Points to Consider

❖ To what regions do you belong?

❖ Why are maps and atlases so important in discussing and studying regions?

❖ What major regions in the world are experiencing change? Which ones seem not to change at all? What are some reasons for the differences?

❖ What regions in the world are experiencing tensions? What are the reasons behind these tensions? How can the tensions be eased?

❖ Why are regions in Africa suffering so greatly?

❖ How will East Asia change as China and Taiwan expand relationships?

❖ Discuss whether or not the nation-state system is an anachronism.

❖ Why is regional study important?

 **Links** www.dushkin.com/online/

13. **AS at UVA Yellow Pages: Regional Studies**
    http://xroads.virginia.edu/~YP/regional/regional.html
14. **Can Cities Save the Future?**
    http://www.huduser.org/publications/econdev/habitat/prep2.html
15. **IISDnet**
    http://iisd1.iisd.ca
16. **NewsPage**
    http://www.individual.com
17. **Telecommuting as an Investment: The Big Picture—John Wolf**
    http://www.svi.org/telework/forums/messages5/48.html
18. **The Urban Environment**
    http://www.geocities.com/RainForest/Vines/6723/urb/index.html
19. **Virtual Seminar in Global Political Economy//Global Cities & Social Movements**
    http://csf.colorado.edu/gpe/gpe95b/resources.html

These sites are annotated on pages 6 and 7.

The Region

The region is one of the most important concepts in geography. The term has special significance for the geographer, and it has been used as a kind of area classification system in the discipline.

Two of the regional types most used in geography are "uniform" and "nodal." A uniform region is one in which a distinct set of features is present. The distinctiveness of the combination of features marks the region as being different from others. These features include climate type, soil type, prominent languages, resource deposits, and virtually any other identifiable phenomenon having a spatial dimension.

The nodal region reflects the zone of influence of a city or other nodal place. Imagine a rural town in which a farm-implement service center is located. Now imagine lines drawn on a map linking this service center with every farm within the area that uses it. Finally, imagine a single line enclosing the entire area in which the individual farms are located. The enclosed area is defined as a nodal region. The nodal region implies interaction. Regions of this type are defined on the basis of banking linkages, newspaper circulation, and telephone traffic, among other things.

This unit presents examples of a number of regional themes. These selections can provide only a hint of the scope and diversity of the region in geography. There is no limit to the number of regions; there are as many as the researcher sets out to define.

"The Rise of the Region State" suggests that the nation-state is an unnatural and even dysfunctional unit for organizing human activity. "Continental Divide" and "Russia's Fragile Union" both deal with changing geopolitical situations. The next two articles consider regional changes in Russia and the Balkans. "Sea Change in the Arctic" analyzes the impact of global warming in the region. The continuing rise of Greenville, South Carolina, is documented in the next article. John Cobourn summarizes flood types in the arid western United States. The hard-pressed Rio Grande River is discussed next. Then, "Does It Matter Where You Are?" considers aspects of geographical location principles in the context of the new global economic systems.

THE RISE OF THE REGION STATE

Kenichi Ohmae

Kenichi Ohmae is Chairman of the offices of McKinsey & Company in Japan.

The Nation State Is Dysfunctional

THE NATION STATE has become an unnatural, even dysfunctional, unit for organizing human activity and managing economic endeavor in a borderless world. It represents no genuine, shared community of economic interests; it defines no meaningful flows of economic activity. In fact, it overlooks the true linkages and synergies that exist among often disparate populations by combining important measures of human activity at the wrong level of analysis.

For example, to think of Italy as a single economic entity ignores the reality of an industrial north and a rural south, each vastly different in its ability to contribute and in its need to receive. Treating Italy as a single economic unit forces one—as a private sector manager or a public sector official—to operate on the basis of false, implausible and nonexistent averages. Italy is a country with great disparities in industry and income across regions.

On the global economic map the lines that now matter are those defining what may be called "region states." The boundaries of the region state are not imposed by political fiat. They are drawn by the deft but invisible hand of the global market for goods and services. They follow, rather than precede, real flows of human activity, creating nothing new but ratifying existing patterns manifest in countless individual decisions. They represent no threat to the political borders of any nation, and they have no call on

20. Rise of the Region State

any taxpayer's money to finance military forces to defend such borders.

Region states are natural economic zones. They may or may not fall within the geographic limits of a particular nation—whether they do is an accident of history. Sometimes these distinct economic units are formed by parts of states, such as those in northern Italy, Wales, Catalonia, Alsace-Lorraine or Baden-Württemberg. At other times they may be formed by economic patterns that overlap existing national boundaries, such as those between San Diego and Tijuana, Hong Kong and southern China, or the "growth triangle" of Singapore and its neighboring Indonesian islands. In today's borderless world these are natural economic zones and what matters is that each possesses, in one or another combination, the key ingredients for successful participation in the global economy.

Look, for example, at what is happening in Southeast Asia. The Hong Kong economy has gradually extended its influence throughout the Pearl River Delta. The radiating effect of these linkages has made Hong Kong, where GNP per capita is $12,000, the driving force of economic life in Shenzhen, boosting the per capital GNP of that city's residents to $5,695, as compared to $317 for China as a whole. These links extend to Zhuhai, Amoy and Guangzhou as well. By the year 2000 this cross-border region state will have raised the living standard of more than 11 million people over the $5,000 level. Meanwhile, Guangdong province, with a population of more than 65 million and its capital at Hong Kong, will emerge as a newly industrialized economy in its own right, even though China's per capita GNP may still hover at about $1,000. Unlike in Eastern Europe, where nations try to convert entire socialist economies over to the market, the Asian model is first to convert limited economic zones—the region states—into free enterprise havens. So far the results have been reassuring.

These developments and others like them are coming just in time for Asia. As Europe perfects its single market and as the United States, Canada and Mexico begin to explore the benefits of the North American Free Trade Agreement (NAFTA), the combined economies of Asia and Japan lag behind those of the other parts of the globe's economic triad by about $2 trillion—roughly the aggregate size of some 20 additional region states. In other words, for Asia to keep pace existing regions must continue to grow at current rates throughout the next decade, giving birth to 20 additional Singapores.

Many of these new region states are already beginning to emerge. China has expanded to 14 other areas—many of them inland—the special economic zones that have worked so well for Shenzhen and Shanghai. One such project at Yunnan will become a cross-border economic zone encompassing parts of Laos and Vietnam. In Vietnam itself Ho Chi Minh City (Saigon) has launched a similar "sepzone" to attract foreign capital. Inspired in part by Singapore's "growth triangle," the governments of Indonesia, Malaysia and Thailand in 1992 unveiled a larger triangle across the Strait of Malacca to link Medan, Penang and Phuket. These developments are not, of course, limited to the developing economies in Asia. In economic terms the United States has never been a single nation. It is a collection of region states: northern and southern California, the "power corridor" along the East Coast between Boston and Washington, the Northeast, the Midwest, the Sun Belt, and so on.

What Makes a Region State

THE PRIMARY linkages of region states tend to be with the global economy and not with their host nations. Region states make such effective points of entry into the global economy because the very characteristics that define them are shaped by the demands of that economy. Region states tend to have between five million and 20 million people. The range is broad, but the extremes are clear: not half a million, not 50 or 100 million. A region state must be small enough for its citizens to share certain economic and consumer interests but of adequate size to justify the infrastructure—communication and transportation links and quality professional services—necessary to participate economically on a global scale.

It must, for example, have at least one international airport and, more than likely, one good harbor with international-class freight-handling facilities. A region state must also be large enough to provide an attractive market for the brand development of leading consumer products. In other words, region states are not defined by their economies of scale in production (which, after all, can be leveraged from a base of any size through exports to the rest of the world) but rather by their having reached efficient economies of scale in their consumption, infrastructure and professional services.

For example, as the reach of television networks expands, advertising becomes more efficient. Although trying to introduce a consumer brand throughout all of Japan or Indonesia may still prove prohibitively expensive, establishing it firmly in the Osaka or Jakarta region is far more affordable—and far more likely to generate handsome returns. Much the same is true with sales and service networks, customer satisfaction programs, market surveys and management information systems: efficient scale is at the regional, not national, level. This fact matters because, on balance, modern marketing techniques and technologies shape the economies of region states.

Where true economies of service exist, religious, ethnic and racial distinctions are not important—or, at least, only as important as human nature requires. Singapore is 70 percent ethnic Chinese, but its 30 percent minority is not much of a problem because commercial prosperity creates sufficient affluence for all. Nor are ethnic differences a source of concern for potential investors looking for consumers.

Indonesia—an archipelago with 500 or so different tribal groups, 18,000 islands and 170 million people—would logically seem to defy effective organization within a single mode of political government. Yet Jakarta has traditionally attempted to impose just such a central control by applying fictional averages to the entire nation. They do not work. If, however, economies of service allowed two or three Singapore-sized region states to be created within Indonesia, they could be managed. And they would ameliorate, rather than exacerbate, the country's internal social divisions. This holds as well for India and Brazil.

The New Multinational Corporation

WHEN VIEWING the globe through the lens of the region state, senior corporate managers think differently about the geographical expansion of their businesses. In the past the primary aspiration of multinational corporations was to create, in effect, clones of the parent organization in each of the dozens of countries in which they operated. The goal of this system was to stick yet another pin in the global map to mark an increasing number of subsidiaries around the world.

More recently, however, when Nestlé and Procter & Gamble wanted to expand their business in Japan from an already strong position, they did not view the effort as just another pin-sticking exercise. Nor did they treat the country as a single coherent market to be gained at once, or try as most Western companies do to establish a foothold first in the Tokyo area, Japan's most tumultuous and overcrowded market. Instead, they wisely focused on the Kansai region around Osaka and Kobe, whose 22 million residents are nearly as affluent as those in Tokyo but where competition is far less intense. Once they had on-the-ground experience on how best to reach the Japanese consumer, they branched out into other regions of the country.

Much of the difficulty Western companies face in trying to enter Japan stems directly from trying to shoulder their way in through Tokyo. This instinct often proves difficult and costly. Even if it works, it may also prove a trap; it is hard to "see" Japan once one is bottled up in the particular dynamics of the Tokyo marketplace. Moreover, entering the country through a different regional doorway has great economic appeal. Measured by aggregate GNP the Kansai region is the seventh-largest economy in the world, just behind the United Kingdom.

Given the variations among local markets and the value of learning through real-world experimentation, an incremental region-based approach to market entry makes excellent sense. And not just in Japan. Building an effective presence across a landmass the size of China is of course a daunting prospect. Serving the people in and around Nagoya City, however, is not.

If one wants a presence in Thailand, why start by building a network over the entire extended landmass? Instead focus, at least initially, on the region around Bangkok, which represents the lion's share of the total potential market. The same strategy applies to the United States. To introduce a new top-of-the-line car into the U.S. market, why replicate up front an exhaustive coast-to-coast dealership network? Of the country's 3,000 statistical metropolitan areas, 80 percent of luxury car buyers can be reached by establishing a presence in only 125 of these.

The Challenges for Government

TRADITIONAL ISSUES of foreign policy, security and defense remain the province of nation states. So, too, are macroeconomic and monetary policies—the taxation and public investment needed to provide the necessary infrastructure and incentives for region-based activities. The government will also remain responsible for the broad requirements of educating and training citizens so that they can participate fully in the global economy.

Governments are likely to resist giving up the power to intervene in the economic realm or to relinquish their impulses for protectionism. The illusion of control is soothing. Yet hard evidence proves the contrary. No manipulation of exchange rates by central bankers or political appointees has ever "corrected" the trade imbalances between the United States and Japan. Nor has any trade talk between the two governments. Whatever cosmetic actions these negotiations may have prompted, they rescued no industry and revived no economic sector. Textiles, semiconductors, autos, consumer electronics—the competitive situation in these industries did not develop according to the whims of policymakers but only in response to the deeper logic of the competitive marketplace. If U.S. market share has dwindled, it is not because government policy failed but because individual consumers decided to buy elsewhere. If U.S. capacity has migrated to Mexico or Asia, it is only because individual managers made decisions about cost and efficiency.

The implications of region states are not welcome news to established seats of political power, be they politicians or lobbyists. Nation states by definition require a domestic political focus, while region states are en-sconced in the global economy. Region states that sit within the frontiers of a particular nation share its political goals and aspirations. However, region states welcome foreign investment and ownership—whatever allows them to employ people productively or to improve the quality of life. They want their people to have access to the best and cheapest products. And they want whatever surplus accrues from these activities to ratchet up the local quality of life still further and not to support distant regions or to prop up distressed industries elsewhere in the name of national interest or sovereignty.

When a region prospers, that prosperity spills over into the adjacent regions within the same political confederation. Industry in the area immediately in and around Bangkok has prompted investors to explore options elsewhere in Thailand. Much the same is true of Kuala Lumpur in Malaysia, Jakarta in Indonesia, or Singapore, which is rapidly becoming the unofficial capital of the Association of Southeast Asian Nations. São Paulo, too, could well emerge as a genuine region state, someday entering the ranks of the Organization of Economic Cooperation and Development. Yet if Brazil's central government does not allow the São Paulo region state finally to enter the global economy, the country as a whole may soon fall off the roster of the newly industrialized economies.

Unlike those at the political center, the leaders of region states—interested chief executive officers, heads of local unions, politicians at city and state levels—often welcome and encourage foreign capital investment. They do not go abroad to attract new plants and factories only to appear back home on television vowing to protect local companies at any cost. These leaders tend to possess an international outlook that can help defuse many of the usual kinds of social tensions arising over issues of "foreign" versus "domestic" inputs to production.

In the United States, for example, the Japanese have already established about 120 "transplant" auto factories throughout the Mississippi Valley. More are on the way. As their share of the U.S. auto industry's production grows, people in that region who look to these plants for their livelihoods and for the tax revenues needed to support local communities will stop caring whether the plants belong to U.S.- or Japanese-based companies. All they will care about are the regional economic benefits of having them there. In effect, as members of the Mississippi Valley region state, they will have leveraged the contribution of these plants to help their region become an active participant in the global economy.

Region states need not be the enemies of central governments. Handled gently, region states can provide the opportunity for eventual prosperity for all areas within a nation's traditional political control. When political and industrial leaders accept and act on these realities, they help build prosperity. When they do not—falling back under the spell of the nationalist economic illusion—they may actually destroy it.

Consider the fate of Silicon Valley, that great early engine of much of America's microelectronics industry. In the beginning it was an extremely open and entrepreneurial environment. Of late, however, it has become notably protectionist—creating industry associations, establishing a polished lobbying presence in Washington and turning to "competitiveness" studies as a way to get more federal funding for research and development. It has also begun to discourage, and even to bar, foreign investment, let alone foreign takeovers. The result is that Boise and Denver now prosper in electronics; Japan is developing a Silicon Island on Kyushu; Taiwan is trying to create a Silicon Island of its own; and Korea is nurturing a Silicon Peninsula. This is the worst of all possible worlds: no new money in California and a host of newly energized and well-funded competitors.

Elsewhere in California, not far from Silicon Valley, the story is quite different. When Hollywood recognized that it faced a severe capital shortage, it did not throw up protectionist barriers against foreign money. Instead, it invited Rupert Murdoch into 20th Century Fox, C. Itoh and Toshiba into Time-Warner, Sony into Columbia, and Matsushita into MCA. The result: a $10 billion infusion of new capital and, equally important, $10 billion less for Japan or anyone else to set up a new Hollywood of their own.

Political leaders, however reluctantly, must adjust to the reality of economic regional entities if they are to nurture real economic flows. Resistant governments will be left to reign over traditional political territories as all meaningful participation in the global economy migrates beyond their well-preserved frontiers.

Canada, as an example, is wrongly focusing on Quebec and national language tensions as its core economic and even political issue. It does so to the point of still wrestling with the teaching of French and English in British Columbia, when that province's economic future is tied to Asia. Furthermore, as NAFTA takes shape the "vertical" relationships between Canadian and U.S. regions—Vancouver and Seattle (the Pacific Northwest region state); Toronto, Detroit and Cleveland (the Great Lakes region state)—will become increasingly important. How Canadian leaders deal with these new entities will be critical to the continuance of Canada as a political nation.

In developing economies, history suggests that when GNP per capita reaches about $5,000, discretionary income crosses an invisible threshold. Above that level people begin wondering whether they have reasonable access to the best and cheapest available products and whether they have an adequate quality of life. More troubling for those in political control, citizens also begin to consider whether their government is doing as well by them as it might.

Such a performance review is likely to be unpleasant. When governments control information—and in large measure because they do—it is all too easy for them to believe that they "own" their people. Governments begin restricting access to certain kinds of goods or services or pricing them far higher than pure economic logic would dictate. If market-driven levels of consumption conflict with a government's pet policy or general desire for control, the obvious response is to restrict consumption. So what if the people would choose otherwise if given the opportunity? Not only does the government withhold that opportunity but it also does not even let the people know that it is being withheld.

Regimes that exercise strong central control either fall on hard times or begin to decompose. In a borderless world the deck is stacked against them. The irony, of course, is that in the name of safeguarding the integrity and identity of the center, they often prove unwilling or unable to give up the illusion of power in order to seek a better quality of life for their people. There is at the center an understandable fear of letting go and losing control. As a result, the center often ends up protecting weak and unproductive industries and then passing along the high costs to its people—precisely the opposite of what a government should do.

The Goal is to Raise Living Standards

THE CLINTON administration faces a stark choice as it organizes itself to address the country's economic issues. It can develop policy within the framework of the badly dated assumption that success in the global economy means pitting one nation's industries against another's. Or it can define policy with the awareness that the economic dynamics of a borderless world do not flow from such contrived head-to-head confrontations, but rather from the participation of specific regions in a global nexus of information, skill, trade and investment.

If the goal is to raise living standards by promoting regional participation in the borderless economy, then the less Washington constrains these regions, the better off they will be. By contrast, the more Washington intervenes, the more citizens will pay for automobiles, steel, semiconductors, white wine, textiles or consumer electronics—all in the name of "protecting" America. Aggregating economic policy at the national level—or worse, at the continent-wide level as in Europe—inevitably results in special interest groups and vote-conscious governments putting their own interests first.

The less Washington interacts with specific regions, however, the less it perceives itself as "representing" them. It does not feel right. When learning to ski, one of the toughest and most counterintuitive principles to accept is that one gains better control by leaning down toward the valley, not back against the hill. Letting go is difficult. For governments region-based participation in the borderless economy is fine, except where it threatens current jobs, industries or interests. In Japan, a nation with plenty of farmers, food is far more expensive than in Hong Kong or Singapore, where there are no farmers. That is because Hong Kong and Singapore are open to what Australia and China can produce far more cheaply than they could themselves. They have opened themselves to the global economy, thrown their weight forward, as it were, and their people have reaped the benefits.

For the Clinton administration, the irony is that Washington today finds itself in the same relation to those region states that lie entirely or partially within its borders as was London with its North American colonies centuries ago. Neither central power could genuinely understand the shape or magnitude of the new flows of information, people and economic activity in the regions nominally under its control. Nor could it understand how counterproductive it would be to try to arrest or distort these flows in the service of nation-defined interests. Now as then, only relaxed central control can allow the flexibility needed to maintain the links to regions gripped by an inexorable drive for prosperity.

Continental Divide

With 13 countries on the applicant list, the European Union confronts the dilemma of expansion in 2000. But are the advantages worth the price of admission? Even as leaders of Bulgaria, Cyprus, the Czech Republic, Estonia, Hungary, Latvia, Lithuania, Malta, Poland, Romania, the Slovak Republic, Slovenia, and Turkey maneuver to bring their policies in line with EU law, there is dissension within member countries over EU standards and the uniformity of their enforcement. And judging from the recent electoral successes of far-right political parties in Western Europe, nationalist sentiment and distaste for open borders will lend controversy to upcoming EU decisions on such hot-button issues as asylum and immigration policy.

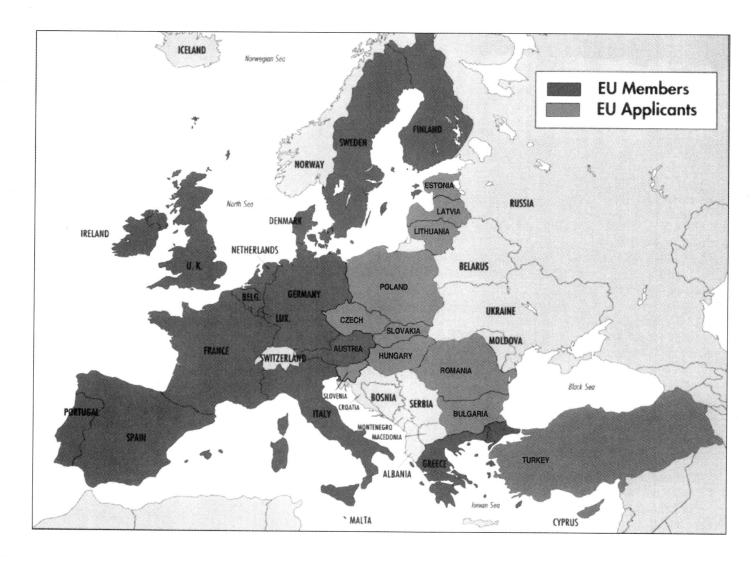

Desperately Seeking Admission

Limitless Growth?

Eighteen months ago the European Union opened admission negotiations with Estonia, Poland, Slovenia, the Czech Republic, Hungary, and Cyprus. In December this group should expand by six additional nations. If Bulgaria, Latvia, Lithuania, Malta, Romania, and Slovakia are included, the 15-member EU will be dealing with 12 candidates. In addition, Turkey, Albania, and the successor states to Yugoslavia have sought admission.

The candidates are pressing for a quick response, but there is no way to say when the EU will act, since it is not clear when the candidates will meet the performance prerequisites. The only exception is Cyprus. The divided island republic faces political hurdles. Economically, Cyprus has long since met all the criteria. For another reason, it is not at all clear that the EU is ready for any expansion.

Even in its present makeup, it finds it very difficult to function. An EU of 20 or even more members would be practically ungovernable. By 2002 the EU wants, therefore, to finish its internal reforms.

But where the European giant ought to be going, just how large the EU ought to be, and what the political and geographic borders of Europe ought to be are unclear. And not by chance. It is only because the future political shape of the Union has been left vague that the integration process has gone as far as it has. Had it, from the beginning, been based on one of the many suggested models—a federative state, a confederation, a union—the European train would have never left the station.

Now, however, with more and more cars hitched on the train, there is increasing pressure for reform. The EU can scarcely continue with the luxury of unanimous decisions and lengthy negotiations preceding any changes. This is true for the first and until now most important pillar of the EU, the internal market, with a common agricultural policy and the economic and currency union. And it will be all the more true for the second pillar, the common foreign and security policy, and the third, domestic and legal policies.

The EU special summit meeting in Tampere, Finland, [in October] made this dilemma clear. It makes sense, in a Europe that is increasingly closely knit, to have a single set of rules governing legal rights, penal codes, border security, and asylum policies. At the same time, this raises the danger that national achievements will be destroyed.

Therefore, it would be reasonable to begin by moving toward a European constitution that would define basic rights and obligations before tackling agreements on the detailed questions. But how should this work, when up till now both the geographical borders and the political form of this entity have been left vague? So, in Tampere there was an agreement to work out a Charter of Basic Rights for Europe by 2000. Long sought by the European left, the charter would in fact represent remarkable progress.

The Council of Sages created by European Commission President Romano Prodi is also aimed at a quasi-constitution for Europe. Belgium's former Prime Minister Jean-Luc Dehaene; Lord Simon, the former British minister to Europe; and former German President Richard von Weizsäcker advise the EU to restructure its legal framework. The Sages' goal is unmistakable: An expanded EU should, and must, abandon the right to veto and replace it with democratic decisions based on majority votes. That would increase the power of the European Parliament. It would also mean that individual nation-states would have to give up even more of their national sovereignty.

But even this proposed reform is a compromise that would consciously leave the future of European integration open. The candidates for membership are of course profiting from this open character—for otherwise the EU would have long since become a closed club.

At the same time, it means living with an undecided and unfinished integration. Deputies to the European Parliament are elected nationally, EU officials are chosen by national governments, and the first concerns of the European Union's heads of government are always the next election back home, and only then European politics. Common interests always take second priority, and when action is taken, fear of what national interests might be affected often leads to stasis or even retreat at the European level.

Europe has put itself under pressure to integrate in order to preserve peace and its own prosperity. That drives the EU onward, but also feeds the tendency of members to keep what they've got. The result is great pressure for conformity. It turns candidates into petitioners who cannot predict the real results of their negotiations or wishes.

The EU's agricultural policy is the best example of this contradiction. It favors intensive farming, aimed at world markets, the kind that damages the environment. To demand of the new Central European nations that they conform to this recipe is hopeless. It would not serve the EU, or Eastern farmers, or the environment, let along Third World farmers.

The way out lies in the reform of the EU agriculture policy. But that flies in the face of narrow-minded yet understandable national interests. The only way out of this dilemma lies in forced democratization of the EU, creating a public receptive to new ideas, and winning supportive majorities. Without stronger citizen participation, the Union will remain a distant, bureaucratic Moloch, or a cash cow that each state will, according to its power, seek to milk without regard for the interests of the entire European Community.

—Torsten Wohlert, *Freitag (leftist weekly),*
Berlin, Oct. 22, 1999.

RUSSIA'S FRAGILE UNION

by Matthew Evangelista

ON NOVEMBER 20, 1998, GALINA Starovoitova, one of Russia's few democratic politicians worthy of the name, was murdered in the entryway of her St. Petersburg apartment. Her death attracted brief notice in the Western press at first. But following an outpouring of anger and sympathy among the Russian public, the media paid attention for a few more days—long enough to cover the demonstrations that coincided with her funeral and the initial investigation into her murder.

Starovoitova, it seems, was the victim of a contract killing of the kind that Russian bankers and businessmen, in the absence of a functioning legal system, typically use to settle economic disputes. But unlike other politicians who had been murdered in similar circumstances, Starovoitova was not known to be involved in any shady deals. She was widely considered honest and incorruptible. A human rights activist and colleague of Andrei Sakharov's during the Soviet period, her work as a deputy of the Russian State Duma had been driven by principle.

Much as Sakharov's death in 1989 demoralized the supporters of the democratic reforms triggered by Mikhail Gorbachev's perestroika, Starovoitova's murder signaled the demise of hope for any way out of the greed, violence, and political bankruptcy of Boris Yeltsin's Russia.

Another death last November went unnoticed by the Western media, but it also served to highlight an important dimension of the Russian crisis. Nearly 12,000 kilometers east of Moscow, in the city of Petropavlovsk-Kamchatsky, 13-year-old Zoya Korshunova was killed when the batteries in a flashlight exploded in her mouth. She had been trying to do her homework in the dark, in a region suffering such an acute shortage of energy that electricity was available only (and briefly) three times a day. Zoya's death received some attention in the Russian media. So did a plea to the United Nations from the local authorities on the Kamchatka peninsula for emergency humanitarian aid.

Russian Prime Minister Yevgeny Primakov, deeply embarrassed, sent a special team to help resolve Kamchatka's energy crisis. Meetings with

November 1998: The funeral of Galina Starovoitova, gunned down at her St. Petersburg home.

Matthew Evangelista teaches international relations at Cornell University. He is the author of the recently published Unarmed Forces: The Transnational Movement to End the Cold War. *Research support for this article was provided by the Smith Richardson Foundation.*

officials from the ministries of Energy, Finance, and Defense came up with a temporary solution: The navy would provide oil from its stockpile.

The circumstances surrounding the two deaths tell us much about today's Russia. Starovoitova's murder is still unsolved, but speculation about it abounds. She had many enemies. For instance, when I met her in Moscow in early November, three weeks before she was killed, Starovoitova had just returned from a Duma session where she had unsuccessfully promoted a motion to censure Gen. Albert Makashov, a notoriously anti-Semitic communist deputy, for remarks inciting violence against Jews. She was already receiving death threats from Makashov's supporters.

It is perhaps more likely, though, that her murder was a local affair unrelated to her role in the Duma. Her concern for her St. Petersburg constituents had led Starovoitova to try to clean up corruption in the municipal elections by running a slate of like-minded candidates. Her efforts annoyed local organized crime figures.

Nevertheless, according to the *St. Petersburg Times*, the investigation of Starovoitova's murder—led by former KGB officials—has focused on digging up dirt on her democratic allies rather than actually solving her murder. "We are going to solve this case in such a way that it buries your democratic movement," one of the investigators told a *Times* journalist who was close to Starovoitova.

A Potemkin village

In the wake of Starovoitova's death, some of the leading "reformers" on the Russian political scene—including Yegor Gaidar, Boris Nemtsov, Anatoly Chubais, and Sergei Kirienko—put aside their internecine squabbles to form a center-right

Russia may not break apart, but the regions will surely take more power from the center.

A three-year-old and her cat in an unheated apartment in the far eastern city of Petropavlovsk, Kamchatka peninsula.

bloc, dubbed "Just Cause," ostensibly under Starovoitova's banner. The efforts of these four politicians—all former prime ministers or deputy prime ministers in Yeltsin's cabinets—only highlighted the advanced state of Russia's political bankruptcy.

In the Gorbachev era, common political discourse identified the orthodox communists as the "right," and supporters of perestroika and democratic socialism as the "left." Now the communists are back on the left and the self-styled reformers are on the right. The words "reform" and "democrat" have become epithets in today's Russia, thanks to Chubais and the others.

During their years in power, many of the reformers preached the virtues of the market while engaging in privatization schemes in which the most valuable state assets were sold at fire-sale prices to their cronies. Just Cause is widely considered the party of the nouveaux riches, a party that follows the dictates of the International Monetary Fund and the U.S. Treasury Department, but which is out of touch with the needs of ordinary Russians.

The reformers are blamed for the devaluation of the Russian ruble last August, the default on Russian loans, and the subsequent collapse of the Russian banking system, when millions lost their savings. That financial collapse precipitated a crisis that continues today. Most important, perhaps, the crisis revealed that the vaunted transition from a command economy to a market economy was a mere Potemkin village—an overused but sadly appropriate metaphor for Russia's reforms.

Much of Russia's economic activity before the August meltdown took place not through normal market mechanisms, but on the basis of sometimes extraordinarily complex systems of barter. The August collapse undermined the Rube Goldberg economy and put the survival of the regions beyond Moscow at risk.

Bartering for heat and light

The story of how Zoya's hometown of Petropavlovsk ended up without fuel oil illustrates how the barter system works—and why it failed in this case. Kamchatskenergo, a partly private, partly state-owned company, supplies most of the energy for the Kamchatka peninsula, which was a key military outpost on the Pacific during the Soviet era. Most of its customers, including the local military base, are agencies of the Russian government.

But the government has been notoriously negligent in paying its bills, as many Russian workers know from the personal experience of going months without receiving wages. In this case, Kamchatskenergo was able to continue supplying heat and energy to government-run schools, hospitals, and other institutions, without payment, in return for tax credits.

As Colin McMahon of the *Chicago Tribune* figured out when he interviewed the director of Kamchatskenergo, the energy company was allowed to trade its tax credits to a profitable Moscow-based oil firm. In turn, the Moscow firm was excused from paying a corresponding amount of taxes to the federal government.

The Moscow company provided crude oil that was sold or bartered on the international market in return for processed fuel oil. In turn, Kamchatskenergo used the fuel oil, acquired at lower-than-market prices, to power its two main energy plants. But when the ruble's value dropped by a third last August, Kamchatskenergo could no longer afford to buy imported oil, which was denominated in dollars.

Meanwhile, Kamchatskenergo's Moscow partner was having trouble selling its crude oil on the glutted international market. All of these events conspired to leave much of the Kamchatka peninsula without heat or electricity as temperatures dropped below zero by mid-November.

The Chechnya debacle

The Russian Federation is formally made up of 89 "subjects," including 21 ethnically identified republics such as Tatarstan, Chechnya, and Bashkortostan, and 68 non-ethnic units of various kinds. All 89 are commonly known as "the regions."

The central government's reaction to the crisis in Kamchatka—and in the regions in general—was revealing. By sending in troubleshooters from Moscow to knock heads together, Prime Minister Primakov reverted to the tried-and-true methods of the Soviet period. But Moscow no longer wields as much power as it once did vis-a-vis the regions. The deal to supply fuel from the navy depot, for instance, unraveled within weeks and Petropavlovsk again ran short of energy.

September 1, 1996: The center of Grozny, a day after Russia's Gen. Alexander Lebed and separatist commander Aslan Maskhadov signed a pact ending the conflict in Chechnya.

Moscow's inability to cope with regional crises has led to speculation that the federation might break apart, much as the Soviet Union did at the end of 1991. This is not the first time such concerns have been raised. When the Soviet Union split into its 15 constituent ethnic republics, many feared that Russia would go the same way.

Boris Yeltsin himself fueled the earliest concerns about a Russian breakup in 1991. As part of his campaign to undermine Soviet President Gorbachev, Yeltsin, then president of Russia, went to Tatarstan's capital city of Kazan and issued a challenge for the regions to "take as much autonomy as you can swallow."

Chechnya and Tatarstan took him at his word and declared formal independence. In 1992 both refused to sign the treaty that the Yeltsin administration promoted as the foundation for relations between the center and the regions. Moscow eventually worked out an uneasy *modus vivendi* with Tatarstan; the republic, acting as if it were a sovereign state, negotiated "treaties" with the federal government to regulate essentially internal matters, such as taxation.

Chechnya did not fare as well. Under the erratic leadership of Dzhokar Dudaev, the republic insisted on a special independent status. In retrospect, many observers believe that if Yeltsin had been willing to meet directly with Dudaev in 1992, the conflict could have been resolved peacefully. Instead Yeltsin thwarted efforts to open negotiations with Dudaev.

In 1992 Galina Starovoitova, then serving as Yeltsin's adviser on ethnic politics, contacted Dudaev in the Chechen capital of Grozny by telephone. Russian hardliners, with Yeltsin's blessing, attacked her efforts at reconciliation.

The next time Starovoitova tried to contact Dudaev, she found that the government phone service to Grozny had been cut. Seeing the writing on the wall in the increasing militarization of Moscow's relations with Chechnya, Starovoitova resigned in protest—well before the Russian armed forces launched an invasion in December 1994.

The war against Chechnya was a low point in Yeltsin's already sagging popularity and a high point in concern about the future of the Russian Federation. At first many of the neighboring regions in the Caucasus expressed solidarity with the Chechens, as citizens from Dagestan and Ingushetia, for example, tried to prevent Russian military forces from crossing their territory.

In the wake of the Chechen war, Andrei Shumikhin, a Russian analyst, expressed a fear shared by many that "a 'brush fire' of drives for independence may pick up elsewhere across Russia, leading to the eventual destruction of Russian territorial integrity."

Starovoitova criticized Moscow's "crude use" of "notorious tools of imperial policy" and predicted that the military intervention in Chechnya would "produce mistrust of the center's policy and centrifugal tendencies in Tatarstan, Bashkortostan," and elsewhere in the Russian Federation.

Gen. Aleksandr Lebed, who led Soviet forces in the disastrous intervention in Afghanistan, warned that an invasion of Chechnya would lead to a similar, protracted guerrilla war, resulting in "a Pyrrhic victory."

Lebed was right. The war in Chechnya dragged on for nearly two years, claimed tens of thousands of casualties, and produced hundreds of thousands of refugees.

But Starovoitova and others who feared that the war would tear apart the federation were wrong. The war ended with a cease-fire brokered by Lebed himself in 1996 and a treaty on future relations signed by Yeltsin and Dudaev's successor Aslan Maskhadov in May 1997.

Creeping autonomy

After the end of the Chechen war, a new—and somewhat contradictory—consensus emerged in the West and in Russia:

■ The Chechen victory would not inspire other regions to attempt to achieve independence. The cost of gaining independence was too high—tens of thousands of lives, mainly civilian, and the total destruction of Grozny, the capital city, as well as many regional centers and villages.

■ The war was also disastrous for Russia, undermining Yeltsin's popularity, demoralizing the armed forces, and draining the federal budget. It was unlikely that Moscow would ever again resort to such a brutal and foolhardy use of violence to resolve a conflict with one of the regions.

■ Finally, common interests would maintain the federation, particularly economic interests, such as trade, distribution of taxes and subsidies, and a common currency; a common transportation network; and a common energy grid.

But the financial crisis of last August undermined such arguments. Perhaps it is too early to speak of a new consensus, but many Russian and Western observers believe a major change in center-regional relations is leading, if not to the breakup of the federation, at least to a substantial decentralization of power and more autonomy for the regions.

A conference in Washington in December, sponsored by the Harvard-based Program on New Approaches to Russian Security, captured the mood among U.S. specialists. Kathryn Stoner-Weiss, a political scientist at Princeton, spoke of "creeping regional autonomy," and Henry Hale, from Harvard's Davis Center for Russian Studies, described the "consolidation of ethnocracy" within the republics and "regional autocratization" throughout the country.

Eva Busza, from the College of William and Mary, focused on the implications of the economic crisis for regionally based military forces. As resources from the center dry up and troops become "reliant on the goodwill of regional leaders for their day-to-day survival," what might those leaders demand in return, particularly as their policies come into conflict with those of the central government?

In many respects the concerns of Western analysts simply reflect fears in Moscow. Primakov put it starkly in a speech to the Duma last August, shortly after being named prime minister: "We are facing a very serious threat of our country being split up." And on Russian television last September, he argued that "89 constituent parts of the federation is too many." He supported plans to consolidate the regions into as few as eight regional economic groups.

In fact, several interregional economic associations already exist, and their horizontal economic links may serve as a "bulwark against disintegration," as Graeme Herd of Aberdeen University has argued.

But Prime Minister Primakov is not satisfied with the status quo. In January, at a meeting of Siberian governors representing the Siberian Accord, an interregional association, Primakov called for the "restoration of the vertical state power structure, where all matters would be solved jointly by the center and local authorities." He also criticized "talk of conflict between the center and the regions" and insisted that separatist trends "must be quelled, liquidated, and uprooted."

The concern of Western analysts about "regional autocratization"—that is, the demise of democratic procedures within the regions—is also heard in Moscow. Many regional governors have manipulated electoral laws to eliminate any challenges to their rule. Galina Starovoitova told me of a particularly blatant, but not unusual, technique employed by the leader of the small ethnic republic of Marii El, located just north of Tatarstan. He sponsored a law requiring that any candidates for the presidency of the republic be fluent in Russian as well as two dialects of the Marii language.

With ethnic Marii people a minority in the republic (43 percent of the population), the law would have eliminated candidates from other groups—such as Russians who make up 48 percent—as well as the many Marii who have lost their native language. In effect the law would have limited the field to the incumbent himself.

Starovoitova made her political reputation in the late 1980s by supporting autonomy for the Armenian enclave of Nagorno-Karabakh in Azerbaijan. But she was not one to offer blind support for ethnic chauvinism at the expense of democratic principles. In one of her last achievements in the service of democracy before her murder, she challenged the Marii El electoral law as unconstitutional and the Russian Constitutional Court ruled in her favor.

Unfortunately, widespread manipulation of local elections often goes unremarked by the Moscow authorities, especially if the regional leaders have supported Yeltsin's own electoral efforts in the past. Primakov favors doing away with regional gubernatorial elections altogether in favor of appointing local rulers directly from Moscow.

More talk; less defiance

Although the concern about antidemocratic tendencies in the regions is well founded—some local politicians have even targeted critical journalists for assassination—the prospects for either successful regional political consolidation or more centralized control from Moscow seem dubious.

Nevertheless, outright disintegration does not appear likely. That impression comes in part from my recent discussions with representatives from some of the regions viewed as most likely to seek greater autonomy—Tatarstan, Bashkortostan, and Dagestan—and one, Chechnya, that has already achieved de facto independence.

Despite the undeniable crisis atmosphere in the regions, there seems to be a strong commitment to working out compromises with the central government. Few regional leaders really want to see Russia go the way of the Soviet Union.

All of the regions have permanent representatives stationed in Moscow who are in daily contact with federal officials to discuss issues ranging from the budget to foreign investment to constitutional disputes. (The constitutions of many regions contradict the federal constitution on key points.)

Even Chechnya, which claims to be independent, has kept its Moscow office open. The most noticeable change is that the staff now answers the phone with a hopeful "Embassy," even though no major country has recognized Chechen sovereignty.

The Chechen office is headed by Vakha Khasanov, a former factory manager from the Nadterechny region of Chechyna, an area that, as Khasanov put it, tried to maintain "neutrality" during the war. Khasanov explained that he was not trained to be either a diplomat or a politician.

His refreshingly outspoken views suggested as much. He criticized the late Chechen leader Dzhokar Dudaev, whose eccentric behavior and megalomania contributed much to the outbreak of the war. He was forthright in admitting that the crime rate in the republic is so high as to discourage any foreign investment. (He told a recent commercial delegation from Malaysia that they would need a small army of security guards even to consider a potential site visit to the republic.) The prerequisite for solving Chechnya's problems, in Khasanov's view, lies in improving relations with Russia.

Further, relations with Moscow are not the only thing that preoccupies the Chechen government. Chechnya teeters on the brink of civil war as the relatively moderate, popularly elected president Aslan Maskhadov—the top military commander during the war—faces challenges from his erstwhile subordinates, including Salman Raduev, whose contributions to the Chechen victory included notorious terrorist attacks and the taking of civilian hostages. Raduev has threatened cross-border attacks on Russian military bases in neighboring Dagestan, putting at risk President Maskhadov's attempts at working out a *modus vivendi* with Moscow.

Tatarstan offers a sharp contrast to Chechnya, in that it was able to gain substantial independence from Moscow without provoking violence. Indeed, the "Tatarstan solution" is often cited as a lost opportunity that Dudaev and Yeltsin should have pursued.

Given its level of de facto independence, one might expect the recent economic crisis to have driven Tatarstan even further away from the central government. To be sure, its authoritarian leader Mintmir Shaimiev has implemented several measures that isolate the republic, including a ban on the export of food stuffs or the import of Russian vodka. At the same time, Tatar officials in the Moscow office are busy trying to reconcile differences with the central government.

Some of their reasons are obvious. "It's too big a price what was paid in Chechnya, and for the Russian Federation," says Mikhail Stoliarov, a deputy head of the Tatarstan delegation in Moscow.

Tatarstan has come a long way since its defiant claims of independence from Russia in the early 1990s. Stoliarov believes the situation then was fraught with the danger of a Chechen-style military conflict, but not anymore. "I'm talking about lessons of Chechnya, which should be a warning to everybody, to all the radicals in Tatarstan as well as in any other republic—and to radicals in Russia, whatever, governmental or non-governmental radicals. So it's a warning that's clear, the Chechen conflict should not be repeated in any form."

Tatarstan and neighboring Bashkortostan are among the richest regions in the Russian Federation, thanks to substantial oil and natural gas deposits and high-technology industries inherited from the Soviet military-industrial sector. They are among the small number of "donor" regions—regions that receive no subsidies from the central budget, but instead contribute through their taxes to the redistribution of income to poorer, "recipient" regions.

In my discussions I probed for some signs of resentment toward the poorer regions, but the reactions of the Tatar and Bashkiri representatives were more mixed. They spoke of pride in their own regions and a desire to protect their achievements. They also seemed to view their economic development as a product of Soviet-era policies, and their abundant resources as a matter of good fortune.

In describing the sources of his republic's wealth, Erik Yumabaevich Ablaev, the representative from Bashkortostan, said: "We understand very well that these were not the achievements of our republic alone. They were developed in the course of 70 years by the whole Soviet people." The vast investments in oil processing were "created by the efforts of the entire Soviet Union."

Ablaev was sensitive to the charge that Bashkortostan's wealth accords it a privileged position vis-a-vis Moscow. But he also stressed the negative side of the republic's development. Industrial pollution poisons the land, water, and air. According to Ablaev, every second child in Bashkortostan is born unhealthy, the victim of an environmental disaster that the central government in Moscow ignores. "They speak of 'privileges,'" he complained. "If we had healthy children, that would be a privilege."

At the other end of the spectrum is Dagestan, the poorest of the 89 regions, except for Chechnya. On economic grounds the republic has no reason to secede, as one of its Moscow representatives readily acknowledged.

Although Chechnya was also poor, it had something Dagestan lacks—a relatively homogeneous population with a strong tradition of opposition to Moscow. Dagestan's two million residents comprise 32 ethnic groups and speak 33 languages. The republic still reflects much of Soviet "internationalist" ideology. Robert Chenciner, the author of a recent book on Dagestan, calls it "a microcosm of the ethnic mosaic of the Soviet Union."

That is pretty much how Mamay Mamayev, the Moscow representative, described his region, as he pointed to it on an old map of the Soviet Union tacked on the wall behind his desk. Most likely, he is not the only Dagestani nostalgic for the stability of Soviet rule. The Chechen war wrought havoc on Dagestan, where some 100,000 refugees fled from the destruction of Grozny and surrounding villages. "Some of them are still living at our house," Mamayev said.

Further, the republic is now awash with weapons. In the capital city of Makhachkala, competing political factions command their own armed militias. Mamayev played down the violence, inviting me to visit the capital, and insisting that only the border area with Chechnya posed any risks. The day after my interview, an American teacher was kidnapped in Makhachkala in broad daylight.

The deeper threat

Despite the crisis atmosphere in much of the country, concern about the imminent disintegration of the Russian Federation seems unwarranted. Relations between the regions and Moscow will not go smoothly. But they will probably continue to be worked out on an ad hoc basis, as they have been for the last several years.

January 1996: Magamed Abdurazakov, Dagestan's internal affairs minister, listens to Chechen rebels engaged in a civil war with Russian troops.

And despite Prime Minister Primakov's ambitious plans for consolidation of the regions, there are few governors or presidents of republics who would willingly relinquish political and economic power to be swallowed up by a larger agglomeration more closely controlled from Moscow.

The real story of the regions is the one represented by the deaths of Galina Starovoitova and Zoya Korshunova. Corruption and political violence are undermining the fragile foundations of Russian democracy. Lack of food, lack of medicine, and lack of heat over the winter months have produced humanitarian catastrophes throughout the country. The Russian government, bulging with more bureaucrats than even Leonid Brezhnev employed, seems utterly incapable of dealing with the various crises, whether large or small.

An official at Radio Free Europe put it just right in January: "It is the collapse of the Russian state, not the breakup of the federation or economic depression, that may in the long run prove the greatest threat to Russian democratic development and international stability."

Russia's regions:
Beyond the Kremlin's walls

President Vladimir Putin is trying to bring Russia's regions back into line

UFA, BASHKORTOSTAN

BUSINESS is what matters; democracy is for later, perhaps. There are elections of a sort, but the president's candidates always win. The media do what they are told. Foreigners are welcome so long as they keep their wallets open and their mouths shut. The place stays afloat thanks to oil and the weak rouble.

A gloomy snapshot of President Vladimir Putin's Russia? The description certainly fits Bashkortostan, a family-run republic in mid-Russia that has become the first target of Mr Putin's attempt to tidy up his country's shambolic internal structure.

The immediate argument is about the republic's constitution, which puts Bashkortostan's laws on an equal footing with Russia's. That has allowed the leadership to ignore federal privatisation programmes and keep the press cowed and elections rigged. Last week, Mr Putin fired off a stiff letter to the speaker of the local parliament, saying that the republic must bring its constitution into line with Russia's.

Mr Putin has launched a wider plan too. He is calling for more power to sack regional bosses and wants to end their automatic right to seats in Russia's upper house of parliament. He also wants to parcel out Russia's 89 component republics and regions into seven new districts, each overseen by a presidential appointee. These governors-general, as the Russian media call them, are to manage the local operations of the "power ministries"—those for defence, the interior, security and justice. All this would severely cramp the style of Bashkortostan and other independent-minded bits of the federation, which have largely taken over the central government's outposts of power.

At stake is the future of Russia as a centrally governed country. Over the past ten years, the centre's hold has weakened sharply, and the differences between regions have grown (see map). The local governments are not an advertisement for federalism: most are by turns thuggish, crony-ridden and plain incompetent, though a handful are now dimly aware of the benefits of foreign trade and investment.

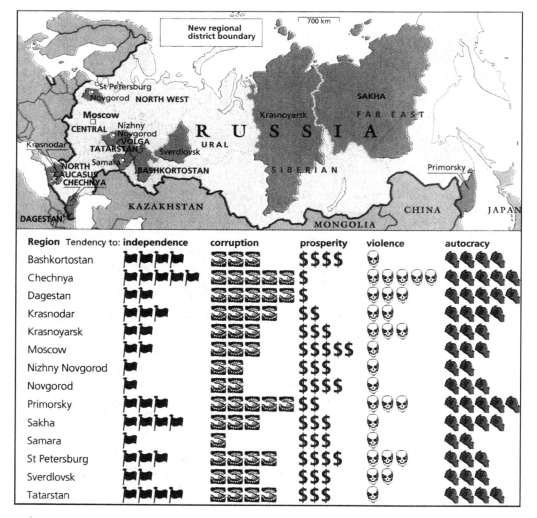

The leadership in Bashkortostan is puzzled rather than panicky. The republic is far from being the worst-run place in Russia, and it loyally supported Mr Putin in the presidential election. Next-door Tatarstan is much more outspoken about political and economic sovereignty. Crime in Bashkortostan is not conspicuous. President Murtaza Rakhimov is an autocrat, but his rule is heavy-handed rather than bloody.

The argument, his friends insist, is really about money, and his government is quite ready to talk about it. Bashkortostan is prosperous by Russian standards, with one of the biggest oil companies in the country (coincidentally run by the president's son, Ural) and one of the strongest banks. "Our republic's status will not be a subject of negotiation," says the republic's speaker. And if Mr Putin differs? "A threat would be counter-productive," he frowns intimidatingly.

The noncommittal welcome given by most regional leaders so far to Mr Putin's plans has worked well in the past. Previous attempts to bring order to the provinces have quickly become bogged down. Under pressure, regional leaders pay lip service to the federal leadership and wait until its attention wanders. The 80-plus presidential representatives in the regions appointed by Boris Yeltsin in an attempt to reestablish his authority have usually become little more than figureheads. Mr Putin's planned new governors-general may well meet the same fate.

"This happens when new people are only just starting work," says the head of another republic in central Russia, in a revealingly patronising tone. "The decree just creates one more bureaucracy with thousands more employees."

All the same, Mr Putin could prove a more serious centraliser than Mr Yeltsin. He worries publicly about separatism, and has strongly backed the war against it in Chechnya. He may be able to muster some serious allies in Moscow. He may, for example, be co-ordinating his plan with the barons of the national oil and gas industries, who would also like to cut down self-important regional leaders—with an eye for the spoils.

Even so, any serious attempt to re-centralise Russia risks changing the regions' current apathy towards Moscow into hostility. And although democrats in places like Bashkortostan rest their hopes, lightly, on Mr Putin, getting rid of bad local leaders will improve things only if the central government for its part starts working properly.

That could yet happen, but the opening weeks of Mr Putin's presidency give little reason for hope. Journalists are panicking about state harassment of the company that owns NTV, the main independent television channel. And there is no sign yet of government support for the increasingly dog-eared economic-reform plans that have been floating around Moscow. Bashkortostan is a useful target for Mr Putin; pessimists fear it could be a useful model too.

EUROPE

The delicate Balkan balance

This autumn sees the biggest round of elections in the Balkans since ex-Yugoslavia broke up eight years ago. This time, voting could genuinely reshape the region. How, in particular, do Bosnia and Kosovo stand?

BANJA LUKA, PRISTINA AND SARAJEVO

THE rosiest Balkan scenario goes something like this. Slobodan Milosevic, the Yugoslav president who has exploited ethnic loyalties so brutally and successfully to stay in power over the past decade, is voted out of office at the presidential poll on September 24th. He is replaced by a doubtless prickly, but nonetheless more decent, Serb who realises that his country can be rebuilt only if he makes terms with the outside world, and accepts the need for a new constitutional settlement for much of the region—within, and perhaps beyond, the rump of Yugoslavia, which still includes the nominally Serbian province of Kosovo.

Thanks to other elections at various levels, most importantly in Bosnia and Kosovo, there is a swing, however small, towards moderates who believe in co-operation rather than confrontation. This matches other moderate successes at local polls in September in Macedonia. There, reasonable people currently run a frail coalition government, which has to fend off a lot of complaints from the country's Slav majority that the ethnic-Albanian minority (at least a quarter of the population) is being allowed to cause too much trouble. In this bright setting, Albania's local elections in October pass off quietly, easing fears of a return to the chaos of the mid-1990s.

The consequent advent of brave new moderate Balkan politics calms tension across the region, enabling economic help to be delivered more effectively and lightening the outside world's task of peacekeeping and protecting minorities.

All too rosy? Very probably, yes. The biggest imponderable is the Yugoslav elections. The government of Montenegro, Serbia's junior partner in the Yugoslav federation, is refusing to co-operate with the poll—and fears are growing that Mr Milosevic will exploit the smaller republic's internal divisions to stir up a new round of killing.

Nor has a convincing candidate emerged to take him on. With two opposition candidates now standing, the anti-Milosevic vote may well anyway be split. Kosovo remains violent, with ruthless ethnic-Albanian nationalists in the ascendant. Meanwhile, in Bosnia, only some of the Muslims—their leaders style themselves Bosniaks—seem truly to believe in a multi-ethnic state.

Still, of late there have been some hopeful advances, especially in Bosnia. Killings in that former charnel-house have been minimal this year. Thanks to greater calm, the NATO-led peacekeeping force (known as SFOR) has been shrunk by a third, to 20,000. The election of a more sensible, westward-looking coalition government in Croatia, under Ivica Racan, has knocked quite a lot of stuffing out of the more bloody-minded of Bosnia's Croats. In April, the harsh nationalists who ruled their respective roosts during the civil war did worse than usual in local elections across Bosnia. From a wretched base, the economy too has begun to improve, with more trade across the "inter-entity boundary" that demarcates Bosnia's constituent bits. Even the three "entity armed forces"—the Croats, Muslims and Serbs—are each on track to trim their numbers by 15% by the end of the year.

Perhaps the most hopeful feature is that more people are returning to their old homes, even in places where a rival ethnic group holds sway. In the first half of this year, there were 20,000 such "minority returns", a threefold increase over the same period of 1999. At that rate, it will still take a long time to reverse the effects of ethnic cleansing. Some 2m Bosnians, about half the pre-war population, were forced to abandon their homes during the early 1990s. Only 300,000 have gone back, and over 1m do not have secure access to their pre-war homes.

But the accelerating rate of minority returns shows that the vicious circle of violently imposed segregation is not unbreakable. In several towns, such as Drvar, in the Muslim-Croat Federation, where many Serbs have returned, the ethnic balance has tipped against the Croats without, yet, provoking the ructions that many feared.

Another notable success, to date, has been the peace prevailing in the strategically sensitive town of Brcko, a once evenly balanced tri-ethnic town at the northern choke-point that links the western and eastern parts of Bosnia's Serb Republic. A decision by outsiders in March to neutralise and demilitarise the town and its surrounding corridor indefinitely did not provoke the violent outrage that Serb militants had promised.

The particular hope of Bosnia's overseers is that, in the forthcoming elections, the Serbs' assorted moderates, led by Milorad Dodik, the prime minister of their "entity" based at Banja Luka, can team up to beat the old Serb Democratic Party, previously led by Radovan Karadzic. It could be: the arch-nationalists did fairly badly in April's local elections. But, as elsewhere in the Balkans, the forging of coalitions and alliances is always fraught with backbiting.

Alija Izetbegovic, the Muslim member of Bosnia's three-person collective presidency, is poised to resign; his stridently Muslim Party of Democratic Action (SDA) has lost ground. One possible successor is Haris Silajdzic, whose moderate group, the Party of Bosnia and Hercegovina, is on the rise.

Yet most Bosnians still see politics in ethnic terms. Leading Bosnian Croats want to dissolve the Muslim-Croat Federation, and give more powers to the ten cantons that make up their hunk of Bosnia; in practice, to each locally dominant ethnic group. Mr Izetbegovic is sympathetic to the idea. The UN's high representative in Bosnia, Wolfgang Petritsch, is said to be pondering it. Similarly, talk of creating a single Bosnian army wins a hearing from some Muslims, but from few Croats or Serbs. "Only the Bosniaks think of themselves as Bosnian," says an SFOR man.

For now, the informal economy seems to be holding the country together as much as anything else. Some two-fifths of people in the Muslim-Croat Federation have no formal job; in the Serb Republic, about half. Most of the old political parties are intricately tied up with organised crime.

Many of those who have returned struggle valiantly to rebuild homes and lives, though many depend on hand-outs. There is a sort of stagnant stability but—despite flickering hopes of an economic revival—no real sign of a state emerging. If SFOR were to clear out, Bosnia would almost certainly relapse into chaos.

And so to Kosovo

If Bosnia is complicated, but with a hint of hope, Kosovo these days has a harsher feel to it. The choices are certainly blunter. Most Albanian Kosovars want all the Serbs to leave—and most have done so. Some 200,000-300,000 Serbs used to live in Kosovo a decade ago. Maybe 100,000 still do, about half of them north of the Ibar river, which cuts through the northern town of Mitrovica. Another 27,000 live in a "Serb crescent" of villages and small towns that curls round the western and southern flanks of Pristina, the province's capital. No more than 700 Serbs remain in the capital itself, out of perhaps 20,000 who were there before NATO's bombing campaign last year.

Apart from the land north of Mitrovica and the Serb crescent, Serbs are few and far between. Kosovo's western valleys, where the ethnic-Albanian insurgency of the Kosovo Liberation Army (KLA) was fiercest before the bombing, are virtually empty of Serbs, except for scattered and dwindling groups huddled under the protection of KFOR, the NATO-led force that is trying to prevent the ethnic Albanians from hounding the Serbs out altogether. British troops around Slivovo, east of Pristina, are trying to consolidate and protect a string of villages where Serbs still survive.

Even within the crescent, few Serbs dare to travel without a KFOR escort. If Serbs cross from Serbia proper, as 100-plus do a day, to visit their relations or test the ethnic climate, they queue up at the border and then set off in convoys, protected by KFOR. In Pristina, Serbs are cooped up in a ghetto of tenement blocks, with KFOR guards on watch 24 hours a day. The Albanians refuse to accept them in Pristina's hospital, so the Serb sick have to be taken, often under KFOR supervision, to a Russian field hospital at Kosovo Polje.

Fewer Serbs are now being killed in Kosovo than during the months immediately after NATO ground troops rolled into the province in mid-1999. Then, by UN estimates, more than 60 people a month were killed, most of them Serbs. In the past three months, some 20-odd have been. But a constant undercurrent of violence and intimidation runs through the province. Life for any Serbs south of Mitrovica—and for non-Serbs north of that city—is wretched.

Yet the Serbs are not without help altogether. Serb paramilitaries are moving around, though mostly lying low, in the crescent near Pristina. North of the Ibar, between Mitrovica and Serbia proper, they hold sway. UN and other agencies barely operate there any more. But, this week, Kosovo's protectors made one of their boldest moves. Under a hail of rocks and abuse from furious Serbs, KFOR seized control of the Zvecan lead-smelter, on the northern side of the divide, citing worries about the appalling level of pollution it was giving rise to. The smelter is part of the huge Trepca complex of mines and factories which communist Yugoslavia regarded as one of its prize economic assets. The UN wants a consortium of French, American and Swedish interests to put the plant in order and reopen it, presumably with a more multi-ethnic workforce.

Farther south, the biggest recent source of unrest has consisted of Albanians attacking Albanians, as politicians and criminals—often one and the

To the polls
Upcoming elections in the Balkans

	Date	Type of election
Macedonia	September 10th	Local
Serbia	September 24th	Local, and Federal Yugoslav parliamentary and presidential
Montenegro	September 24th	Federal Yugoslav parliamentary and presidential*
Albania	October 1st	Local
Slovenia	October 15th	Parliamentary
Kosovo	October 28th	Local
Bosnia	November 11th	Parliamentary elections in Muslim-Croat Federation and Serb Republic

*Montenegrin government pledged to boycott election

same—joust for supremacy in the run-up to elections for Kosovo's 30 municipalities.

After NATO's arrival, the 20,000-plus guerrillas of the KLA were disbanded, and were supposed to hand in their arms. Their force was reinvented, under the UN's aegis, as the much smaller Kosovo Protection Corps (KPC). The KPC is meant to do civilian neighbourly good works—putting out fires, building bridges and so on. But in practice the force sees itself as the core of an armed elite that will one day be the vanguard of the army of an independent Kosovo.

Its commander, Agim Ceku, was the KLA commander. It answers, unofficially, to Hashim Thaci, leader of the Democratic Party of Kosovo (PDK), which emerged from the KLA. A lot of the intra-Albanian violence is perpetrated by the PDK against the other main Kosovar factions, especially Ibrahim Rugova's more moderate Democratic League of Kosovo (LDK).

Last month, one of Mr Rugova's advisers was kidnapped and killed. Recently, his party leader in Podujevo just escaped being assassinated. KFOR simply does not fully control these ex-KLA men. The Americans, a central part of KFOR, are loth to take them on, for fear that the perception of NATO troops will turn from that of liberators to oppressors—and that the force will start taking casualties at the hands of the Kosovars. Another ethnic-Albanian guerrilla force, known as UCPMB, operates almost at will in an Albanian-inhabited strip of Serbia proper, with the declared aim of seizing control of three Serbian towns just to the east of Kosovo. The guerrillas, most of them former KLA men, claim they number 500; KFOR guesses 150. They want this bit of Serbia to belong to their would-be Albanian state. The idea of swapping it for the chunk above Mitrovica is in the air—though discounted by western mediators, fearful where this might lead with tricky borders elsewhere.

By all informed accounts, Mr Rugova's lot is far more popular than the young thugs who look to Mr Thaci. But the more militant PDK has more of the local media behind it, more cash, and more criminal and business links. Somewhere between the two stands a new player, Ramush Haradinaj.

Several of Kosovo's most prominent businessmen, such as Mustafa Remi in Podujevo and Saved Geci, are leading lights in Kosovar politics. Of the 50-odd Albanian parties which have registered for the municipal elections, many call for a Greater Albania, embracing a big slice of Macedonia and Albania proper. One even demands the Greek island of Corfu.

Just as KFOR cannot decide how robustly to handle the former KLA people, politicians in the West have little idea how to resolve Kosovo's eventual constitutional status. The judicial system is still weak; the UN's police force, which was meant to number more than 4,700, has yet to reach full strength: not surprisingly, given that it has 42 nationalities, it cannot cope. Besides, a proper criminal investigation unit is badly needed. There is no proper prison. Bernard Kouchner, the French doctor who administers the province in the name of the UN, is struggling to bring in foreign judges to help oversee law and order.

The main immediate hope of Dr Kouchner and the KFOR commanders in charge of some 20,000 overstretched troops is that Mr Rugova, and perhaps Mr Haradinaj, will do well at the elections. Dr Kouchner could then appoint additional people, including Serbs, who are boycotting the polls, to local councils. KFOR might then have more authority to enable it to bring KPD miscreants to book.

In the longer run, however, it may be that the only way to settle the problem of Kosovo would be to hold a new Balkan conference, where the entire gamut of territorial and constitutional issues across the region could be discussed in the round. Might, for instance, Montenegro, Kosovo and Serbia proper be reshaped as three republics linked in one federation? Might Bosnia's constitution be rewritten? Might there be referendums to decide under whose flag the inhabitants of contested areas want to live?

The trouble is that no such bold ideas are worth discussing until Mr Milosevic is out of the way. A defeat for him in the September election would hugely boost the prospects for Balkan peace. Jacques Chirac, president of France, which holds the current six-month presidency of the European Union, has contemplated holding such a grand Balkan summit in the autumn.

But Mr Milosevic could well stay in power for quite some time yet. If he does, the West will simply have to grit its teeth, in Kosovo perhaps more doggedly even than in Bosnia, and prepare itself for a long haul as an unloved policeman in a wretched part of the world. For if SFOR and KFOR were to leave, or to draw down their forces too fast, a Balkan bonfire, in Kosovo, Bosnia and elsewhere, could easily reignite.

Sea Change in the Arctic

An oceanful of clues points to climatic warming in the far North

By RICHARD MONASTERSKY

The first sign that something was wrong came just a few days after the icebreaker Des Groseilliers left Tuktoyaktuk, a port on the Arctic coast of western Canada. Steaming north with a full load of 20 scientists and provisions for 18 months, the Canadian coast guard ship was heading for a date with the Arctic pack ice—the perennially frozen layer that covers the top of the globe like a white skullcap.

Old hands at polar research, such as oceanographer Miles G. McPhee, expected to meet the ice between 71° and 72°N. That's where the southern edge of the pack showed up in the 1970s, when McPhee made several trips far into the Arctic. Two decades later, the Canadian icebreaker cruised past 72°N at full speed, with no ice in the water to slow its progress. McPhee, who runs a research company out of Naches, Wash., wondered silently, "Oh my God, where did all the ice go?"

The case of the shrinking pack ice is only one of many climatic conundrums troubling scientists who study the Arctic. In recent years, researchers have discovered myriad signs of changes in the far North, affecting everything from the ocean currents flowing 1,000 meters beneath the ice pack to the howling winds at the top of Earth's atmosphere. "I think the changes are verging on what could be called dramatic," says John M. Wallace, an atmospheric scientist at the University of Washington in Seattle.

Some aspects of the shifting Arctic weather resemble the patterns expected from greenhouse warming, leading scientists to wonder whether they are witnessing early warning signs of conditions to come during the next century. The polar regions, especially the Arctic, are regarded as the climatic equivalent of canaries in a coal mine. The poles are the places on Earth most sensitive to global warming. With the canaries looking distinctly wan, the search is now on to determine what has caused their decline.

These are some of the scientific issues that propelled McPhee and his colleagues northward on the *Des Groseilliers* in the fall of 1997. Following the plans of the $19.5 million international research project, the Canadian icebreaker finally plowed into the pack ice, cut its way to the middle of a large floe, and then stopped its engines at 75°N.

For an entire year, the ship remained frozen amid the ice, dragged 2,800 kilometers on a roundabout course northwest across the Arctic Ocean, a landless expanse of water and ice. Over that time, planes ferried a total of 170 researchers out to the ship in shifts of several weeks to months. The scientific teams fanned out across the nearby ice to collect measurements on how heat shuttles among the ocean, the sea ice, and the atmosphere through the different seasons. They called the project SHEBA, an acronym for Surface Heat Budget of the Arctic.

From the beginning, SHEBA taught investigators to question the prevailing ideas about the Arctic. Project planners had expected to park the icebreaker amid floes measuring 3 m thick, the kind of ice seen in the 1970s during the last major U.S. Arctic initiative. The SHEBA crew, however, was dismayed by what it encountered in 1997.

"When we went up there, the first problem we had was trying to find a floe that was thick enough. The thickest ice we could find was 1.5 to 2 m," says SHEBA chief scientist Donald K. Perovich of the U.S. Army Cold Regions Research and Engineering Laboratory in Hanover, N.H.

Once the crew set up the ice station, another startling fact came to light. During the first few weeks of work, McPhee and his colleagues found that the shallow waters of the Arctic Ocean were warmer and less salty than they had been 22 years earlier. This observation implies that a significant quantity of ice had melted during the previous summer, says McPhee's team, which published the SHEBA finding in the May 15, 1998 GEOPHYSICAL RESEARCH LETTERS.

More shocks came toward the end of the project in October 1998. "The big surprise of our work was that the ice ended up at the end of the year thinner than when we had started," says Perovich. The warm winter and long summer of 1988 had shaved off about one-third of a meter from the already thin ice.

SHEBA investigators suggest that some of the changes they saw could have stemmed from the 1997–1998 El Niño in the equatorial Pacific, although this temporary warming can't explain all the findings. Arctic sea ice has been declining for many years.

In 1997, researchers from NASA's Goddard Space Flight Center in Greenbelt, Md., reported that the area covered by sea ice decreased by more than 5 percent between 1978 and 1996. Other satellite measurements of the sea ice have revealed evidence of increased summertime melting since 1979 (SN:2/21/98, p. 116).

Sea ice is only a thin skin over the Arctic Ocean, but in many ways it serves as the linchpin of the region's climate and perhaps that of the whole globe. If the pack diminished substantially, climate models predict big changes in the ocean and atmosphere. The reason stems mainly from something scientists call the ice-albedo feedback.

Because ice is so bright, it reflects more than half the sunlight that hits it during summer. The dark water, by contrast, absorbs 90 percent of the incident sunlight, says Perovich. The existence of the sea ice, therefore, keeps the Arctic Ocean cool by shielding it from solar energy.

If the ice starts to disappear, according to theory, the ocean will rapidly warm and melt more ice—a potentially runaway process that could strip the Arctic of its protective cap and allow solar energy to stream into the ocean unhindered. This ice-albedo feedback would greatly amplify the effects of greenhouse gas polution, causing the Arctic to warm much more than the rest of the globe, according to climate models. The Arctic shift would have a domino effect, rerouting ocean currents and weather patterns further south.

Until recently, scientists had little access to the Arctic Ocean, which served as the sparring ring for superpower submarines. With

Three sets of submerged mountain ridges (black) split the Arctic Ocean into separate basins. The SHEBA project studied conditions on the Alaskan side of the Arctic.

the end of the Cold War and the withdrawal of military forces, scientists started invading the region. Researchers who once sat on opposite sides of the Iron Curtain began sharing data and collaborating on projects. Since 1993, the U.S. Navy has invited oceanographers along on several submarine cruises beneath the ice pack.

In a strange confluence of climate and current events, the thaw in political tensions a decade ago coincided with an apparent shift toward warmer conditions in the North. Some of the biggest changes have occurred within the Arctic Ocean, an oblong body of water nearly one and a half times the size of the United States. The ocean is divided into four unequal basins separated by three roughly parallel mountain ranges on the seafloor—an arrangement that looks somewhat like an empty TV dinner tray.

Much of the water in the Arctic comes from the Atlantic Ocean, coursing north as a vestigial extension of the Gulf Stream. On the opposite side of the pole, water from the Pacific Ocean enters the Arctic via the Bering Strait. As the Pacific water passes through the shallow strait, it loses much of its heat to the atmosphere and consequently grows colder than the Atlantic water. The two distinct types of water meet midway in the Arctic, creating a front much like the kind that separates warm and cold air masses in the atmosphere.

Water measurements made by Russian and Western scientists indicate that the front between these two types of water bisected the Arctic Ocean for most of the past half-century. From 1949 through the late 1980s, the front generally lined up with a submerged mountain range called the Lomonosov Ridge, which runs from northern Greenland toward eastern Siberia, passing close to the North Pole.

During a scientific submarine cruise in 1993, researchers on board the USS *Pargo* found that the front had shifted substantially. The warmer, saltier Atlantic waters had pushed further into the Arctic, causing a retreat of the Pacific waters, says James H. Morison of the University of Washington, who participated in the *Pargo* cruise.

At the same time, the Atlantic-water layer over the Lomonosov Ridge warmed by 1°C, reaching a temperature not seen in the data going back to 1949, according to a January 1998 report by Morison and his colleagues in DEEP-SEA RESEARCH PART I.

Other shifts have altered the layered arrangement of water in the Arctic. To understand these divisions, imagine lowering a thermometer and a salt meter off

25. Sea Change in the Arctic

a ship in the middle of the Arctic. The uppermost layer of the ocean would be extremely fresh with a temperature near the freezing point. Below that, the instruments would pass through a layer called the halocline, where the water remains cold but grows saltier with depth. Lower still would be a thick region of relatively warm water, so dense with salt that it remains stuck on the bottom, trapped below the fresher, colder layer above.

In a series of three submarine cruises during the 1990s, oceanographers witnessed the halocline weakening in the central part of the Arctic. Early in the decade, a particularly cold sheet of halocline water straddled the Lomonosov Ridge. By 1995, however, the sheet had disappeared from the European side of this range, according to Michael Steele of the University of Washington. He and colleague Timothy Boyd of Oregon State University in Corvallis reported their finding in the May 15, 1998 JOURNAL OF GEOPHYSICAL RESEARCH.

The retreat of this layer could have profound effects, says Steele, because the halocline helps stratify the Arctic Ocean, keeping the warm, deep waters from reaching the surface. "Because there is decreased stratification, there should be enhanced heat transfer from the warm layer up to the surface, and hence thinner ice," says Steele.

Data collected by British submarines show that ice has been thinning in this region for some time. In 1990, researchers with the Scott Polar Research Institute at Cambridge University in England reported that ice thickness on the European side of the Lomonsov Ridge had decreased by 15 percent between 1976 and 1987. Preliminary analysis of measurements made in 1996 now suggest that the downward trend has continued, says Cambridge's Norman Davis.

With so many sigus of disruptions in the Arctic climate, scientists are looking skyward for answers. The ice pack and even the water beneath it, they suspect, are taking their cues from the pattern of winds that swirl around the edges of the

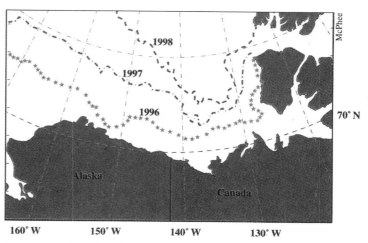

Frozen retreat: The edge of summertime sea ice in the Beaufort Sea pulled northward between 1996 and 1998, according to satellite data. Oceanographers are watching closely to see whether the ice withdrawal continues this summer.

Arctic. "Really, that is where it all starts," says Steele.

What's going on in the Arctic may fit into a larger picture that includes much of the globe's northern half. According to a theory proposed last year, atmospheric pressure tends to fluctuate in seesaw fashion over large parts of the Northern Hemisphere. When pressure increases over the Arctic, it decreases in a donut-shaped ring at the latitude of Washington, D.C. During the reverse part of the irregular cycle, air pressure drops in the Arctic and rebounds in the midlatitudes. Wallace and David W. J. Thompson of the University of Washington called this pattern the Arctic oscillation in the May 1, 1998 GEOPHYSICAL RESEARCH LETTERS.

Atmospheric temperatures in northern Europe and Asia dance in step with this Arctic oscillation. When the atmospheric pressure sags, air temperatures tend to rise and vice versa, says Wallace, who has tracked data going back to 1900.

For most of the century, the Arctic oscillation shifted randomly from month to month and year to year, with no distinct preference for either extreme. Recently, however, the oscillation has leaned strongly toward low Arctic air pressure and high temperatures in northern Eurasia. The shift started in the early 1970s and intensified in the late 1980s, bringing atmospheric conditions unseen in the century of measurements.

"That's the link to perhaps some of the other things going on in the Arctic," says Wallace. As the air pressure drops, the westerly winds encircling the Arctic grow stronger. The winds weaken the ice pack by pulling it apart and opening up watery channels between individual ice floes, says Wallace.

The souped-up westerlies also alter conditions in heavily populated parts of the globe. As they scream over the North Atlantic Ocean, the winds carry more heat downwind onto land. This translates into warmer winters in northern Europe and drier winters in southern Europe. "We've seen a strong warming in high latitudes over land in large areas of Russia," says Wallace. "We're seeing changes in the circulation that are big enough in winter that people in Europe are noticing them."

The key question is, What has caused all this change? In a study using computer models of the atmosphere, a team of Russian researchers has found evidence that thinning of the Arctic ozone layer could be driving the shift in the atmosphere. As chlorofluorocarbons and other chemicals eat away ozone, the polar stratosphere cools off. In the Russian model, this cooling effect enhances the westerlies that circle the Arctic and hence reduces the atmospheric pressure there.

Greenhouse gases could also have a hand in the Arctic shift. Carbon dioxide and other heat-trapping pollutants tend to warm the lowest layer of the atmosphere—the troposphere—while cooling off the stratosphere above. Researchers at NASA's Goddard Institute for Space Studies in New York City used a computer model to investigate this split effect of greenhouse gases. They found that the combination speeds up the westerly polar winds, pushing the Arctic oscillation in the direction seen during the past 30 years, says NASA's Drew T. Shindell.

Wallace thinks there is a good chance that one or both of these human influences could have precipitated the Arctic changes. Yet he cautions against drawing any conclusions. "We have to allow some possibility that this is of natural origin," as part of a long-term cycle that will eventually reverse, he says. To rule out a natural cause, the Arctic oscillation would have to keep on its present course for another 5 to 10 years, he says.

If that indeed transpires, researchers may not be able to repeat an experiment like SHEBA. With sustained warming and stronger westerlies, the ice-albedo feedback could kick in and rapidly rid the Arctic of its white cover during summer. McPhee and his colleagues raised this possibility last year in their paper in GEOPHYSICAL RESEARCH LETTERS. In the title, they asked, "Is perennial sea ice disappearing?"

Early in his career, McPhee would have found such a question unthinkable. Now, he says, "I'm starting to wonder whether we're not going to see it happen."

GREENVILLE FROM BACK COUNTRY to FOREFRONT

Eugene A. Kennedy

What factors are crucial in determining the success or failure of an area? This article explores the past and present and glimpses what may be the future of one area which is experiencing great success. The success story of Greenville County, S.C. is no longer a secret. This article seeks to find the factors which led to its success and whether they will provide a type of yardstick to measure the future.

The physical geography of this area is explored, as well as the economic factors, history, transportation, energy costs, labor costs and new incentive packages designed to lure new industries and company headquarters to the area.

Physical geography: advantageous

Greenville County is situated in the northwest corner of South Carolina on the upper edge of the Piedmont region. The land consists of a rolling landscape butted against the foothills of the Appalachian Mountains. Monadnocks, extremely hard rock structures which have resisted millions of years of erosion, rise above the surface in many places indicating that the surface level was once much higher than today. Rivers run across the Piedmont carving valleys between the plateaus. The cities, farms, highways and rail lines are located on the broad, flat tops of the rolling hills.

Climatologically, the area is in a transition zone between the humid coastal plains and the cooler temperatures of the mountains, resulting in a relatively mild climate with a long agricultural growing season. The average annual precipitation for Greenville County is 50.53 inches at an altitude of 1040 feet above sea level. The soil is classified as being a Utisoil. This type of soil has a high clay base and is usually found to be a reddish color due to the thousands of years of erosion which has leached many of the minerals out of the soil, leaving a reddish residue of iron oxide. This soil will produce good crops if lime and fertilizer containing the eroded minerals are added. Without fertilizers, these soils could sustain crops on freshly cleared areas for only two to three years before the nutrients were exhausted and new fields were needed. This kept large plantations from being created in the Greenville area. The climate and land are such that nearly anything could be cultivated with the proper soil modification. Physical potential, although a limiting factor, is not the only determining factor in the success of an area.

PHOTO: E. KENNEDY
An open courtyard off Main Street, downtown Greenville, S.C.

3 ❖ THE REGION

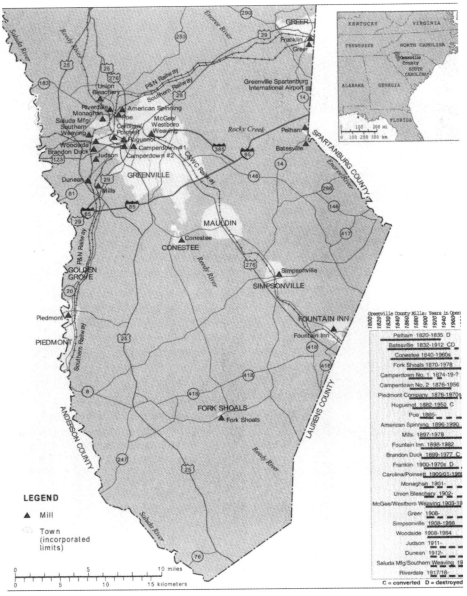

KAREN SEVERUD COOK, UNIVERSITY OF KANSAS MAP ASSOCIATES

As Preston James, one of the fathers of modern geography pointed out, the culture of the population which comes to inhabit the area greatly determines the response to that particular physical environment.

From European settlement through the textile era

An Englishman named Richard Pearis was the first to begin to recognize the potential of what was then known as the "Back Country." The area was off limits to white settlers through a treaty between the British and the Indians. In order to get around the law, Pearis married a native American and opened a trading post in 1768 at the falls of the Reedy River. He soon built a grist mill and used the waterfalls for power. Pearis prospered until the end of the Revolutionary War. He had remained loyal to the British, lost all his property when the new nation was established and left the country.

Others soon realized the potential of what was to become Greenville County. Isaac Green built a grist mill and became the area's most prominent citizen. In 1786, the area became a county in South Carolina and was named for Isaac Green. The Saluda, Reedy and Enoree Rivers along with several smaller streams had many waterfalls, making them excellent locations for mills during the era of water power. During the Antebellum period, Greenville County's 789 square miles was inhabited by immigrants with small farms and also served as a resort area for Low Country planters who sought to escape the intense summer heat and the disease carrying mosquitoes which flourished in the flooded rice fields and swampy low country of the coastal plains. Most of the permanent residents farmed and a few mills were built to process the grains grown in the area.

Although very little cotton was ever actually grown in Greenville County, cotton became the driving force behind its early industrialization. Beginning with William Bates in 1820, entrepreneurs saw this area's plentiful rivers and waterfalls as a potential source of energy to harness. William Bates built the first textile mill in the county sometime between 1830 and 1832 on Rocky Creek near the Enoree River. This was known as the Batesville Mill. Water power dictated the location of the early southern textile mills, patterned after the mills built in New England. Mill owners purchased the cotton from farms and hauled it to their mills, but lack of easy transportation severely limited their efforts until 1852. That year, the Columbia and Greenville Railroad finally reached Greenville County. Only the interruption of the Civil War kept the local textile industry from becoming a national force during the 1850s and 1860s.

The area missed most of the fighting of the Civil War and escaped relatively unscathed. This provided the area with an advantage over those whose mills and facilities had been destroyed during the war. The 1870 census reported a total $351,875 in textiles produced in the county. This success encouraged others to locate in Greenville. Ten years later, with numerous mills being added each year, the total reached $1,413,556.11. William Bates' son-in-law, Colonel H. P. Hammett, was owner of the Piedmont Company which was the county's largest producer. Shortly after the construction of the water powered Huguenot Mill in the downtown area of the City of Greenville County in 1882, the manufacture of cotton yarn would no longer be controlled by the geography of water power.

The development of the steam engine created a revolution in the textile industry. No longer was the location of the mill tied to a fast moving stream, to turn a wheel that moved machinery. Large amounts of water were still needed but the dependence upon the waterfalls was severed. Between 1890 and 1920, four textile plants were built in the county outside the current city limits of the City of Greenville. At least thirteen large mills were built near the city to take advantage of the rail system, as shown on the map, "Greenville County Mills." Thus, with cheaper and more efficient steam

26. Greenville

PHOTO: E. KENNEDY

F.W. Poe Manufacturing Co., Old Buncombe Road, Greenville, S.C. Built in 1895, purchased by Burlington Industries and closed in 1997. Palmetto State Dyeing and Finishing Co. opened in 1987. The company employs approximately 110 people.

power, transportation costs became a deciding factor. These mills built large boiler rooms adjacent to their plants and dug holding ponds for water.

Another drastic change took place in the textile industry around 1900. This change would provide even greater flexibility for the mill owners. A hydroelectric dam was constructed on the Saluda River, five miles west of the city of Greenville. It was completed in 1902 and would provide cheap electricity for the county. John Woodside, a local mill owner who foresaw electricity as the next step in the evolution of the industry, built what was then the largest textile mill in the world in the city of Greenville that same year. He located it further from a water source than previously thought acceptable. However, John had done some primitive locational analysis and chose the new site well. It was located just beyond the city boundary to limit his tax liability and directly between the lines of two competing rail companies—the Piedmont Railroad and the Norfolk and Western (now known as Norfolk and Southern). John Woodside's mill proved to be a tremendous success. With water no longer a key factor of location, the owners identified transportation as the key factor of location. Others began to build near rail lines.

The textile industry made Greenville County very prosperous. The mills needed workers and shortly outstripped the area's available labor supply. Also, many did not want to work in the hot, poorly ventilated, dangerous conditions found in the mills. When most of the mills were still built of wood, the cotton fibers floating in the air made fire a very real danger. Many businesses sprang up to service the needs of the workers and the textile mill owners. Farmers, sharecroppers, former slaves and children of former slaves were recruited to work in the mills. Housing soon became scarce and the infrastructure wasn't equipped to handle the influx of new workers. To alleviate the problem, the mill owners built housing for their workers. These were very similar to the coal camps of Appalachia and other factory owned housing in the north. They were very simple dwellings built close to the mill so the workers could easily walk to and from work. They also provided company-owned stores, doctors and organized recreational activities for their employees, creating mill communities. Many people who worked for the mills would have told you they lived at Poe Mill or Woodside, the names of their mill communities, rather than Greenville.

In the 1960s, rail transportation of textiles was a cost the owners wished to lower. They found a cheaper, more versatile form of transportation in the trucking industry. The interstate highway system was now well developed and provided a means of keeping costs down for the operators. In the 1970s, owners began to identify wages and benefits as a major factor in their cost of operation and many firms relocated in foreign countries, which offered workers at a fraction of the wages paid in the United States and requiring few if any benefits.

Meeting the challenge of economic diversification

Greenville County used its natural physical advantage to become the "Textile Capital of the World." Many of the other businesses were tied directly or indirectly to the textile industry. These ranged from engineering companies who designed and built textile machinery to companies which cleaned or repaired textile machines. Employment in the textile industry in Greenville County peaked in 1954 with 18,964 workers directly employed in the mills. As the industry began to decline, the leaders of the industry along with local and state leaders showed great foresight by combining their efforts into an aggressive move to transform Greenville County into a production and headquarters oriented economy. A state sponsored system of technical schools greatly facilitated this effort. Workers could get the training they needed to pursue almost any vocation at these centers. This system still is a factor in Greenville County's success.

Table 1
CORPORATE HEADQUARTERS IN GREENVILLE COUNTY

1. American Leprosy Mission International
2. American Federal Bank
3. Baby Superstores, Inc.
4. Bowater Inc.
5. Builder Marts of America Inc.
6. Carolina First Bank
7. Delta Woodside Industries Inc.
8. Ellcon National Inc.
9. First Savings Bank
10. Heckler Manufacturing and Investment Group
11. Henderson Advertising
12. Herbert-Yeargin, Inc.
13. JPS Textile Group Inc.
14. Kemet Electronics Corp.
15. Leslie Advertising
16. Liberty Corp.
17. Mount Vernon Mills Inc.
18. Multimedia Inc.
19. Ryan's Family Steakhouses
20. Span America
21. Steel Heddle Manufacturing Co.
22. Stone Manufacturing Co.
23. Stone International.
24. TNS Mills Inc.
25. Woven Electronics Corp.

3 ❖ THE REGION

> **What ultimately swayed the automaker to choose Greenville? One of the main reasons was physical location.**

The group emphasized the ability to make a profit in Greenville County. The focus of their efforts was turned to creating a sound technical education network along with the flexibility to negotiate packages of incentives to lure large employers. Incentives included negotiable tax and utility rates, plus a strong record of worker reliability due to South Carolina's nonunion tradition, with very few work stoppages. The foresight of this group has paid off handsomely. The majority of the textile mills which provided the backbone of the economy of Greenville County are no longer in business. Many of the old buildings still stand. Ten of the mills built before 1920 now are used in other capacities. American Spinning was built in 1896 and now is used as a warehouse, office space and light manufacturing all under one roof. Most of the mills are used for warehouse space or light manufacturing such as the Brandon Duck Mills, which operated between 1899 and 1977 as a cotton mill. It now houses two small factories which assemble golf clubs and part of the mill is used as a distribution center. The low lease cost (from $1 to $15 per square foot) is an enticement for other businesses to locate in these old buildings.

The old Huguenot Mill, the last water powered mill built in the county, was recently gutted and has been rebuilt as offices for the new 35 million dollar Peace Center entertainment complex in downtown Greenville. The Batesville Mill, the first in the county, was built of wood. It burned and was rebuilt in brick in 1881. It closed its doors in 1912 because the water-powered mill was not competitive. The mill was purchased by a husband and wife in 1983, converted into a restaurant, and was the cornerstone and headquarters of a chain of FATZ Restaurants until it burned again in 1997. So, in considering diversification, one of the first steps was to look for other uses for the facilities which already existed.

Other efforts also met with great success. As businesses began to look south during the 1970s for relocation sites, Greenville began to use its natural advantages to gather some impressive companies into its list of residents. By 1992, the combination of these efforts made Greenville County the wealthiest county in the state of South Carolina. Twenty-five companies have their corporate headquarters in the county, as shown in Table 1.

Forty-nine others have divisional headquarters in the county, as shown in Table 2. This constitutes a sizable investment for the area, yet even this list does not include a 150 million dollar investment by G. E. Gas Turbines in 1992 for expansion of their facility. This was the largest recent investment until 1993.

Along with American companies, foreign investment was sought as well. Companies such as Lucas, Bosch, Michelin, Mita and Hitachi have made major investments in the county. Great effort has been put into reshaping the face of Main Street in Greenville as well. The city is trying to make a place where people want to live and shop. Many specialty stores have opened replacing empty buildings left by such long time mainstays as Woolworths. The Plaza Bergamo was created to encourage people to spend time downtown. The Peace Center Complex provides an array of entertainment choices not usually found in a city the size of Greenville. The Memorial Auditorium, which provided everything from basketball games, to rodeo, concerts, high school graduations and truck pulls has closed its doors and was demolished in 1997 to make way for a new 15,000 seat complex which will be named for its corporate sponsor. It will be called the Bi-Lo Center. Bi-Lo is a grocery store chain and a division of the Dutch Company, Ahold. A new parking garage is being built for this center and two other garages have recently been added to improve the infrastructure of the city. City leaders have traveled to cities such as Portland, Oregon to study how they have handled and managed growth and yet kept the city friendly to its inhabitants.

The largest gamble for Greenville County came in early 1989. The automaker BMW announced that it was considering building a factory in the United States. Greenville County and the state of South Carolina competed against several other sites in the midwest and southeast for nearly two years. On June 23, 1992, the German automaker chose to locate in the Greenville-Spartanburg area. Although the plant is located in Spartanburg County, the headquarters are in Greenville and both counties will profit greatly. When the announcement was made, the ques-

Panorama of downtown Greenville, S.C.

PHOTO: E. KENNEDY

Table 2
DIVISIONAL HEADQUARTERS IN GREENVILLE COUNTY

1. Ahold, Bi-Lo.
2. BB&T, BB&T of South Carolina.
3. Bell Atlantic Mobile
4. Canal Insurance Company
5. Coats and Clark, Consumer Sewing Products Division.
6. Cryovac—Div. of W. R. Grace & Co.
7. Dana Corp., Mobile Fluid Products Division.
8. DataStream Systems, Inc.
9. Dodge Reliance Electric
10. Dunlop Slazenger International, Dunlop Slazenger Corp.
11. EuroKera North America, Inc.
12. Fluor Daniel Inc.
13. Fulfillment of America
14. Frank Nott Co.
15. Gates/Arrow Inc.
16. General Nutrition Inc., General Nutrition Products Corp.
17. Gerber Products Co., Gerber Childrenswear, Inc.
18. GMAC
19. Goddard Technology Corp.
20. Greenville Glass
21. Hitachi, Hitachi Electronic Devices (USA)
22. Holzstoff Holding, Fiberweb North America Inc.
23. IBANK Systems, Inc.
24. Insignia Financial Group, Inc.
25. Jacobs-Sirrine Engineering
26. Kaepa, Inc.
27. Kvaerner, John Brown Engineering Corp.
28. Lawrence and Allen
29. LCI Communications
30. Lockheed Martin Aircraft Logistics Center Inc.
31. Manhattan Bagel Co.
32. Mariplast North America, Inc.
33. Michelin Group, Michelin North America.
34. Mita South Carolina, Inc.
35. Moovies, Inc.
36. Munaco Packing and Rubber Company.
37. National Electrical Carbon Corp.
38. O'Neal Engineering
39. Personal Communication Services Dev.
40. Phillips and Goot
41. Pierburg
42. Rust Environment and Infrastructure
43. SC Teleco Federal Credit Union
44. Sodotel
45. South Trust Bank
46. Sterling Diagnostic Imaging, Inc.
47. Umbro, Inc.
48. United Parcel Service
49. Walter Alfmeier GmbH & Co.

tion was: what ultimately swayed the automaker to choose Greenville? One of the main reasons was physical location. The site is only a four hour drive from the deepwater harbor of Charleston, SC. Interstates 26 and 85 are close by for easy transportation of parts to the assembly plant. The Greenville-Spartanburg Airport is being upgraded so that BMW can send fully loaded Boeing 747 cargo planes and have them land within five miles of the factory. Plus, the airport is already designated as a U.S. Customs Port of Entry and the flights from Germany can fly directly to Greenville without having to stop at Customs when entering the country. Other incentives in the form of tax breaks, negotiated utility rates, worker training and state purchased land helped BMW choose the 900 acre site where it will build automobiles.

The large incentive packages might appear self-defeating but BMW's initial investment was scheduled to be between 350 and 400 million dollars. The majority of the companies supplying parts for BMW also looked for sites close enough to satisfy BMW's just-in-time manufacturing needs. The fact that Michelin already made tires for their cars here, Bosch could supply brake and electrical parts from factories already here and J.P. Stevens and others could supply fabrics for automobile carpets and other needs readily from a few miles away also was a factor.

One major BMW supplier, Magna International, which makes body parts for the BMW Roadster and parts for other car manufacturers, located its stamping plant in Greenville County. Magna invested $50 million and will invest $35 million more as BMW expands. Magna needed 100 acres of flat land without any wetlands and large rock formations. This land needed to be close enough to provide delivery to BMW. After studying several sites, Magna chose South Donaldson Industrial Park, formerly an Air Force base, just south of the city of Greenville. The county and state will help prepare the location for their newest employer.

Road improvements, the addition of a rail spur and an updating of water and sewer facilities will all be provided to Magna in this agreement. Also, Magna will receive a reduced 20 year fixed tax rate along with other incentives for each worker hired. These incentive packages may seem unreasonable but they have proven to be necessary in the 1990s

PHOTO: E. KENNEDY

Huguenot Mill (lower left). Built in 1882, on Broad Street, Greenville, S.C., is the last waterpowered mill in Greenville County. It is being refurbished to become part of the Peace Center Complex at right.

3 ❖ THE REGION

when large organizations are deciding where to locate.

The future: location, location, and location

From the time of the earliest European settlers, the natural advantages of Greenville County helped bring it to prosperity. The cultural background of the settlers was one of industry and a propensity for changing the physical environment to maximize its industrial potential. Nature provided the swift running rivers and beautiful waterfalls. The cultural background of the settlers caused them to look at these natural resources and see economic potential.

The people worked together to create an environment which led Greenville County to be given the title of "Textile Center of the World" in the 1920s. Then, again taking advantage of transportation opportunities and economic advantages, the area retained its textile center longer than the majority of textile centers.

Today, after 30 years of diversification, economic factors now are normally the deciding factor in the location of a new business or industry. Greenville County with its availability of land, reasonable housing costs, low taxes, willingness to negotiate incentive packages, and positive history of labor relations helped make it a desirable location for business. Proximity to interstate transportation, rail and air transport availability help keep costs low. The county's physical location about half way between the mega-growth centers of Charlotte, North Carolina and Atlanta, Georgia places it

> A combination of physical, environmental, and cultural factors greatly influence the location of businesses.

in what many experts call the mega-growth center of the next two decades. Now, with BMW as a cornerstone industry for the 1990s and beyond, Greenville County looks to be one of the areas with tremendous growth potential.

Thus, a combination of physical, environmental and cultural factors greatly influence the location of businesses. Transportation costs, wage and benefit packages and technical education availability are all interconnected.

The newest variable involves incentive packages of tax, utility reduction, worker training, site leasing and state and local investment into improving the infrastructure for attracting employers. The equation grows more and more complex with no one factor outweighing another; however, economic costs of plant or office facilities, wages and benefits and transportation seem to be paramount. Greenville County is blessed with everything it needs for success. It will definitely be one the places "to be" in the coming years.

Eugene A. Kennedy is a native of West Virginia who attended Bluefield State College, Bluefield, West Virginia, and received an M.A. in geography from Marshall University in Huntington. He is currently a public educator in the Greenville County School system, Greenville S.C. He was awarded a "Golden Apple" by Greenville television station WYFF in 1997, has been a presenter at the South Carolina Science Conference, and a consultant to the South Carolina State Department of Education. He can be reached at GEOGEAK@aol.com

References and further readings

DuPlessis, Jim. 1991. Many Mills Standing 60 Years After Textile Heydays. *Greenville News-Piedmont* July 8. pp. 1c–2c.

Greater Greenville Chamber of Commerce, 1990. *1990 Guide to Greenville.*

Greenville News-Piedmont. 1991–1993. *Fact Book 1991; 1992; 1993.*

Patterson, J. H. 1989. *North America.* Oxford University Press: N.Y. Eighth edition.

Scott, Robert. 1993. Upstate Business. *Greenville-News Piedmont.* 15 August. pp. 2–3.

Shaw, Martha Angelette. 1964. *The Textile Industry in Greenville County.* University of Tennessee Master's Thesis.

Strahler, Arthur. 1989. *Elements of Physical Geography.* John Wiley and Sons: N.Y. Fourth edition.

Flood Hazards and Planning in the Arid West

Can we somehow marry flood control or flood planning with water quality and best management practices? This would require truly integrated watershed management.

By John Cobourn

IN Nevada and other areas in the west, many urbanizing areas have not been inhabited for very long. In these areas, there is a short period of record regarding local flood hazards. Therefore, people commonly lack awareness about flood hazards. One question of particular interest is, "What kind of watershed approach is suitable to address flooding?" In this day-and-age, dams are not necessarily the most popular solution for flooding. What about non-structural techniques? What about best management practices? How can we link flood control or flood management with watershed concerns like water quality and habitat?

In the arid west, there are two main types of floods that are completely different. They are like night and day. There is winter flooding, also called "river valley bottom" flooding. It comes down a river like the Truckee or Carson River from the Sierra Nevada or another mountain range. It produces tremendous flooding in the valleys. Most valley bottoms are Zone A flood zones. The other type of flooding is called alluvial fan flooding or "flash" flooding. These happen in the summer and are called dry mantle events because they happen when the soil is not saturated. Because desert soils often have poor infiltration capacity, if a cloudburst occurs that lasts even an hour or two, it can cause a flash flood to rush down a normally dry wash. These creeks are normally bone-dry 365 days a year, unless there is a flash flood. The water might come down at 500, 1,000 or 2,000 cfs for 30 minutes or an hour and damage a subdivision on an alluvial fan.

River Valley Floods

Six feet of snow fell in the Sierra Nevada near Lake Tahoe just before Christmas of 1996, and on December 27–28, another storm dumped two to three more feet of snow. On New Year's Eve and New Year's Day a huge rainstorm hit the area. The weather changed from an Alaskan snowstorm to a Hawaiian rainstorm.

As a result, in Nevada and California, major river valley floods occurred on the Truckee, Carson and Walker Rivers in January of 1997. They surprised many people. A rain-on-snow even at high elevations of the Sierra Nevada caused large amounts of runoff to come down the rivers into the local valleys. There was five feet of water in some casinos in downtown Reno. There was also terrible flood damage to the Reno-Tahoe International Airport, to the large industrial section of Sparks, which is next to the river in the floodplain, and to a residential neighborhood in Gardnerville.

The Carson River parallels the Truckee, coming down from the Sierra and going out through Fallon to its terminus, the Stillwater Wildlife Refuge. The Truckee River also fails to reach the sea, as its terminus is the large desert lake, Pyramid Lake, owned by the Paiute Tribe. The watersheds are almost the same size with almost the same water flow. The two rivers are an interesting contrast.

The Truckee River goes through Reno and has many reservoirs on the upper watershed, including Lake Tahoe, which has a six-foot dam on it. The Carson River has very few reservoirs.

The Truckee River has a tremendous urban area right in the middle of the watershed. Reno and Sparks have a population of over 250,000, with much of

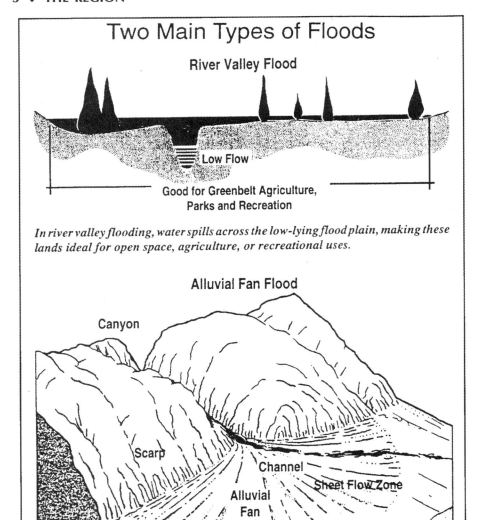

In river valley flooding, water spills across the low-lying flood plain, making these lands ideal for open space, agriculture, or recreational uses.

In a flash flood on an alluvial fan, the old wash sometimes fills with sediment, causing the channel to shift to a new location, disturbing new areas.

their commercial-industrial area situated in the floodplain.

There is not a great deal of urbanization in the floodplain of the Carson River. There is a great deal of agricultural land still in the flood plain. However, the Carson Valley is one of the fastest growing population centers in Nevada. The Carson River flows through the Carson Valley and normally, in late summer, has dwindled to a small stream. In the flood of January 1997 however, it created a temporary lake that covered Highway 395, the main highway between that area and Reno, and much of the agricultural land on the valley floor.

The population of the Carson Valley is now about 30,000 and is projected to approach 60,000 in the next 20 years or so. There is debate about whether there should be more urbanization in the current floodplain. Should Planners repeat in the Carson Valley what has been done in Reno and Sparks? It is a big question, and it was discussed at two major conferences on flooding and watershed management that were held in Nevada in 1997 and 1998. Hydrologists and geomorphologists are urging county planners and local ranchers to leave agriculture in the valley floodplains rather than subdivide them.

In the 1997 flood, most Zone A flood zones were inundated in both watersheds. Luckily, there was not a great deal of flooding of homes. One levee did break and flooded about 50 homes in upper Gardnerville.

Water Quality Concerns

The 1997 winter flood scoured and widened the channels of many creeks in the low elevations around Reno. A large amount of sediment washed down many urban tributaries to the Truckee River.

Can we somehow marry flood control or flood planning with water quality and best management practices? This would require truly integrated watershed management.

Steamboat Creek is the largest tributary to the Truckee River in the Reno area. Cooperative Extension and citizen volunteers are working with the Washoe-Storey Conservation District and Nevada Division of Environmental Protection to clean it up, not through structures like dams or levees, but through implementation of best management practices. Nevada is one of the fastest growing states in the nation, so there is a large potential for erosion in this area on new construction sites. In a big event, there is a huge flush of pollution as sediment and nutrients flow downstream.

During a rainstorm, the flow of a creek will increase much more gradually, with a lower flood crest, if there are many wetlands in the watershed than if there are few or no wetlands. The wetlands serve as holding ponds and capture some of the rain before it flows into the river. One golf course in Reno incorporated wetlands as part of the golf course design. The course is located on Steamboat Creek, so some upstream sediment is filtered out at the golf course. When valley floors are used for agriculture, a flooding river can usually flow out of its banks onto its floodplain. When rivers have access to their floodplains, the erosive power of the flood water is dissipated, so that less sedimentation occurs.

Alluvial Fan Flash Flooding

Let's consider alluvial fan flooding. It is easy to see how sediment coming down a canyon can create a fan-shaped deposit. Along almost all of western mountain ranges, at the bottom of even small canyons that are dry now, there are alluvial fans. In many parts of the southwest, alluvial fans are being urbanized. Over half of the developable land in Nevada is on alluvial fans, which are subject to flooding. Many people do not recognize that they live on alluvial fans.

One area in Douglas County has had five flash floods between 1990 and 1994. This alluvial fan is one of the fastest growing areas in Douglas County. In the master plan, it is slated for extensive and rapid urban growth. When a thunderstorm occurs, this area may receive intense rain and flash floods.

When subdivisions are built in Nevada, developers are required to put in detention ponds so that the runoff water from the rooftops and pavements in the subdivision can be gathered and possibly create some wildlife habitat. Residents in one subdivision decided after about five years that they didn't like the way their detention pond was working. The pond had a spillway that was high enough so that, when a flash flood occurred, the water it detained backed up into adjacent streets. There was eight inches of water in the street, and the residents didn't like it. The subdivision brought in some bulldozers and lowered the spillway. The spillway still retains some water, but more of the water will flow downstream to where another subdivision will be built someday. It could flood residents of that subdivision in the future.

Many residents and county planners in the west have not yet realized the flash flood dangers of the region's numerous alluvial fans. With their gently sloping topography, they seem to be ideal locations for homesites and subdivisions. Yet the Federal Emergency Management Agency (FEMA) has identified "a critical need to provide guidance to communities, developers and citizens on how to safely accommodate growth while protecting life and property from flood hazards on alluvial fans." (FEMA 165, 1989)

The flash floods that occurred in Douglas County, Nevada in the early 1990s were warning floods. They have not been close in size to a 100-year event. If five similar floods occurred in five years, by definition they couldn't be 100-year events. These floods caused some damage, but many people think, "So, we have had five floods in five years. I'm not going to worry about it. They didn't flood my house." These people don't realize that they haven't seen a true 100-year event yet. The planners and developers keep developing for flood rates of 600 cfs because that is what they have seen in the last five years. They haven't seen the 2,000-cfs flood event yet. Due in part to the recent floods, some of these washes are being restudied by FEMA now.

In 1974, a flood occurred in El Dorado Canyon, located in southern Nevada. A series of warning floods had gone through a national park campground causing damage to picnic tables, and some campers had gotten wet. The Park Service wanted to move the campground because it was in a flood area. Many of the people who had been using the park area for 10 years signed a petition and said they loved that campground. "It is right on the lake and we can bring our boats here. Don't move the campground." Two years later, a flood wiped out the entire campground, demolished every structure, and killed nine people. If people had paid more attention to the small "warning floods", they might have been out of harm's way when the true 100-year event struck the campsite.

One of the things Cooperative Extension is trying to do in Nevada is to educate people about this kind of problem through conferences and publications. In Douglas County, a citizens group called the Buckbrush Wash Flood Safety Coalition has formed. The people there have said, "Let's see if we can work with our county government and local government." These citizens have been using public education and working with their local search and rescue unit to educate the children and their parents about the flash flood areas. They are educating the children not to play in the washes.

This citizens group has also come up with a design for a diversion structure. In Nevada, a diversion structure for flood water is listed as a best management practice. There are several new houses that have been built in the Zone A flood zone near the apex of this fan. The diversion structure, which the Coalition designed with help from NRCS in Nevada, could divert the peak flow from a major flood into an uninhabited basin with no houses or roads. The structure would cost $1.4 million. The county's previous estimate for reducing the flash flood potential was $5 million. The citizens were looking for an innovative way to solve this problem, and they came up with a relatively inexpensive proposal that would also be much better for water quality because it would prevent flood peaks from washing through urban neighborhoods.

There is not much appetite around the nation for dams on rivers. However, with urbanized alluvial fans, because the land surface is convex, non-structural techniques alone will not solve the problem. That is because on most alluvial fans, the entire fan surface below its apex constitutes the floodplain. If you live in an area where people are urbanizing alluvial fans, it would be interesting to see if PL-566 funds or other public grant funds could be used to help create structures to safeguard lives and property.

Conclusion

According to the National Weather Service, floods and flash floods are the number one weather related killer in the United States. To plan appropriately for floods, drought, water quality, urban growth and a sustainable ecosystem, managers of diverse agencies and private concerns need to come together and collaborate on integrated watershed management. When flood problems are viewed as interconnected with pollution, infrastructure, and habitat problems, we will begin the uphill journey to better solutions. If we can learn to cooperate and help each other, we can make substantial progress toward creating safer communities and protecting water quality.

For more information, contact John Cobourn, University of Nevada Cooperative Extension, P.O. Box 8208, Incline Village, NV 89452-8208, (775) 832-4150, fax (775) 832-4139.

The Río Grande

Beloved river faces rough waters ahead

By Steve Larese

Corrales farmer Gus Wagner allows himself just a minute to admire a peach and honey sunrise pouring over the Sandía Mountains, its patchy light filtering through cottonwoods and bobbing on the Río Grande. Turning back to his work, his callused hands turn the wheel of his *compuerta* (floodgate), and the river's paced water soon sluices through the *acequia* to flood his blossoming apple orchard.

"The river's beautiful, *que no?*" he asks, stealing another moment from his busy day. "I thank God everyday for the Río Grande and how it allows my family to live here. The river is our lifeblood."

It's safe to say there's not a Land of Enchanter around who would disagree with Wagner. Three states and two countries lay claim to the Río Grande, but only New Mexico claims it as our soul. To New Mexicans past and present, the Río Grande has been a gentle miracle in a harsh land.

But our beloved river was recently named the seventh most endangered in the nation by the respected American Rivers, a national conservation organization (www.Amrivers.org). "The Río Grande is really on its deathbed," says Betsy Otto, American Rivers' director of river restoration finance. "Increased population, outdated irrigation techniques and misguided engineering are drying the river up. But that being said, the situation isn't hopeless. Rivers are resilient if you let them be rivers. But too many people think of the Río Grande only as an irrigation ditch, drinking fountain or a sewer."

Beginning life as a twist of alpine streams 12,500 feet above sea level in the San Juan Mountains of Colorado, the Río Grande courses through Colorado, New Mexico and forms the 1,000-mile border between Texas and Mexico. Its life ends 1,887 miles later when it merges with salt water in the Gulf of Mexico. Today, the river appears fairly linear through New Mexico, lined by the nation's largest continuous cottonwood forest called the *bosque*, or woods. But this hasn't always been the case, says Dr. Cliff Crawford, a University of New Mexico research professor emeritus who has extensively studied the Río Grande and its ecosystem.

"The river and *bosque* as we see them today look quite different from how they appeared just a few decades ago," he says.

Crawford explains that until about the mid-'50s, the Río Grande was a braided river, meaning several sizable rivulets would snake over one another with cottonwood islands between them. Instead of perfectly lining the river, cottonwoods, coyote willows and other native flora created a mosaic pattern, their seeds deposited on the whim of the river. Depending on the year's runoff, the rivulets would swell to make one river worthy of the name Río Grande.

This seasonal flooding would establish new groves of trees, clean away debris and keep in check non-native species such as salt cedar and Russian olive. During heavy runoff years, the river would occasionally spread past its average bounds and devastate villages and pueblos built within the flood plain. The force of such flooding could reroute the very course of the river, cutting new channels through former farm fields and irrigation ditches.

28. Rio Grande

STEVE LARESE

LAURENCE PARENT

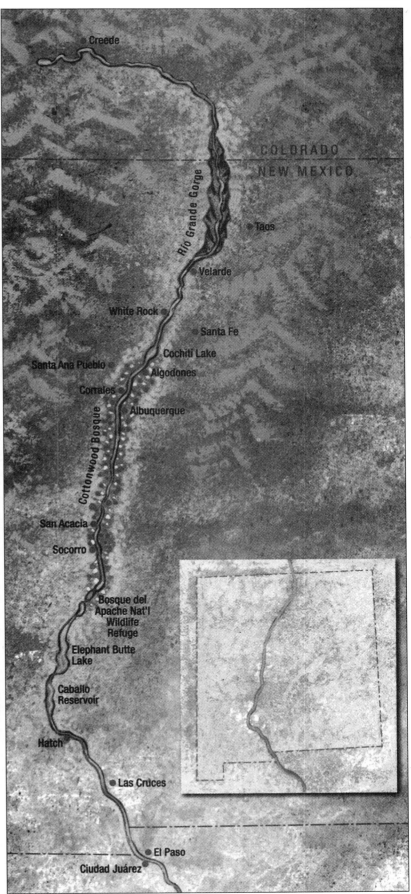

Above—The river courses through the Río Grande Gorge, an impressive feature it carved over millions of years through the volcanic basalt of the Taos Plateau. The canyon starts as a ditch in Colorado's San Juan Hills and eventually plummets to depths of 1,000 feet. **Below**—The Río Grande as it leaves New Mexico and forms the Texas/Mexico border. **Right**—Map of New Mexico's portion of the Great River.

STEVE LARESE

The Middle Río Grande supports the largest continuous cottonwood forest left in the United States, seen here north of Albuquerque. Scientists, citizens and environmentalists worry the forest (bosque) is dying out, partly because the river hasn't been able to flood in the past half century. Flooding is crucial in the establishment of new cottonwoods. Also, deadwood has been allowed to build up, which creates a severe fire hazard. The Catch 22 is that if flooding is allowed, there may not be enough water to meet water commitments to Texas, Chihuahua, Coahuila, Nuevo Leon and Tamaulipas, Mexico. As New Mexico enters an expected period of even drier weather, the growing cities of Albuquerque, El Paso and Ciudad Juárez are also becoming more dependent on the river.

Ciénegas (marshes) and oxbow lakes remained after the flood subsided. For the most part, New Mexicans accepted this fact of life, and many understood it was necessary. Like Egypt's Nile, with the floods came rich silt that was deposited on farmland. The infrequent inconvenience of rebuilding waterlogged adobes was worth the annual gifts provided by the river.

But New Mexico's dynamics were changing. As more people moved to the state and communities—especially Albuquerque—grew, the price of flooding increased. Finally, after Albuquerque in part had to be evacuated because of the floods of 1941 and '42, state and federal governments decided the Great River needed to be controlled.

In the 1950s, the Southwest suffered through a devastating drought. The Río Grande dried up, and New Mexico was unable to meet its water obligations to Texas under the Río Grande Compact, a water-rights agreement signed by Colorado, New Mexico and Texas and approved by Congress in 1939. New Mexico avoided a lawsuit by aggressively pursuing channelization of the river to maximize flow downstream. Levees were built along the river's banks to contain flooding. "Jetty Jacks," the same type of crossed metal structures used to deter amphibious assaults during World War II, made the Río Grande look like Normandy Beach. Jetty Jacks lined the desired channels with other structures directing the water flow into the channels. The result was to trap silt, sand and debris, building up the banks and further confining the river. By narrowing the channel, more water was being delivered downstream, which also greatly reduced flooding.

The efforts worked perfectly. But altering the natural tendencies of the river on such a grand scale couldn't happen without affecting its very nature. Today, Crawford says, we are seeing those effects.

"The changes made to the river were very justified in people's minds at the time," he says. "But the hydrology of the system has completely changed. Unless we allow the system to come back to some level of how its ecosystem used to work, Albuquerque's *bosque* at least will be lost."

"You can't blame people for not wanting their houses washed away or their crops destroyed," says Rob Yaksich, an interpretive ranger at the Río Grande Nature Center State Park in Albuquerque. "But a relatively short time later, we're seeing the cottonwoods are certainly losing ground. Without flooding, they aren't regenerating. Most of the youngest trees we

28. Rio Grande

CLAY MARTIN

STEVE LARESE

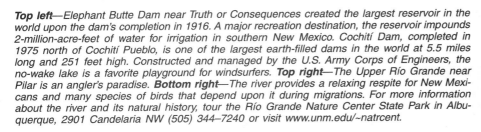

Top left—Elephant Butte Dam near Truth or Consequences created the largest reservoir in the world upon the dam's completion in 1916. A major recreation destination, the reservoir impounds 2-million-acre-feet of water for irrigation in southern New Mexico. Cochití Dam, completed in 1975 north of Cochití Pueblo, is one of the largest earth-filled dams in the world at 5.5 miles long and 251 feet high. Constructed and managed by the U.S. Army Corps of Engineers, the no-wake lake is a favorite playground for windsurfers. *Top right*—The Upper Río Grande near Pilar is an angler's paradise. *Bottom right*—The river provides a relaxing respite for New Mexicans and many species of birds that depend upon it during migrations. For more information about the river and its natural history, tour the Río Grande Nature Center State Park in Albuquerque, 2901 Candelaria NW (505) 344-7240 or visit www.unm.edu/~natrcent.

STEVE LARESE

have were established in the floods of the '40s. Without some type of flooding, when they're gone in 40 or so years, that'll be it for the *bosque* as we know it."

Early last century, salt cedar and Russian olive trees were introduced to New Mexico as ornamental vegetation. Some were planted along the river to further stabilize its banks. These newcomers have done extremely well, and have choked out native trees such as cottonwoods in areas. Salt cedars have increased the salinity of the soil, and they have also blocked the sun that young cottonwoods need, says Crawford. "New cottonwoods need to be established in the open with lots of sun and silt," he says. "But it's a matter of who wins the shade race."

Realizing the threat, several governments and organizations have begun to take action to maintain and restore New Mexico's *bosque*.

Todd Caplan, department of natural resources director at Santa Ana Pueblo, watches as his crew plants new cottonwoods along part of the six miles of river that crosses the 74,000-acre pueblo. "What we're doing is restoring 7,000 acres of floodplain here," Caplan says. "The river through here used to average 1,200 feet wide. Now it's 300 feet. Cochití Dam has certainly reduced flooding, but now the *bosque* needs help."

Using a combination of state, federal and tribal funds, the Pueblo has painstakingly ripped out hun-

3 ❖ THE REGION

Above—Having escaped the drought and hardships that befell northwestern New Mexico, many groups of ancestral Pueblo people eventually settled along the Río Grande, creating the oldest communities in the United States. Here, a pet helps out during San Juan Pueblo's deer dance. **Above Right**—Union and Confederate troops clashed along the Río Grande during the Civil War battle of Valverde near Socorro, which is re-enacted every February. Victorious Confederate troops rebuffed Union attacks from Fort Craig, captured six cannons and continued north to take Albuquerque and Santa Fe. For more information, log on to www2.cr.nps.gov/abpp/battles/nm001.htm. **Below**—Glowing red in the late afternoon sun, salt cedar lines much of the river through New Mexico. The non-native, thirsty shrub was introduced from Eurasia as an ornamental plant early last century. Also called tamarisk, it is considered a problem along the Río Grande because it uses much water and chokes out native trees like cottonwoods and coyote willows.

28. Rio Grande

STEVE LARESE

STEVE LARESE

Above—This section of river north of Cochití Dam demonstrates the braided nature and oxbows once common for much of New Mexico's Río Grande. By damming and bank reinforcement, the river below Cochití Dam is now fairly straight. This has eliminated flooding and improved water availability downstream, but it has also destroyed much of the habitat needed by endangered species such as the silvery minnow and Southwest willow flycatcher.

Top left—Ron J. Sarracino places a beaver cage around a newly planted cottonwood as part of Santa Ana Pueblo's Río Grande Bosque Rehabilitation Project. **Middle left**—Todd Caplan, Santa Ana's director of natural resources, explains how his employees have cleared out non-native salt cedar and Russian olive trees from the Pueblo's bosque, leaving cottonwoods and returning the area to how it appeared before the introduction of non-native species. **Bottom left**—Rafters take advantage of the annual spring snow runoff that turns the Upper Río Grande into a national destination for thrill seekers.

STEVE LARESE

dreds of acres of salt cedar and Russian olive that overran traditional cottonwood and willow habitat. The felled trees are cut and delivered to tribal elders for firewood. Towering piles of disassembled Jetty Jacks await new lives as fence posts. Below what will be the tribe's new Hyatt Regency Tamaya golf resort, 115 acres of native salt grass will be planted, and miles of nature trails will wind through the restored *bosque*, Caplan says. Already, the change is apparent.

"The best feeling I can have is when the elders come down and take a look at what we've done so far and say, 'This is what it looked like when we were growing up and playing down here.' What we're doing is for the health, recreation and enjoyment of the pueblo, but it will also positively impact the river downstream. The river is a living creature, and what is good for a part is good for the whole."

Bosque del Apache National Wildlife Refuge is also aggressively maintaining and restoring its 57,191 acres by tearing out salt cedar, planting cottonwoods and conducting controlled flooding this month, says ecologist Gina Dello Russo. "We have one of the nicest examples of what the river used to look like," she says. "We've had great success in bringing back native habitat."

But conditions look like they're going to get harder before they get easier. A major drought is being predicted for New Mexico in the near future. The El Niño weather pattern gets much of the blame, but as old-timers, historians and scientists say, periods of unusually dry conditions aren't so unusual.

"The issues we're seeing with the Río Grande have always, always, always been true," says Steve Hansen, a Bureau of Reclamation hydrologist. "The Río Grande has always been in danger of drying up. It has a long history of water poverty."

True, historic accounts are filled with tales of drought and deluge. Coronado's conquistadores gave the river its impressive name after seeing what was described as a body of water many leagues wide. After seeing a trickle of water pick its way through a dusty riverbed, Will Rogers declared the Río Grande was the only river he'd ever seen in need of irrigation. Both accounts summarize the Great River, which has always been a paradox. It's the country's third-longest river next to the Mississippi and Missouri, yet some years you can step across it.

"We're coming out of a 20-year wet period," Hansen says. "It's going to get drier, which is normal. What has changed is that there's a lot more people depending on the Río Grande now, and we care about the environment a lot more. Fact is, we need more water but there's going to be less to go around. Everybody is going to have a compromise and be neighbors and partners in this."

Does it matter where you are?

The cliché of the information age is that instantaneous global telecommunications, television and computer networks will soon overthrow the ancient tyrannies of time and space. Companies will need no headquarters, workers will toil as effectively from home, car or beach as they could in the offices that need no longer exist, and events half a world away will be seen, heard and felt with the same immediacy as events across the street—if indeed streets still have any point.

There is something in this. Software for American companies is already written by Indians in Bangalore and transmitted to Silicon Valley by satellite. Foreign-exchange markets have long been running 24 hours a day. At least one California company literally has no headquarters: its officers live where they like, its salesmen are always on the road, and everybody keeps in touch via modems and e-mail.

Yet such developments have made hardly a dent in the way people think and feel about things. Look, for example, at newspapers or news broadcasts anywhere on earth, and you find them overwhelmingly dominated by stories about what is going on in the vicinity of their place of publication. Much has been made of the impact on western public opinion of televised scenes of suffering in such places as Ethiopia, Bosnia and Somalia. Impact, maybe, but a featherweight's worth.

World television graphically displayed first the slaughter of hundreds of thousands of people in Rwanda and then the flight of more than a million Rwandans to Zaire. Not until France belatedly, and for mixed motives, sent in a couple of thousand soldiers did anyone in the West lift so much as a finger to stop the killing; nor, once the refugees had suddenly poured out, did western governments do more than sluggishly bestir themselves to try to contain a catastrophe.

Rwanda, of course, is small (population maybe 8m before the killings began). More important, it is far away. Had it been Flemings killing Walloons in Belgium (population 10m) instead of Hutus slaying Tutsi in Rwanda, European news companies would have vastly increased their coverage, and European governments would have intervened in force. Likewise, the only reason the Clinton administration is even thinking about invading Haiti is that it lies a few hundred miles from American shores. What your neighbours (or your kith and kin) do affects you. The rest is voyeurism.

The conceit that advanced technology can erase the contingencies of place and time ranges widely. Many armchair strategists predicted during the Gulf war that ballistic missiles and smart weapons would make the task of capturing and holding territory irrelevant. They were as wrong as the earlier seers who predicted America could win the Vietnam war from the air.

In business, too, the efforts to break free of space and time have had qualified success at best. American multinationals going global have discovered that—for all their world products, world advertising, and world communications and control—an office in, say, New York cannot except in the most general sense manage the company's Asian operations. Global strengths must be matched by a local feel—and a jet-lagged visit of a few days every so often does not provide one.

Most telling of all, even the newest industries are obeying an old rule of geographical concentration. From the start of the industrial age, the companies in a fast-growing new field have tended to cluster in a small region. Thus, in examples given by Paul Krugman, an American economist, all but one of the top 20 American carpetmakers are located in or near the town of Dalton, Georgia; and, before 1930, the American tire industry consisted almost entirely of the 100 or so firms carrying on that business in Akron, Ohio. Modern technology has not changed the pattern. This is why the world got Silicon Valley in California in the 1960s. It is also why tradable services stay surprisingly concentrated—futures trading (in Chicago), insurance (Hartford, Connecticut), movies (Los Angeles) and currency trading (London).

History's Heavy Hand

This offends not just techno-enthusiasts but also neo-classical economics: for both, the world should tend towards a smooth dispersion of people, skills and economic competence, not towards their concentration. Save for transport costs, it should not matter where a tradable good or service is produced.

The reality is otherwise. Some economists have explained this by pointing to increasing returns to scale (in labour as well as capital markets), geographically uneven patterns of demand and transport costs. The main reason is that history counts: where you are depends very much on where you started from.

The new technologies will overturn some of this, but not much. The most advanced use so far of the Internet, the greatest of the world's computer networks, has not been to found a global village but to strengthen the local business and social ties among people and companies in the heart of Silicon Valley. As computer and communications power grows and its cost falls, people will create different sorts of space and communities from those that exist in nature. But these modern creations will supplement, not displace, the original creation; and they may even reinforce it. Companies that have gone furthest towards linking their global operations electronically report an increase, not a decline in the face-to-face contact needed to keep the firms running well: with old methods of command in ruins, the social glue of personal relations matters more than ever.

The reason lies in the same fact of life that makes it impossible really to understand from statistics alone how exciting, say, China's economic growth is unless you have physically been there to feel it. People are not thinking machines (they absorb at least as much information from sight, smell and emotion as they do from abstract symbols), and the world is not immaterial: "virtual" reality is no reality at all; cyberspace is a pretence at circumventing true space, not a genuine replacement for it. The weight on mankind of time and space, of physical surroundings and history—in short, of geography—is bigger than any earthbound technology is ever likely to lift.

Unit 4

Unit Selections

30. **Transportation and Urban Growth: The Shaping of the American Metropolis,** Peter O. Muller
31. **GIS Technology Reigns Supreme in Ellis Island Case,** Richard G. Castagna, Lawrence L. Thornton, and John M. Tyrawaski
32. **Teaching About Karst Using U.S. Geological Survey Resources,** Joseph J. Kerski
33. **Gaining Perspective,** Molly O'Meara Sheehan
34. **Counties With Cash,** John Fetto
35. **Do We Still Need Skyscrapers?** William J. Mitchell
36. **Bio Invasion,** Janet Ginsburg

Key Points to Consider

❖ Describe the spatial form of the place in which you live. Do you live in a rural area, a town, or a city, and why was that particular location chosen?

❖ How does your hometown interact with its surrounding region? With other places in the state? With other states? With other places in the world?

❖ How are places "brought closer together" when transportation systems are improved?

❖ What problems occur when transportation systems are overloaded?

❖ How will public transportation be different in the future? Will there be more or fewer private autos in the next 25 years? Defend your answer.

❖ How good a map reader are you? Why are maps useful in studying a place?

 **Links** www.dushkin.com/online/

20. **Edinburgh Geographical Information Systems**
    http://www.geo.ed.ac.uk/home/gishome.html
21. **GIS Frequently Asked Questions and General Information**
    http://www.census.gov/ftp/pub/geo/www/faq-index.html
22. **International Map Trade Association**
    http://www.maptrade.org
23. **PSC Publications**
    http://www.psc.lsa.umich.edu/pubs/abs/abs94-319.html
24. **U.S. Geological Survey**
    http://www.usgs.gov/research/gis/title.html

These sites are annotated on pages 6 and 7.

Spatial Interaction and Mapping

Geography is the study not only of places in their own right but also of the ways in which places interact. Places are connected by highways, airline routes, telecommunication systems, and even thoughts. These forms of spatial interaction are an important part of the work of geographers.

In "Transportation and Urban Growth: The Shaping of the American Metropolis," Peter Muller considers transportation systems, analyzing their impact on the growth of American cities. The next article explores GIS as the technology used to solve a 160-year-old boundary controversy involving Ellis Island. Next, Joseph Kerski reviews the advantages of using USGS maps and other resources in teaching about landscapes. An extensive analysis of satellite imagery and its applications follows. The next article illustrates the power of the choropleth map to tell its story. "Do We Still Need Skyscrapers?" questions the need for high-density structures in the new era of extensive communications. The last article documents the diffusion of diseases into the United States, resulting from higher rates of immigration and animal importation.

It is essential that geographers be able to describe the detailed spatial patterns of the world. Neither photographs nor words could do the job adequately, because they literally capture too much of the detail of a place. Therefore, maps seem to be the best way to present many of the topics analyzed in geography. Maps and geography go hand in hand. Although maps are used in other disciplines, their association with geography is the most highly developed.

A map is a graphic that presents a generalized and scaled-down view of particular occurrences or themes in an area. If a picture is worth a thousand words, then a map is worth a thousand (or more!) pictures. There is simply no better way to "view" a portion of Earth's surface or an associated pattern than with a map. With computer cartography and satellite imagery, the world of maps and mapping has grown even bigger.

Transportation and Urban Growth

The shaping of the American metropolis

Peter O. Muller

In his monumental new work on the historical geography of transportation, James Vance states that geographic mobility is crucial to the successful functioning of any population cluster, and that "shifts in the availability of mobility provide, in all likelihood, the most powerful single process at work in transforming and evolving the human half of geography." Any adult urbanite who has watched the American metropolis turn inside-out over the past quarter-century can readily appreciate the significance of that maxim. In truth, the nation's largest single urban concentration today is not represented by the seven-plus million who agglomerate in New York City but rather by the 14 million who have settled in Gotham's vast, curvilinear outer city—a 50-mile-wide suburban band that stretches across Long Island, southwestern Connecticut, the Hudson Valley as far north as West Point, and most of New Jersey north of a line drawn from Trenton to Asbury Park. This latest episode of intrametropolitan deconcentration was fueled by the modern automobile and the interstate expressway. It is, however, merely the most recent of a series of evolutionary stages dating back to colonial times, wherein breakthroughs in transport technology unleashed forces that produced significant restructuring of the urban spatial form.

The emerging form and structure of the American metropolis has been traced within a framework of four transportation-related eras. Each successive growth stage is dominated by a particular movement technology and transport-network expansion process that shaped a distinctive pattern of intraurban spatial organization. The stages are the Walking/Horsecar Era (pre-1800–1890), the Electric Streetcar Era (1890–1920), the Recreational Automobile Era (1920–1945), and the Freeway Era (1945–present). As with all generalized models of this kind, there is a risk of oversimplification because the building processes of several simultaneously developing cities do not always fall into neat time-space compartments. Chicago's growth over the past 150 years, for example, reveals numerous irregularities, suggesting that the overall metropolitan growth pattern is more complex than a simple, continuous outward thrust. Yet even after developmental ebb and flow, leapfrogging, backfilling, and other departures from the idealized scheme are considered, there still remains an acceptable correspondence between the model and reality.

Before 1850 the American city was a highly compact settlement in which the dominant means of getting about was on foot, requiring people and activities to tightly agglomerate in close proximity to one another. This usually meant less than a 30-minute walk from the center of town to any given urban point—an accessibility radius later extended to 45 minutes when the pressures of industrial growth intensified after 1830. Within this pedestrian city, recognizable activity concentrations materialized as well as the beginnings of income-based residential congregations. The latter was particularly characteristic of the wealthy, who not only walled themselves off in their large homes near the city center but also took to the privacy of horse-drawn carriages for moving about town. Those of means also sought to escape the city's noise and frequent epidemics resulting from the lack of sanitary conditions. Horse-and-carriage transportation enabled the wealthy to reside in the nearby countryside for the disease-prone summer months. The arrival of the railroad in the 1830s provided the opportunity for year-round daily commuting, and by 1840 hundreds of affluent businessmen in Boston, New York, and Philadelphia were making round trips from exclusive new trackside suburbs every weekday.

As industrialization and its teeming concentrations of working-class housing increasingly engulfed the mid-nineteenth century city, the deteriorating physical and social environment reinforced the desires of middle-income residents to suburbanize as well. They were unable, however, to afford the cost and time of commuting by steam train, and with the walking city now stretched to its morphological limit, their aspirations intensified the pressures to improve intraurban transport technology. Early attempts involving stagecoach-like omnibuses, cablecar systems, and steam railroads proved impractical, but by 1852 the first meaningful transit breakthrough was finally introduced in Manhattan in the form of the horse-drawn trolley. Light street rails were easy to install, overcame the problems of muddy, unpaved roadways, and enabled horsecars to be hauled along them at speeds slightly (about five mph) faster than those of pedestrians. This modest improvement in mobility permitted the opening of a narrow belt of land at the city's edge for new home construction. Middle-income urbanites flocked to these "horsecar suburbs," which multiplied rapidly after the Civil War. Radial routes were the first to spawn such peripheral development, but the relentless demand for housing necessitated the building of cross-town horsecar lines, thereby filling in the interstices and preserving the generally circular shape of the city.

The less affluent majority of the urban population, however, was confined to the old pedestrian city and its bleak, high-density industrial appendages. With the massive immigration of unskilled laborers, (mostly of European origin after 1870) huge blue-collar communities sprang up around the factories. Because these newcomers to the city settled in the order in which they arrived—thereby denying them the small luxury of living in the immediate company of their fellow ethnics—social stress and conflict were repeatedly generated. With the immigrant tide continuing to pour into the nearly bursting industrial city throughout the late nineteenth century, pressures redoubled to further improve intraurban transit and open up more of the adjacent countryside. By the late 1880s that urgently needed mobility revolution was at last in the making, and when it came it swiftly transformed the compact city and its suburban periphery into the modern metropolis.

(Library of the Boston Athenaeum)

Horse-drawn trolleys in downtown Boston, circa 1885.

The key to this urban transport revolution was the invention by Frank Sprague of the electric traction motor, an often overlooked innovation that surely ranks among the most important in American history. The first electrified trolley line opened in Richmond in 1888, was adopted by two dozen other big cities within a year, and by the early 1890s swept across the na-

(Library of the Boston Athenaeum)

Electric streetcar lines radiated outward from central cities, giving rise to star-shaped metropolises. Boston, circa 1915.

tion to become the dominant mode of intraurban transit. The rapidity of this innovation's diffusion was enhanced by the immediate recognition of its ability to resolve the urban transportation problem of the day: motors could be attached to existing horsecars, converting them into self-propelled vehicles powered by easily constructed overhead wires. The tripling of average speeds (to over 15 mph) that resulted from this invention brought a large band of open land beyond the city's perimeter into trolley-commuting range.

The most dramatic geographic change of the Electric Streetcar Era was the swift residential development of those urban fringes, which transformed the emerging metropolis into a decidedly star-shaped spatial entity. This pattern was produced by radial streetcar corridors extending several miles beyond the compact city's limits. With so much new space available for homebuilding within walking distance of the trolley lines, there was no need to extend trackage laterally, and so the interstices remained undeveloped.

Before 1850 the American city was a highly compact settlement in which the dominant means of getting about was on foot, requiring people and activities to tightly agglomerate in close proximity to one another.

The typical streetcar suburb of the turn of this century was a continuous axial corridor whose backbone was the road carrying the trolley line (usually lined with stores and other local commercial facilities), from which gridded residential streets fanned out for several blocks on both sides of the tracks. In general, the quality of housing and prosperity of streetcar subdivisions increased with distance from the edge of the central city. These suburban corridors were populated by the emerging, highly mobile middle class, which was already stratifying itself according to a plethora of minor income and status differences. With frequent upward (and local geographic) mobility the norm, community formation became an elusive goal, a process further retarded by the grid-settlement morphology and the reliance on the distant downtown for employment and most shopping.

Within the city, too, the streetcar sparked a spatial transformation. The ready availability and low fare of the electric trolley now provided every resident with access to the intracity circulatory system, thereby introducing truly "mass" transit to urban America in the final years of the nineteenth century. For nonresidential activities this new ease of movement among the city's various parts quickly triggered the emergence of specialized land-use districts for commerce, manufacturing, and transportation, as well as the continued growth of the multipurpose central business district (CBD) that had formed after mid-century. But the greatest impact of the streetcar was on the central city's social geography, because it made possible the congregation of ethnic groups in their own neighborhoods. No longer were these moderate-income masses forced to reside in the heterogeneous jumble of row-houses and tenements that ringed the factories. The trolley brought them the opportunity to "live with their own kind," allowing the sorting of discrete groups into their own inner-city social territories within convenient and inexpensive traveling distance of the workplace.

By World War I, the electric trolleys had transformed the tracked city into a full-fledged metropolis whose streetcar suburbs, in the larger cases, spread out more than 20 miles from the metropolitan center. It was at this point in time that intrametropolitan transportation achieved its greatest level of efficiency—that the bustling industrial city really "worked." How much closer the American metropolis might have approached optimal workability for all its residents, however, will never be known because the next urban transport revolution was already beginning to assert itself through the increasingly popular automobile. Americans took to cars as wholeheartedly as anything in the nation's long cultural history. Although Lewis Mumford and other scholars vilified the car as the destroyer of the city, more balanced assessments of the role of the automobile recognize its overwhelming acceptance for what it was—the long-awaited attainment of private mass transportation that offered users the freedom to travel whenever and wherever they chose. As cars came to the metropolis in ever greater numbers throughout the interwar decades, their major influence was twofold: to accelerate the deconcentration of population through the development of interstices bypassed during the streetcar era, and to push the suburban frontier farther into the countryside, again producing a compact, regular-shaped urban entity.

While it certainly produced a dramatic impact on the urban fabric by the eve of World War II, the introduction of the automobile into the American metropolis during the 1920s and 1930s came at a leisurely pace. The earliest flurry of auto adoptions had been in rural areas, where farmers badly needed better access to local service centers. In the cities, cars were initially used for weekend outings—hence the term "Recreational

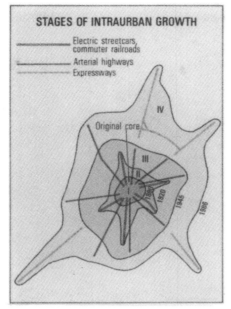

Cartography Lab. Dept. of Geography Univ. of Minnesota

Auto Era"—and some of the earliest paved roadways were landscaped parkways along scenic water routes, such as New York's pioneering Bronx River Parkway and Chicago's Lake Shore Drive. But it was into the suburbs, where growth rates were now for the first time overtaking

The ready availability and low fare of the electric trolley now provided every resident with access to the intracity circulatory system, thereby introducing truly "mass" transit to urban America.

those of the central cities, that cars made a decisive penetration throughout the prosperous 1920s. In fact, the rapid expansion of automobile suburbia by 1930 so adversely affected the

30. Transportation and Urban Growth

metropolitan public transportation system that, through significant diversions of streetcar and commuter-rail passengers, the large cities began to feel the negative effects of the car years before the auto's actual arrival in the urban center. By facilitating the opening of unbuilt areas lying between suburban rail axes, the automobile effectively lured residential developers away from densely populated traction-line corridors into the suddenly accessible interstices. Thus, the suburban homebuilding industry no longer found it necessary to subsidize privately-owned streetcar companies to provide low-fare access to trolley-line housing tracts. Without this financial underpinning, the modern urban transit crisis quickly began to surface.

The new recreational motorways also helped to intensify the decentralization of the population. Most were radial highways that penetrated deeply into the suburban ring and provided weekend motorists with easy access to this urban countryside. There they obviously were impressed by what they saw, and they soon responded in massive numbers to the sales pitches of suburban subdivision developers. The residential development of automobile suburbia followed a simple formula that was devised in the prewar years and greatly magnified in scale after 1945. The leading motivation was developer profit from the quick turnover of land, which was acquired in large parcels, subdivided, and auctioned off. Understandably, developers much preferred open areas at the metropolitan fringe, where large packages of cheap land could readily be assembled. Silently approving and underwriting this uncontrolled spread of residential suburbia were public policies at all levels of government: financing road construction, obligating lending institutions to invest in new homebuilding, insuring individual mortgages, and providing low-interest loans to FHA and VA clients.

Because automobility removed most of the pre-existing movement

4 ❖ SPATIAL INTERACTION AND MAPPING

(Boston Public Library)

Afternoon commuters converge at the tunnel leading out of central Boston, 1948.

constraints, suburban social geography now became dominated by locally homogeneous income-group clusters that isolated themselves from dissimilar neighbors. Gone was the highly localized stratification of streetcar suburbia. In its place arose a far more dispersed, increasingly fragmented residential mosaic to which builders were only too eager to cater, helping shape a kaleidoscopic settlement pattern by shrewdly constructing the most expensive houses that could be sold in each locality. The continued partitioning of suburban society was further legitimized by the widespread adoption of zoning (legalized in 1916), which gave municipalities control over lot and building standards that, in turn, assured dwelling prices that would only attract newcomers whose incomes at least equaled those of the existing local population. Among the middle class, particularly, these exclusionary economic practices were enthusiastically

Americans took to cars as wholeheartedly as anything in the nation's long cultural history.

supported, because such devices extended to them the ability of upper-income groups to maintain their social distance from people of lower socioeconomic status.

Nonresidential activities were also suburbanizing at an increasing rate during the Recreational Auto Era. Indeed, many large-scale manufacturers had decentralized during the streetcar era, choosing locations in suburban freight-rail corridors. These corridors rapidly spawned surrounding working-class towns that became important satellites of the central city in the emerging metropolitan constellation. During the interwar period, industrial employers accelerated their intraurban deconcentration, as more efficient horizontal fabrication methods replaced older techniques requiring multistoried plants-thereby generating greater space needs that were too expensive to satisfy in the high-density central city. Newly suburbaniz-

Central City-Focused Rail Transit

The widely dispersed distribution of people and activities in today's metropolis makes rail transit that focuses in the central business district (CBD) an obsolete solution to the urban transportation problem. To be successful, any rail line must link places where travel origins and destinations are highly clustered. Even more important is the need to connect places where people really want to go, which in the metropolitan America of the late twentieth century means suburban shopping centers, freeway-oriented office complexes, and the airport. Yet a brief look at the rail systems that have been built in the last 20 years shows that transit planners cannot—or will not—recognize those travel demands, and insist on designing CBD-oriented systems as if we all still lived in the 1920s.

One of the newest urban transit systems is Metrorail in Miami and surrounding Dade County, Florida. It has been a resounding failure since its opening in 1984. The northern leg of this line connects downtown Miami to a number of low- and moderate-income black and Hispanic neighborhoods, yet it carries only about the same number of passengers that used to ride on parallel bus lines. The reason is that the high-skill, service economy of Miami's CBD is about as mismatched as it could possibly be to the modest employment skills and training levels possessed by residents of that Metrorail corridor. To the south, the prospects seemed far brighter because of the possibility of connecting the system to Coral Gables and Dadeland, two leading suburban activity centers. However, both central Coral Gables and the nearby International Airport complex were bypassed in favor of a cheaply available, abandoned railroad corridor alongside U.S. 1. Station locations were poorly planned, particularly at the University of Miami and at Dadeland—where terminal location necessitates a dangerous walk across a six-lane highway from the region's largest shopping mall. Not surprisingly, ridership levels have been shockingly below projections, averaging only about 21,000 trips per day in early 1986. While Dade County's worried officials will soon be called upon to decide the future of the system, the federal government is using the Miami experience as an excuse to withdraw from financially supporting all construction of new urban heavy-rail systems. Unfortunately, we will not be able to discover if a well-planned, high-speed rail system that is congruent with the travel demands of today's polycentric metropolis is capable of solving traffic congestion problems. Hopefully, transportation policy-makers across the nation will heed the lessons of Miami's textbook example of how not to plan a hub-and-spoke public transportation network in an urban era dominated by the multi-centered city.

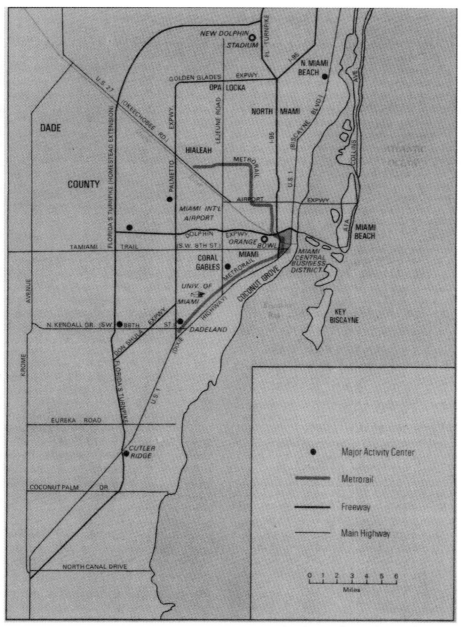

Cartography Lab. Dept. of Geography Univ. of Minnesota

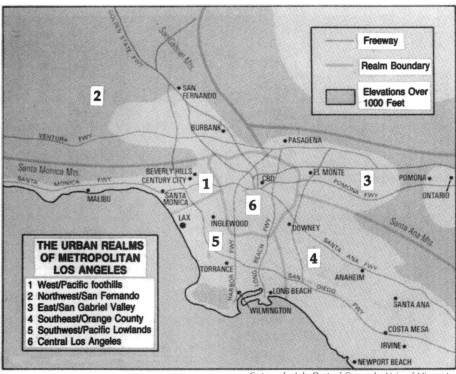

Cartography Lab. Dept. of Geography Univ. of Minnesota

ing manufacturers, however, continued their affiliation with intercity freight-rail corridors, because motor trucks were not yet able to operate with their present-day efficiencies and because the highway network of the outer ring remained inadequate until the 1950s.

The other major nonresidential activity of interwar suburbia was retailing. Clusters of automobile-oriented stores had first appeared in the urban fringes before World War I. By the early 1920s the roadside commercial strip had become a common sight in many southern California suburbs. Retail activities were also featured in dozens of planned automobile suburbs that sprang up after World War I—most notably in Kansas City's Country Club District, where the nation's first complete shopping center was opened in 1922. But these diversified retail centers spread slowly before the suburban highway improvements of the 1950s.

Unlike the two preceding eras, the postwar Freeway Era was not sparked by a revolution in urban transportation. Rather, it represented the coming of age of the now pervasive automobile culture, which coincided with the emergence of the U.S. from 15 years of economic depression and

> *Retail activities were featured in dozens of planned automobile suburbs that sprang up after World War I—most notably in Kansas City's Country Club District, where the nation's first complete shopping center was opened in 1922.*

war. Suddenly the automobile was no longer a luxury or a recreational diversion: overnight it had become a necessity for commuting, shopping, and socializing, essential to the successful realization of personal opportunities for a rapidly expanding majority of the metropolitan population. People snapped up cars as fast as the reviving peacetime automobile industry could roll them off the assembly lines, and a prodigious highway-building effort was launched, spearheaded by high-speed, limited-access expressways. Given impetus by the 1956 Interstate Highway Act, these new freeways would soon reshape every corner of urban America, as the more distant suburbs they engendered represented nothing less than the turning inside-out of the historic metropolitan city.

The snowballing effect of these changes is expressed geographically in the sprawling metropolis of the postwar era. Most striking is the enormous band of growth that was added between 1945 and the 1980s, with freeway sectors pushing the metropolitan frontier deeply into the urban-rural fringe. By the late 1960s, the maturing expressway system began to underwrite a new suburban co-equality with the central city, because it was eliminating the metropolitanwide centrality advantage of the CBD. Now any location on the freeway network could easily be reached by motor vehicle, and intraurban accessibility had become a ubiquitous spatial good. Ironically, large cities had encouraged the construction of radial expressways in the 1950s and 1960s because they appeared to enable the downtown to remain accessible to the swiftly dispersing suburban population. However, as one economic activity after another discovered its new locational flexibility within the freeway metropolis, nonresidential deconcentration sharply accelerated in the 1970s and 1980s. Moreover, as expressways expanded the radius of commuting to encompass the entire dispersed metropolis, residential location constraints relaxed as well. No longer were most urbanites required to live within a short distance of their job: the workplace had now become a locus of opportunity offering access to the best possible residence that an individual could afford anywhere in the urbanized area. Thus, the overall pattern of locally uniform, income-based clusters that had emerged in prewar automobile suburbia was greatly magnified in the Freeway Era, and such new social variables as age and lifestyle produced an ever more balkanized population mosaic.

The revolutionary changes in movement and accessibility introduced during the four decades of the Freeway Era have resulted in nothing less than the complete geographic restructuring of the metropolis. The single-center urban structure of the past has been transformed into a polycentric metropolitan form in which several outlying activity concentrations rival the CBD. These new "suburban downtowns," consisting of vast orchestrations of retailing, office-based business, and light industry, have become common features near the highway interchanges that now encircle every large central city. As these emerging metropolitan-level cores achieve economic and geographic parity with each other, as well as with the CBD of the nearby central city, they provide the totality of urban goods and services to their surrounding populations. Thus each metropolitan sector becomes a self-sufficient functional entity, or *realm*. The application of this model to the Los Angeles region reveals six broad realms. Competition among several new suburban downtowns for dominance in the five outer realms is still occurring. In wealthy Orange County, for example, this rivalry is especially fierce, but Costa Mesa's burgeoning South

The new freeways would soon reshape every corner of urban America, as the more distant suburbs they engendered represented nothing less than the turning inside-out of the historic metropolitan city.

Coast Metro is winning out as of early 1986.

The legacy of more than two centuries of intraurban transportation innovations, and the development patterns they helped stamp on the landscape of metropolitan America, is suburbanization—the growth of the edges of the urbanized area at a rate faster than in the already-developed interior. Since the geographic extent of the built-up urban areas has, throughout history, exhibited a remarkably constant radius of about 45 minutes of travel from the center, each breakthrough in higher-speed transport technology extended that radius into a new outer zone of suburban residential opportunity. In the nineteenth century, commuter railroads, horse-drawn trolleys, and electric streetcars each created their own suburbs—and thereby also created the large industrial city, which could not have been formed without incorporating these new suburbs into the pre-existing compact urban center. But the suburbs that materialized in the early twentieth century began to assert their independence from the central cities, which were ever more perceived as undesirable. As the automobile greatly reinforced the dispersal trend of the metropolitan population, the distinction between central city and suburban ring grew as well. And as freeways eventually eliminated the friction effects of intrametropolitan distance for most urban functions, nonresidential activities deconcentrated to such an extent that by 1980 the emerging outer suburban city had become co-equal with the central city that spawned it.

As the transition to an information-dominated, postindustrial economy is completed, today's intraurban movement problems may be mitigated by the increasing substitution of communication for the physical movement of people. Thus, the city of the future is likely to be the "wired metropolis." Such a development would portend further deconcentration because activity centers would potentially be able to locate at any site offering access to global computer and satellite networks.

Further Reading

Jackson, Kenneth T. 1985. *Crabgrass Frontier: The Suburbanization of the United States.* New York: Oxford University Press.

Muller, Peter O. 1981. *Contemporary Suburban America.* Englewood Cliffs, N.J.: Prentice-Hall.

Schaeffer, K. H. and Sclar, Elliot. 1975. *Access for All: Transportation and Urban Growth.* Baltimore: Penguin Books.

GIS Technology Reigns Supreme in Ellis Island Case

Richard G. Castagna, Lawrence L. Thornton and John M. Tyrawski

In 1998, the U.S. Supreme Court ruled on a territorial squabble between the states of New Jersey and New York over Ellis Island that had been brewing for more than 160 years. GIS technology as implemented by the New Jersey Department of Environmental Protection (NJDEP) was instrumental in effecting the outcome. Here's how.

Early agreements between New York and New Jersey set the state boundary line as the middle of the Hudson river down through New York Bay, but the actual boundary around historic Ellis Island was never officially determined. Although the boundary dispute predates the Revolution, the story officially begins with an 1834 compact between the states, which was approved by the U.S. Congress. This compact set the boundary for Ellis Island as follows: *all non-submerged lands of the island belong to New York, and all submerged lands surrounding the island belong to New Jersey.* At that time, the non-submerged area of the island occupied approximately three acres.

Enlargement of Non-submerged Area

In the intervening years, the boundary question became more muddled. The United States, which owned the island since 1808, used the island as a fort in the early 19th century and as a powder magazine in the mid-19th century. In 1890, the federal government decided to use Ellis Island as an immigration station. Requiring more space, the federal government began filling the submerged lands around the island. By 1934, the island was enlarged tenfold by successive landfills, from its original 2.75 acres to 27.5 acres. Since 1890, New Jersey has contended that the filling to enlarge and develop the island was done on New Jersey territory.

The NJDEP mapped the natural island as part of the riparian mapping program undertaken by the Bureau of Tidelands in

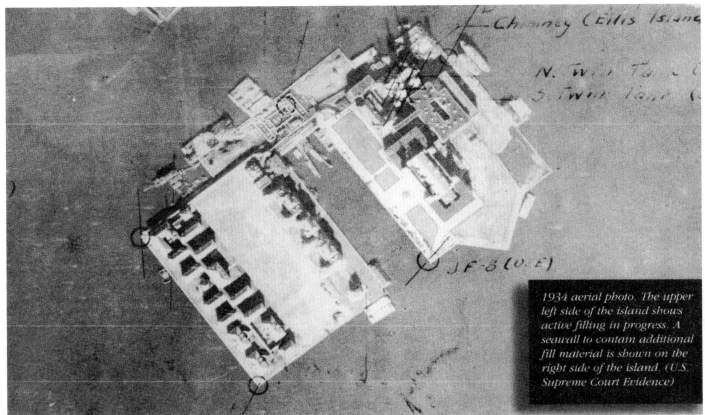

1934 aerial photo. The upper left side of the island shows active filling in progress. A seawall to contain additional fill material is shown on the right side of the island. (U.S. Supreme Court Evidence)

Source: New Jersey Department of Environmental Protection, Aerial Photo Library.

31. GIS Technology Reigns Supreme

1980. Using the riparian map as a preliminary claim, New Jersey officially asked the U.S. Supreme Court in 1993 to adjudicate the boundary dispute. Although both states and several congressional mandates had reviewed the boundary question over the years, the U.S. Supreme Court has sole jurisdiction to resolve boundary issues between states.

State Boundary Case Procedures

The process by which a state boundary case is presented to the Supreme Court is an involved one. The Court first names a Special Master, who reviews all pertinent information submitted by the states to support their claims. The Special Master then submits a report to the full Supreme Court summarizing the evidence and making a recommendation on the settlement. The high court then makes its determination on the final boundary settlement.

In preparation for the trial, the NJDEP Bureau of Tidelands, assisted by the NJDEP GIS Unit, the NJDEP Land Use Regulation Program and the Geodetic section of the New Jersey Department of Transportation, prepared several maps on the DEP's GIS showing the proposed boundary line based

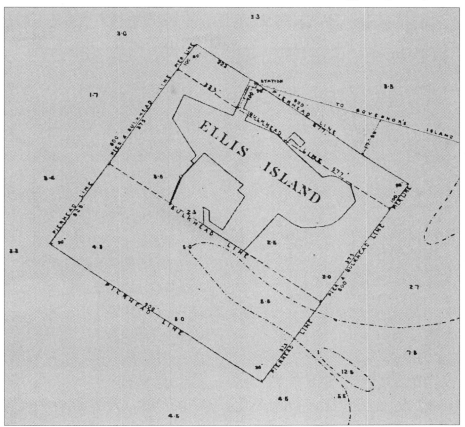

Source: National Archives, June 1890.

(below) "Plan of Ellis' Island 1870," prepared by the Bureau of Ordinance, Navy Department. This map clearly shows the two angles in the Fort Gibson wall used in the 1995 GPS survey. (U.S. Supreme Court Evidence)

(above) This map is titled "Pierhead & Bulkhead Lines for Ellis Island, New Jersey, New York Harbor as recommended by the New York Harbor Line Board." Special Master Paul Verkuil noted, "The significance of this map is that it was approved by the Secretary of War, Elihu Root, and produced over his signature. His signature . . . with the designation Ellis Island, New Jersey, makes this weighty evidence." (U.S. Supreme Court Evidence)

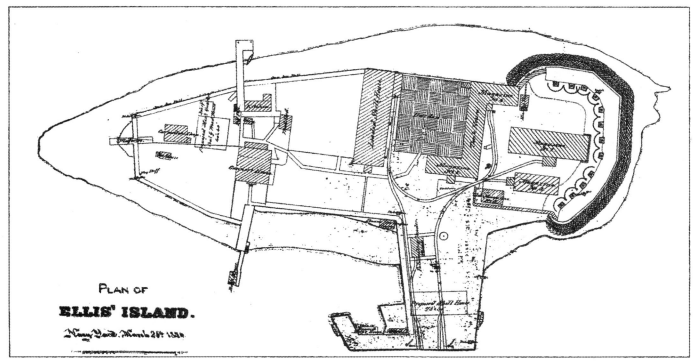

Source: National Archives

4 ❖ SPATIAL INTERACTION AND MAPPING

Ellis Island in 1995 and in 1857 Showing the Low Water Line

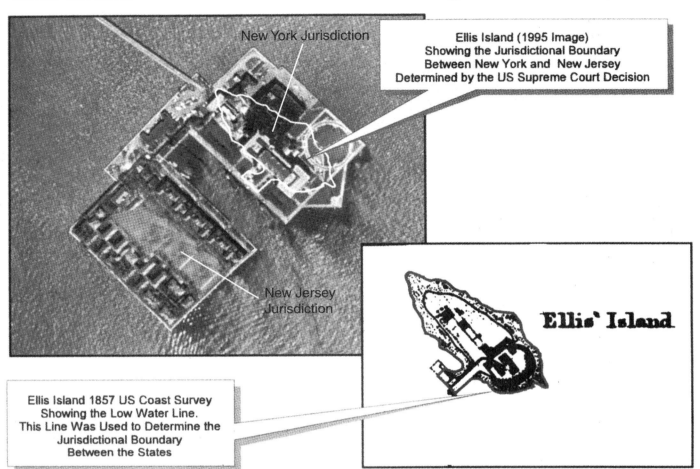

(above) 1857 U.S. Coast Survey map and the jurisdictional boundary line from the 1857 map superimposed on a 1995 aerial photograph. This exhibit was prepared after the trial and was not used as evidence. *(right)* Aerial photograph dated about 1993. Note that the open area that straddles the left side of the circular "Wall of Honor" is the exposed wall from Fort Gibson. Two angle points in the excavated sections of the wall were used in a 1995 GPS survey. The wall location was a crucial part of the Supreme Court case. This photo was not used as evidence.

on the 1857 U.S. Coast Survey map. Also mapped were the perimeter of the existing island as determined using GPS, as well as the locations of a portion of the historic fort built on Ellis Island before 1812. The location of the original wall (which has been excavated in part) was helpful in supporting the alignment of the historical maps used to define New Jersey's claim.

1857 Map Chosen to Define Boundary

When the Special Master reviewed all of the evidence, he determined that the 1857 U.S. Coast Survey map should define the boundary between the two states. However, the Special Master disagreed with New Jersey's

31. GIS Technology Reigns Supreme

use of the mean high water line on the island as the boundary and directed the state to use the low water line. In May of 1998, after oral arguments were presented by the states, the Supreme Court determined that the 1857 low water line of the natural island should be used to delineate the jurisdictional boundary.

The NJDEP was directed to prepare the Ellis Island boundary based on the Special Master's recommendation. The mapping was completed using GIS. The 1857 U.S. Coast Survey map was scanned and captured as a TIFF image file by the New Jersey Geological Survey. The image file was then brought into ArcView, and the low water line was captured as an edit function. The line depicting low water was represented on the 1857 map by a series of dots. The center of each dot was used to enter each point used to define the low water boundary. Once completed, the points and the line they describe were given geographic referencing by first converting the shape files to coverages—creating tic files for each—and then transforming and projecting the coverages to New Jersey State Plane Feet, NAD83. With this referencing, the line could be plotted on NJDEP's 1995 digital imagery and be integrated into the outer boundary survey, which was completed with the GPS done by NJDOT. The line was then un-generated and sent to NJDOT to prepare a final hard copy map.

New York Officials Approve Map

The map was then presented to New York state officials, who subsequently approved the delineation after several minor changes. The final step in the process is the Special Master's approval of the delineation.

While possibly not the first use of GIS to solve a boundary dispute, this may be a first use of GIS to present and solve a boundary dispute before the U.S. Supreme Court. The success and power of GIS in the Ellis Island case suggests that it will not be the last. After more than 160 years, the jurisdictional fight over Ellis Island is finally over. The high court issued its final decree and approved the boundary line on May 17, 1999. New Jersey was granted sovereign authority over 22.80 acres, and New York was granted authority over the remaining 4.68 acres.

RICHARD G. CASTAGNA *is a regional supervisor with the Bureau of Tidelands, NJDEP. He testified as an expert witness before the U.S. Supreme Court in the Ellis Island case on behalf of the State of New Jersey, describing physical changes to the island from the 18th century to the present.*

LAWRENCE L. THORNTON *is manager of the GIS Unit for NJDEP in Trenton, New Jersey. He delineated a claims line around Ellis Island for the state's riparian claim in 1980 and assisted in the delineation of the low water line, implementing the Supreme Court's decision in 1998.*

JOHN M. TYRAWSKI *is currently a research scientist with the GIS Unit of NJDEP. For the Ellis Island case, he assisted in the digital creation of the historic 1857 shoreline and in the development of the exhibits used to present the case for the State of New Jersey.*

Teaching About Karst Using U.S. Geological Survey Resources

Joseph J. Kerski

All of the resources referred to in this article can be purchased from the USGS by calling 1–800–HELP–MAP, by sending an E-mail to infoservices@usgs.gov, or by using the Global Land Information System (GLIS) at http://edcwww.cr.usgs.gov/webglis/. Each map costs $4, with a volume discount for teachers, plus a $3.50 handling fee per order. The prices of other products, such as professional papers, digital data, and aerial photographs, vary according to the length and the type of media. Most USGS products are also available from the 2,600 authorized USGS map dealers around the world, accessible at http://mapping.usgs.gov/esic/usimage/dealers.html.

What is karst?

Karst can be broadly defined as all land-forms that are produced primarily by the dissolution of rocks, mainly limestone and dolomite. Landforms in karst terrain include closed surface depressions called sinkholes, disrupted surface drainage, and underground drainage networks that include openings formed from the solution of calcium carbonate in water. These openings range in size from enlarged cracks to large caves. This surface drainage may contain streams that run along the surface for a certain distance and then disappear to follow an underground channel. They may reappear further downstream.

> **Students can learn to visualize any landscape from analyzing the spacing, elevation, and pattern of contour lines.**

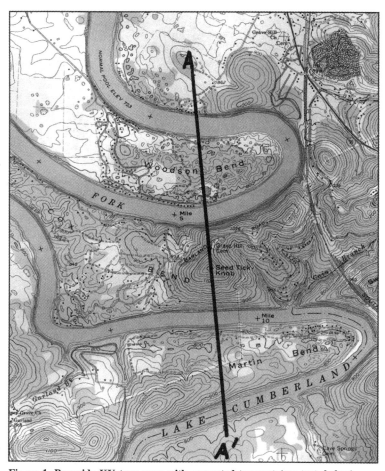

Figure 1. Burnside KY topo map with suggested transect (see text, below).

Why use USGS maps?

Because some of these features are shown on USGS topographic maps, the maps are excellent tools for teaching about the imprint of limestone on the Earth. The predominant scales of USGS topographic maps are 1:24,000, 1:100,000, and 1:250,000, supplemented in some states by a county series at 1:50,000 scale. In Alaska and Hawaii, other scales such as 1:63,360 and 1:25,000 are available. Sinkholes are symbolized by depression contours with tic marks.

Students can learn to visualize any landscape by analyzing the spacing, elevation, and pattern of contour lines.

32. Teaching About Karst

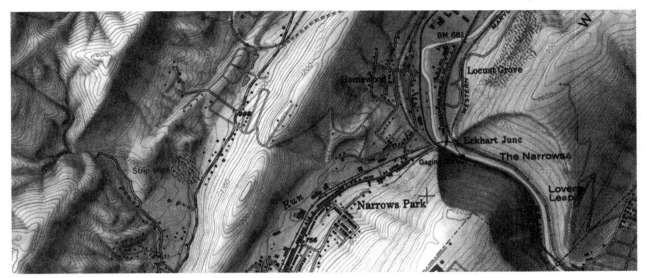

Figure 2. Shaded-relief map of Cumberland, Maryland.

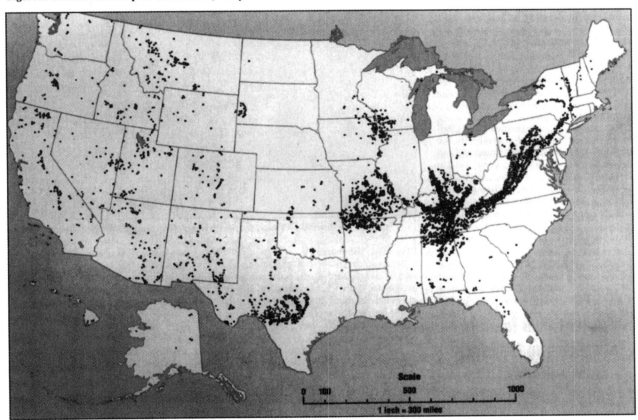

Figure 3. USGS Map of cave locations, from the "Exploring Caves" teachers packet.

Karst landscapes are no exception. Drawing cross-sections of selected transects across the landscape or modeling the landscape with clay, foam, or plaster helps students to understand the three-dimensional shape of the earth's surface and its distinguishing features. For example, a cross-section of transect A to A′ across the topographic map of Burnside, Kentucky, reveals a landscape dissected by river valleys and pitted by holes (figure 1).

The use of a shaded relief map, available for selected areas, most often at the state scale, allows students to better visualize the land surface, as illustrated by this map of Cumberland, Maryland. Students can test a transect that they derived from a topographic map on to a shaded relief map (figure 2).

One resource for discovering the location of karst terrain in the United States is the free USGS teacher's packet named "Exploring Caves." It includes the map (figure 3) which provides a pattern of cave locations, including ones formed by coastal erosion, lava, and limestone.

Free state map indexes can then be used to select topographic maps covering the areas that students will investigate. Topographic maps include famous examples of karst, such as Mammoth Cave, Kentucky (figure 4).

Karst landscapes may include streams that disappear from the surface because the water follows a crack or pit in the

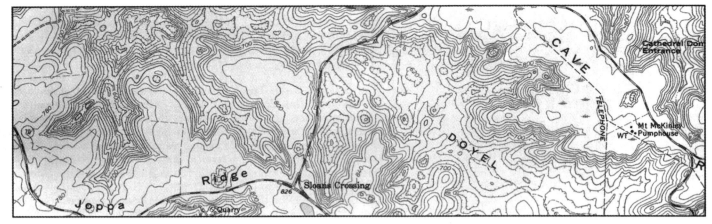

Figure 4. Topographic map of Mammoth Cave, Kentucky.

limestone. These streams may appear on the surface again further downstream (figures 5 and 6).

What can be learned about karst from these maps?

Ways to teach about karst from a map such as this could include posing the following questions: "What percentage of sinks are filled with water?" "Are these sinks perennial or intermittent, and why?" On the basis of their observations, students could estimate the elevation of the water table and identify the places most likely to contain springs. Some springs are shown on USGS topographic maps. Students could also determine the primary direction of water flow in the area. Flow direction observations can be linked with an analysis of drainage basins on the Hydrologic Unit Code maps of each state.

Karst landscapes may include streams that "disappear" from the surface

The effect of limestone on the landscape may be less evident than the effects of coastal erosion, tectonics, or glaciation. Karst can therefore be difficult to detect, even on maps with a scale of 1:24,000, but particularly on maps with a coarser scale and a larger contour interval. When investigating karst, students can view themselves as detectives, uncovering the mysteries of the landscape.

Furthermore, depending on the climate of the area, limestone can act as an erosion-resistant caprock or as an easily-erodible material found on valley floors. Students can investigate maps and data on climate and rainfall for an area to determine if the limestone will act as a caprock (predominantly in arid or semiarid regions) or as a weak layer (in regions with a greater amount of rainfall). Students could then be guided in a consideration of the different types of weathering. Karst landscapes are pri-

Figure 5. Stream disappearing into a sinkhole in karst terrain in Texas. (Photograph by Jon Gilhousen, USGS).

marily the result of chemical, rather than mechanical, weathering, which explains some of the subtlety of the resultant landforms. Carlsbad Caverns, New Mexico, lies at the top of a Permian-age reef that has been exposed on the surface as a high ridge (figure 8).

32. Teaching About Karst

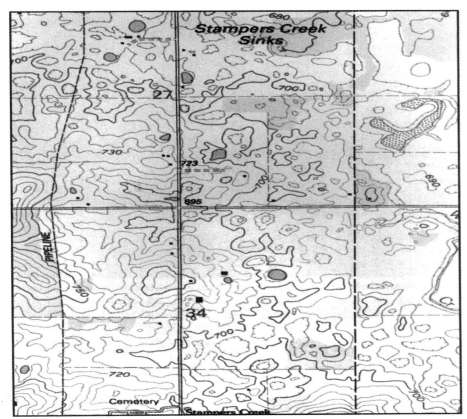

Figure 6. On the Paoli, Indiana topographic map, both Stampers Creek and Wolf Creek disappear below the surface.

soluble limestone beneath. Most land surface depressions containing lakes in this region are formed when the surface deposits slump into sinkholes that have formed in the limestone.

Human impact on karst landscapes can be examined by using prints made from historical editions of topographic maps. These historical editions, available from the USGS Earth Science Information Centers, make population growth and land use change evident. Students can be encouraged to consider how future change might affect cave formations, animal life in caves, and the water quality of the region.

It should be emphasized that the correct interpretation of geomorphic forces operating in a region under study may require the use of additional maps or reports to supplement the topographic map. For instance, in one region, the presence of pits in the landscape may be due to the presence of sand hills or lava covering the surface, rather than from water percolating through limestone.

In that example, a USGS geologic map can aid students in landform study by indicating the ages and types of rock in the region. Geologic maps exist for each state at 1:500,000 or similar scale. In addition, numerous geologic quadrangles exist for selected areas at 1:250,000 or 1:24,000 scale. Most geologic maps also show fault lines, which can be analyzed by the student to determine how they act as controls on landforms. In an arid or semiarid region,

Other available USGS resources

Aerial photographs from the USGS can also aid in teaching about karst. The USGS's PhotoFinder, at http://edcwww.cr.usgs.gov/webglis, allows the user to browse data about flight heights, cloud cover, and image types. Aerial photographs are custom-made products that can be ordered from the USGS Earth Science Information Centers. Photographs can be ordered on a variety of media, such as diazo paper or photographic paper. The most common and up-to-date photograph is usually a 1:40,000-scale National Aerial Photography Program (NAPP) product.

Using an aerial photograph, students can consider all surface features, rather than only those features included on topographic maps. For example, they can investigate how karst topography affects vegetation. What kinds of vegetation grow on north-versus south-facing slopes, and in river bottomlands? This orthophotoquad (figure 9) of Carlsbad Caverns, New Mexico, made from aerial photographs, is the same scale as the 1:24,000-scale topographic map.

Satellite images can also aid in karst studies and are available both on paper and in digital form. The section of the South Florida satellite image map shows a mantled karst region (figure 10). Here, unconsolidated deposits overlie the highly

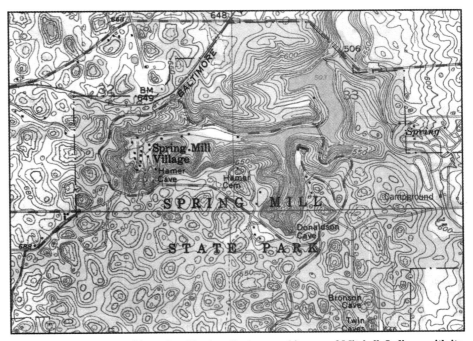

Figure 7. The presence of karst is evident on the topographic map of Mitchell, Indiana, with its numerous caves, springs, and sinkholes.

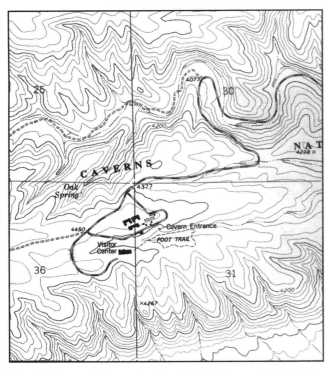

Figure 8. Topographic map showing Carlsbad Caverns, New Mexico.

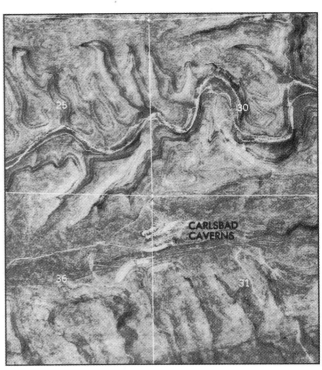

Figure 9. Orthophotoquadrangle showing Carlsbad Caverns, New Mexico.

streams may appear to vanish below the surface, when actually evaporation has caused them to disappear. Climatic maps and data indicating rainfall and temperature can aid in deciphering the most important forces upon the landscape.

Learning about cultural geography and water resources from maps

The study of place names on the landscape incorporates history and human geography into physical geography. The USGS Geographic Names Information System (GNIS) lists the location and type of feature for the name of every stream, town, canyon, sinkhole, and any other feature found on any topographic map for the USA. More than two million names are in the GNIS, available in CD-ROM format or on the Internet at http://mapping.usgs.gov/www/gnis/ Terms such as karst, limestone, cave, or hidden can be input into the system. Due to the Federal Cave Resources Protection Act, the caves and their locations are not listed. However, GNIS displays 988 other features, such as hills, towns, and rivers, that include the word cave and 351 features that include the word sink. The resulting latitude-longitude coordinates for the features can be mapped with stickers, markers, or push-pins on a map of the country, or they can be digitally plotted with a geographic information system.

The use of karst studies in geography and environmental science can be an excellent means of introducing water resources education. Students can learn what a watershed is and the boundaries of the watershed in which they live. Through studying and appreciating the fragile nature of karst ecology, students can learn the route of pollutants and their effects on the ecosystem. The USGS map "Surface Water and Related Land Resources Development in the United States and Puerto Rico" emphasizes river networks, allowing students to derive watershed boundaries. A series of water education posters and publications such as "Ground Water and the Rural Homeowner" are other examples of water resource materials that could be useful in teaching about rivers and drainage basins.

Publications; paper and digital availability

Circulars, professional papers, bulletins, and open file reports are other USGS resources that could be useful in teaching about karst. For example, Bulletin 1673 is entitled "Selected Caves and Lava-tube Systems in and near Lava Beds National Monument, California." Circular 1139 contains photographs, diagrams, and explanatory text concerning "Ground Water and Surface Water: A Single Resource."

Thematic maps are also useful for karst studies. One of the best examples is Miscellaneous Field Investigations Map 2262 (figure 12), entitled "Sinkholes and Karst-related features of the Shenandoah Valley in the Winchester 30X60-Minute Quadrangle, Virginia and West Virginia."

All USGS topographic maps are available as digital raster graphics (DRGs) which can be viewed on a computer. Digital raster graphics are tagged image format files (tiff) files that can be im-

> **Students can learn the route and effect of pollutants to the ecosystem.**

32. Teaching About Karst

ported into a school's geographic information system. To find out more about them, visit the site http://mcm-cweb.er.usgs.gov/drg/. Digital orthophoto- quadrangles (DOQs) are computer versions of aerial photographs that can also be viewed on a computer. Their spatial resolution of 1 meter on the ground allows for a detailed analysis of karst landforms. To find out more about DOQs, visit the site http://mapping.usgs.gov/www/ndop/. To discover more USGS resources on caves and karst, use the web-based publications search engine at http://www.usgs.gov/pubprod.

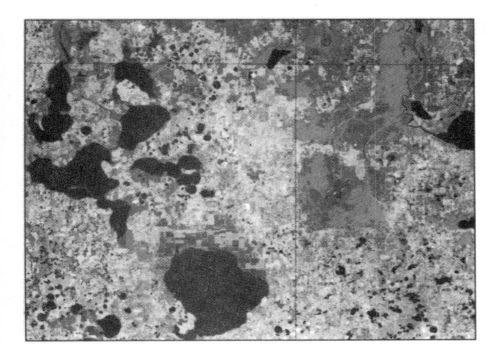

Figure 10. South Florida satellite image map.

Figure 11. Although the unnamed lake in this aerial photograph taken near Sebring, Florida, has a nearly perfect circular shape, indicating its sinkhole origin, many solution-formed lakes do not have this shape (Photograph by E.P. Simonds, USGS).

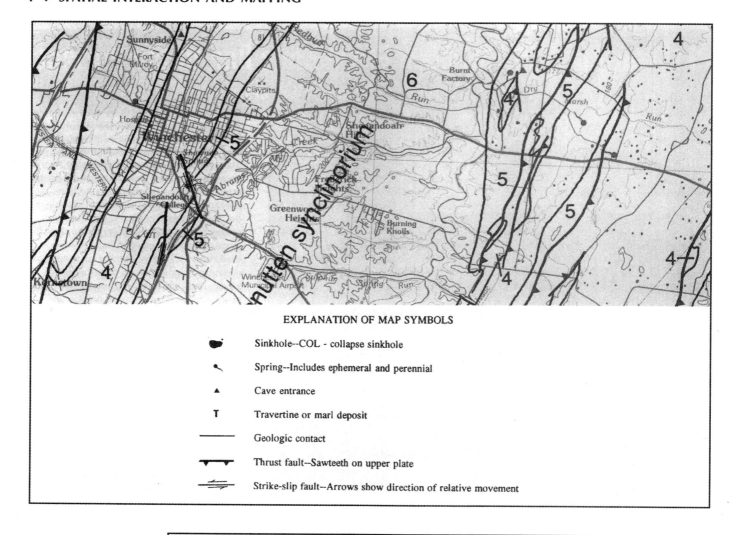

Figure 12. Section of Shenandoah Valley karst map.

GAINING PERSPECTIVE

The proliferation of satellite technology, from spy-quality photos to low-resolution radar images, is giving us new, more meaningful ways to envision complex information about the Earth. But whether we will act on the picture of ecological destruction this technology is cobbling together—from global climate change to wholesale clearing of forests—remains to be seen.

by Molly O'Meara Sheehan

During the last few decades of the 20th century it became evident that tropical rainforests were endangered not only by road-building, timber-cutting, and other incursions of the bulldozer and saw, but by thousands of wildfires. Historically, relatively small fires have been set by slash-and-burn farmers trying to clear patches of jungle for farm land. But starting in 1997, fires in the world's tropical forests from Brazil to Papua New Guinea raged on a scale never recorded before. The causes of these huge conflagrations raised questions, because tropical rainforests rarely burn naturally.

In the wake of several haze-induced accidents and public health warnings in smoke-covered Indonesia, the need for a clear answer to these questions was given legal significance when President Suharto, under pressure from neighboring countries, passed a decree making it illegal to set forest fires. The politically connected timber industry had managed to direct most of the blame for forest fires on the small-scale, slash-and-burn farmers, but Indonesia's rogue environment minister Sarwono Kusumaatmadja employed a relatively new intelligence-gathering technology to get a clear picture of the situation: he downloaded satellite images of burning Indonesian rainforests from a U.S. government website and compared them to timber concession maps. The satellite images confirmed that many of the blazes were being set in areas the timber companies wanted to clear for plantations. With the satellite evidence in hand, Kusnmaatmadja got his government to revoke the licenses of 29 timber companies.

HIGH SPEED INTELLIGENCE

The environment minister's quick work on the rainforest issue is just one of many recent cases involving environmental questions in which satellite surveillance has been used to provide answers that might otherwise not have been known for years, if ever. The images captured by cameras circling high above the planet are proving effective not only because they scan far more extensively than ground observers can, but because they can be far faster than traditional information-gathering methods.

Pre-satellite studies of the oceans, for example, had to be done from boats, which can only reach a tiny fraction of the oceanic surface in any given month or year. And even after centuries of nautical exploration, most of the information scientists have gathered about winds, currents, and temperatures comes from the commercial trade routes of the North Atlantic between the United States and Europe. Satellites don't replace on-the-water research, as they can't collect water samples, but for some kinds of data collection they can do in minutes what might take boats centuries to do. Satellites can, in principle, watch the whole of the world's oceans, providing almost immediate assessments of environmental conditions everywhere.

Similarly quick surveillance is available for many parts of the biosphere that are otherwise difficult to reach—the polar ice, dense forest interiors, and atmosphere. As a result, says Claire Parkinson of the U.S. National Aeronautics and Space Administration (NASA), "theory and explanations no longer have a database restricted to areas and times where humans have physically [gone] and made observations or left instruments to record the measurements." The speed of environmental research has taken a quantum leap.

Speed isn't only a matter of technical capability, however. In practice, it's also a matter of access. Spy satellites began circling the globe soon after Russia's Sputnik went into orbit in 1957. But the information they relayed to Soviet and U.S. intelligence agencies was kept sequestered. The difference now is that satellites are increasingly being used for purposes other than espionage or military intelligence. The great majority are for telecommunications. However, more than 45—many owned by governments, but a growing number of them privately owned—are being used for monitoring various phenomena on the ground, on the water, or in the atmosphere. In addition, more than 70 launches are planned during the next 15 years by civil space agencies and private companies. How these instruments are used, and by whom, will have enormous consequences for the world.

THE RACE AGAINST TIME

If incidents like the Indonesian forest fire intervention are any indication, environmental monitoring by orbiting cameras could play a critical role in reversing the global trends of deforestation and ecological collapse that now threaten the long-term viability of civilization. Denis Hayes, the former Worldwatch Institute researcher who is the chairman of Earth Day 2000 asked a few years ago, in a speech, "How can we have won so many environmental battles, yet be so close to losing the war?" Since then, we have edged still closer. Clearly, the number of battles being won is too small, and the time it takes to win them is too long.

Another way of posing Hayes's now famous question might be to ask whether the processes of information-gathering and dissemination essential to changing human behavior can be speeded up enough to accelerate the environmental movement. Telecommunications satellites began providing part of the answer several decades ago by facilitating the formation of an active international environmental community that can mobilize quickly—whether to protest a dam on the Narmada River of India or to stop the use of genetically modified organisms in food production in Europe.

But while activism gained momentum, field work remained ominously slow—biologists slogging about in boots and rowboats, while the forces they were trying to understand raced over the Earth on the wings of global commerce, or ripped into it with the blades of industrial agriculture and resource extraction. However, satellite monitoring has begun to help researchers to more quickly assemble the data needed to bring decisive change. Remotely sensed images are contributing to critical areas of environmental research and management, including:

- **Weather:** The first meteorological satellites were launched in the early 1960s, and quickly became a key part of the U.N. World Meteorological Organization's World Weather Watch. In addition to the satellite data, virtually all nations contribute surface measurements of temperature, precipitation, and wind to this program to aid weather prediction, which has enormous social and economic benefits. In recent years, optical sensors that collect data on sea surface temperature and radar sensors that estimate ocean height have proven useful in understanding and predicting El Niño events, which can damage fisheries and agriculture by bringing warmth and wetness to much of the west coasts of South and North America,

FLOODING IN BANGLADESH

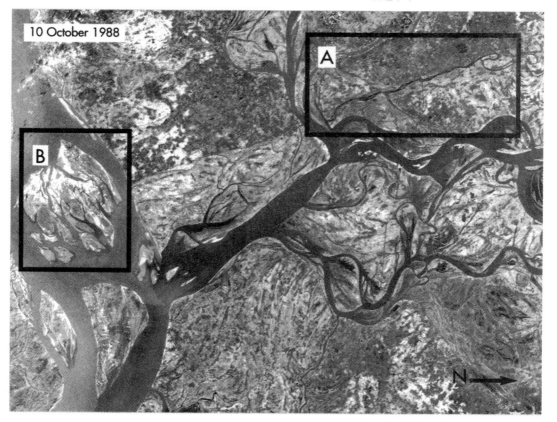

This medium resolution SPOT image of the confluence of the Meghna and Ganges Rivers and nearby Dhaka, the capital of Bangladesh, depicts a kind of collision between population growth and the ebbs and flows of the powerful rivers that feed the Ganges Delta. The land here is some of the most fertile—and heavily cultivated—in the world, replenished with rich silt washed down from the Himalayas by annual floods. The sprawling city of Dhaka and a couple of roads are perceptible in the northwest. And a closer view shows considerable development and cultivation throughout the image. (Even though the area is densely populated, this is difficult to see as most of the people live in small towns or villages.)

SPOT SATELLITE IMAGERY: © CNES 2000. COURTESY SPOT IMAGE CORPORATION, WWW.SPOT.COM.

and drought to Southeast Asia, Australia, and parts of Africa.

- **Climate:** In the 1990s, researchers began to delve into satellite archives to study longer-term climate patterns. For instance, satellite images have helped reveal a decrease in snow in the Northern Hemisphere, a lengthening growing season in northern latitudes, and the breakup of major ice sheets. Radar sensors have been used to construct topographical maps of the ocean bottom, which in turn provide better understanding of the ocean currents, tides, and temperatures that affect climate. However, it was not until recently that space agencies began to design satellite systems dedicated specifically to climate research. In 1999, the United States launched Terra, which carries five different sensors for recording climatic variables such as radiative energy fluxes, clouds, water vapor, snow cover, land use, and the biological productivity of oceans. It is to be the first in a series of satellites that will create a consistent data set for at least 18 years.

- **Coastal boundary changes:** Whether as a result of warmer temperatures contributing to sea-level rise or irrigation projects shrinking lakes, coastal configurations change over time—sometimes dramatically. Scientists have compared declassified images from covert U.S. military satellites pointed at Antarctica in 1963 to recent images of the same regions, for example, to reveal changes in the continent's ice cover. Other comparisons show the extent to which central Asia's Aral Sea and Africa's Lake Chad have diminished in size.

- **Habitat Protection:** The destruction of habitats as a result of human expansion has been identified as the largest single cause of biodiversity loss. Satellite images have proved quite effective in revealing large-scale forest destruction, whether by fire or clearcutting, not only in Indonesia but in the Amazon and other biological hotspots. New, more detailed imagery may reveal small-scale habitat niches. In Australia, for example, the Australia Koala Foundation plans to use detailed satellite images to identify individual eucalyptus trees. Researchers will compare these images to field data to determine what this species of tree looks like from above, then use the information to more quickly map individual trees or groves than would be possible from the ground. In the oceans, the same satellite-generated maps of undersea topography and sea surface temperature used to study weather and climate can be used to track the upwellings of nutrient-rich water that help to sustain fisheries.

- **Environmental law enforcement:** International organizations and national governments can use remote imaging to put more teeth in environmental laws and treaties. One of the leading fishing nations, Peru, is monitoring its coastal waters to prevent the kind of heavy overfishing that has so often caused fisheries to collapse. In Italy, the city of Ancona plans to buy satellite images to detect illegal waste dumps. Within the next decade, large-scale use of this technology could give urgently needed new effectiveness to such agreements as the Kyoto Protocol to the Climate Convention, the Biodiversity Convention, the

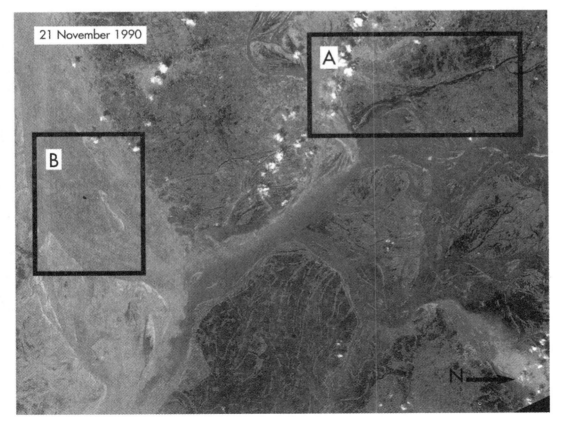

Bangladesh is one of the most densely populated countries in the world, and one of the most low-lying—the majority of the country is barely above sea level—so even the smallest of the annual floods can cause considerable damage. This image, in which darker areas are water-covered, was recorded during a devastating period of flooding, which took thousands of lives. Dhaka (A) is waterlogged and the once-cultivated islands in the Ganges, in the south of the image (B), are completely inundated.

A SAMPLING OF ENVIRONMENTAL APPLICATIONS OF SATELLITES

Spy-quality, high-resolution satellites (Ikonos, Orb View-3 and -4, QuickBird):

- Map urban development for transportation planning.
- Monitor arms control and humanitarian emergencies.
- Track refugee movements and estimated numbers.
- Provide maps for urban planners, farmers, foresters, electric utilities, and mineral surveyors.
- Give detailed assessments of crop and forest health.
- Help map biodiversity hotspots and sensitive bioregions to promote wildlife conservation.
- Track down illegal waste dumps.

Medium-resolution satellites (Landsat, SPOT, etc.):

- Measure vegetation health and land cover change, such as deforestation and flooding.
- Track urban growth, the effects of oil spills and nuclear accidents, and changes caused by dams and river diversions.
- Record the location and extent of damage caused by forest fires.
- Track coastal boundary changes such as melting ice cover or shifting barrier islands.
- Monitor status of fisheries to prevent overfishing.

Low-resolution satellites (AVHRR, OrbView-1, etc.):

- Detect meteorological phenomena and climate patterns such as El Niño.
- Map large areas of vegetation to reveal large-scale patterns of forest destruction.
- Analyze ocean circulation and geological formations.

Radar satellites (ERS-1 and -2, Radarsat, etc.):

- Help ships navigate ice fields and around icebergs.
- Create topographical maps of the ocean floor, which provide information about ocean currents, tides, and temperatures that affect the climate.
- Track nutrient-rich waters that sustain fisheries.

may, when seen from many kilometers above, resolve themselves into startling pictures.

The first pictures from space were photographs made from film, by astronauts aboard the first manned flights to the moon in the 1960s. These photos, of a fragile blue planet suspended in the vast blackness of space, helped to inspire the nascent environmental movement—one of them becoming the emblem of the first Earth Day in 1970.

In later surveillance from satellites, the imaging was digitized so that the data could be sent down in continuous streams and in much larger quantities than would be possible with film. Although one Russian satellite still uses regular camera film that is dropped to Earth in a canister and retrieved from the North Sea, most remote sensing satellites now use digital electronic sensors. The binary data they send down can be reconstructed into visual images by ground-based computers.

The amount of detail varies with the type of sensor (see table, "Selected Satellite Systems"). For instance, an image of a 1,000 square kilometer tract of land obtained by a fairly low-resolution satellite such as AVHRR, which is used in continental and global studies of land and ocean, might contain 1,000 picture elements—or "pixels" (one piece of data per square kilometer). In contrast, an image of the same tract from the new, high-resolution Ikonos satellite would have 1 *billion* pixels (one per square meter). Between the broad perspective of satellites like AVHRR and the telescopic imaging of those like the new spy-quality satellite Ikonos, are medium-resolution sensors, such as those aboard the Landsat and SPOT satellites, in which one pixel represents a piece of land that is between 30 and 10 meters across. However, the level of detail is also limited by the size of the medium on which an image is displayed. For instance, if an Ikonos image with 1 billion pixels were reproduced in a magazine image one-quarter the size of this page, it would be reduced to 350,000 pixels.

Different tasks require different levels of detail. The value of high resolution lies in its enabling the viewer to hone in on a much smaller piece of the ground and see it in a kind of detail that the lower resolution camera could not capture. For a larger area, a lower resolution would suffice to give the human eye and brain as clear a pattern as it can recognize. Whereas the wide coverage provided by lower-resolution satellites has proved useful in understanding large-scale natural features such as geo-

Convention on Illegal Trade in Endangered Species (CITES), or the Law of the Sea.

Making Sense of Nonsense

Look closely at a small detail of a newspaper or magazine photo—put it under a magnifying glass—and it may make no sense. The dots don't form any recognizable image. But stand back and see the photo as a whole, and it snaps into focus. Satellite images do the same thing, only on a vastly larger scale. Bits of information that might make no meaningful pattern when seen from the ground

33. Gaining Perspective

logic formations and ocean circulation, very detailed imagery may be best able to reveal niche habitats—such as the individual treetops that are home to the koala—and manmade structures, such as buildings, tanks, weapons, and refugee camps.

But satellite surveillance can do much more than provide huge volumes of sharp visual detail of the kind recorded by conventional optical cameras. The orbital industry also deploys a range of sensors that pick up information outside the range of the human eye, which can then be translated into visual form:

• Near-infrared emissions from the ground can be used to assess the health of plant growth, either in agriculture or in natural ecosystems, because healthy green vegetation reflects most of the near-infrared radiation it receives;

• Thermal radiation can reveal fires that would otherwise be obscured by smoke;

• Microwave emissions can provide information about soil moisture, wind speed, and rainfall over the oceans;

• Radar—short bursts of microwaves transmitted from the satellite—can penetrate the atmosphere in all conditions, and thus can "see" in the dark and through haze, clouds, or smoke. Radar sensors launched by European, Japanese, and Canadian agencies in the 1990s have been used mainly to detect changes in the freezing of sea ice in dark, northern latitudes, helping ships to navigate ice fields and steer clear of icebergs. Radar is what enabled satellites to map the ocean bottom, which would otherwise be obscured.

While satellites have the technical capability to monitor the Earth's entire surface—day or night, cloud-covered or clear, on the ground or underwater—this doesn't mean we now have updated global maps of whatever we want. Aside from the World Weather Watch, there is no process for coordinating a worldwide, long-term time series of comparable data from Earth observations. Rather, individual scientists collect data to answer specific questions for their own projects. In recent years, national space admini-

THE NILE DELTA, EGYPT

SPOT SATELLITE IMAGERY: © CNES 2000. COURTESY SPOT IMAGE CORPORATION, WWW.SPOT.COM.

This medium-resolution SPOT image of the Nile Delta shows the steady march of irrigation (indicated by the dark shades and crop circles) out into the desert near Cairo. While the Nile Delta and its floodplain have been farmed for thousands of years, only within the past 30 years have crops been planted intensively out in the desert. And with good reason: Cairo's population has grown from 5 million in 1970 to more than 11 million today.

strations have teamed up with research funding agencies and two international research programs to support an Integrated Global Observing Strategy that would create a framework for uniting environmental observations. Researchers are now trying to demonstrate the viability of this approach with a suite of projects, including one on forests and another on oceans.

In addition, to make sense of remotely sensed data requires comparison with field observations. An important element of the weather program's success, for instance, is the multitude of observations from both sky and land. And satellite estimations of sea-surface temperature can't generate El Niño forecasts automatically, but must be combined with other data sources, including readings from a network of ocean buoys that monitor wind speed and a satellite altimeter that measures water height. Similarly, the radar scans used to make maps of the ocean floor are calibrated and augmented by sounding surveys conducted by ships. Even a task as straightforward as the location of eucalyptus trees for the Australia koala project requires initial field observations to confirm that the typical visual pattern being searched out from above is indeed that of the eucalyptus, and not of some other kind of tree.

Satellite imagery has become even more useful with the advent of geographic information systems (GIS), which allow users to combine satellite images with other data in a computer to create maps and model changes over time. In much the same way that old medical encyclopedias depict human anatomy, with transparencies of the skeleton, circulatory system, nervous system, and organs that can be laid over a picture of the body, a GIS stores multiple layers of geographically referenced information. The data layers might include satellite images, topography, political boundaries, rivers, highways, utility lines, sources of pollution, and wildlife habitat.

Maps that are stored in a GIS allow people to exploit the data storage capacity and calculating power of computers. Thus when geographically referenced data are entered into a GIS, the computer can be harnessed to look at changes over time, to identify relationships between different data layers, to change variables in order to ask "what if" questions, and to explore various alternatives for future action.

Because human perception can often identify patterns more easily on maps than in written text or numbers, maps can help people understand and analyze problems in ways that other types of information cannot. The Washington, D.C.-based World Resources Institute (WRI), for example, has used GIS to analyze threats to natural resources on a global scale. Researchers have combined ground and satellite data on forests with information about wilderness areas and roads to map the world's remaining large, intact "frontier" forests and identify hot-spots of deforestation. A similar WRI study, investigating threats to coral reefs, assembled information from 14 global data sets, local studies of 800 sites, and scientific expertise to conclude that 58 percent of the world's reefs are at risk from development.

With advances in computing power, some GIS software packages can now be run on desktop computers, allowing more people to take advantage of them. In fact, the number of people using GIS is swelling by roughly 20 percent each year, and the leading software company, ESRI, grew from fewer than 50,000 clients in 1990 to more than 220,000 in 1999.

A related technology spurring the market for geographic information is the Global Positioning System—a network of 24 navigation satellites operated by the U.S. Department of Defense. A GPS receiver on the ground uses signals from different satellites to triangulate position. (For security reasons, the Defense Department purposefully introduces a distortion into the signal so that the location is correct only to within 100 meters.) As GPS receivers have become miniaturized, their cost has come down. The technology is now built into some farm machines, cars, and laptop computers. Researchers can take air or water samples and feed the data directly into a GIS, with latitude and longitude coordinates supplied by the GPS receiver in their computers.

The relatively new field of "precision agriculture" demonstrates how satellite imagery, GIS, and GPS systems can all be used to show farmers precisely how their crops are growing. Conditions in every crop row can be monitored when a farmer walks into the field—or when a satellite flies over—and recorded for analysis in a GIS. During the growing season, satellite monitoring can track crop conditions, such as the amount of pest damage or water stress, and allow farmers to attend to affected areas. The central component of a precision agriculture operation is a yield monitor, which is a sensor in a harvesting combine that receives GPS coordinates. As the combine harvests a crop such as corn, the sensor records the quality and quantity of the harvest from each section of the field. This detailed information provides indicators about the soil quality and irrigation needs of different parts of the field, and allows the farmer to apply water, fertilizer, and pesticides more accurately the following season.

Meanwhile, the growth of the Internet is allowing satellite images and GIS data to be more easily distributed. In late 1997, Microsoft Corporation teamed up with the Russian space agency Sovinformsputnik and image providers such as Aerial Images, Inc. and the U.S. Geological Survey to create TerraServer, the first website to allow people to view, download, and purchase satellite images. Some satellite operators have begun to offer catalogs of their images on the Internet.

Whose Picture Is It?

Throughout history, people have fought for possession of pieces of the Earth's land and water. It was not until the advent of Earth observation satellites that ownership of the images of those places was seriously debated. To legitimize its satellite program, the United States argued strongly that the light reflected off the oceans or mountains, like the air, should be in the public domain—a claim now accepted by many other countries. But for the last two decades, the United States has also promoted the involvement of private U.S. companies in earth observation. Until the end of the Cold War, companies were reluctant to enter the satellite remote sensing business for fear of restrictions related to national security concerns. But in September 1999, a U.S.-based firm called Space Imaging launched the first high-resolution commercial satellite; this year, two other U.S. enterprises, Orbimage and Earth-Watch, plan to launch similar instruments.

This trend raises questions about the tension between public and private interests in exploiting space. On the one hand, the widespread availability of detailed images means greater openness in human conduct. With satellite imagery, it is impossible to hide (or not find out about) such harmful or threatening activities as Chernobyl-scale nuclear accidents, or widespread forest clearing, or major troop movements. On the other hand, whether the information will be used to its full potential is up to governments and citizens.

There are obvious benefits to commercially available high-resolution images. Governments wary of revealing secrets have traditionally restricted the circulation of detailed satellite information. Now, images of the Earth that were once available to a select few intelligence agencies are accessible to anyone with a credit card. The information may be valuable to many non-military enterprises—public utilities, transportation planners, telecommunications firms, foresters, and others who already rely on up-to-date maps for routine operations.

In the world of NGOs, the impact of high-resolution imagery may be most dramatic for groups that keep an eye on arms control agreements and government military activities. "When one-meter black-and-white pictures hit the market, a well-endowed non-governmental organiza-

SELECTED SATELLITE SYSTEMS PRODUCING COMMERCIALLY AVAILABLE IMAGERY

Satellite	Launch Date	Owner	Spatial Resolution	Spectral Range	Price per Square Mile
Landsat series	1972	NASA,	30–120m,	visible light (red, green, blue); near-, short-wave, and thermal infrared	$.02–.03
Landsat-7	1999	(U.S. space agency)	15–60m		
Terra	1999	NASA	15m–22km	visible, near-, short-wave, mid-, and thermal infrared	N/A
SPOT series	1986	CNES	10–30m	visible, near- and short-wave infrared	$1–3
SPOT-4	1997	(French space agency)			
AVHRR	1979	NOAA (U.S. agency)	1.1km	visible, thermal infrared	$.08–80 per 10,000 square miles
IRS-1D	1997	Indian remote sensing agency	6m	visible, near- and short-wave infrared	$1.30–6.20
Ikonos	1999	Space Imaging Corp.	1–4m	visible; near-infrared	$75–250
Orb View		Orbimage Corp.			
OrbView-1	1995		10 km	visible; near-infrared	
OrbView-2	1997		1 km	visible; near-infrared	$.0003
OrbView-3	2000		1–8m	visible; near-infrared	N/A
OrbView-4	2000–01		1–8m	visible; near-infrared	N/A
QuickBird	2000	Earthwatch Corp.	1–4m	visible; near-infrared	N/A
ERS SAR					
ERS-1	1991	European space agencies	25m	radar; C-band	
ERS-2	1995				
JERS-1	1992	Japanese space agency	18m	radar; L-band	
Radarsat					
Radarsat-1	1995	Canadian space agency	8–100m	radar; C-band	$.04–5.40

tion will be able to have pictures better than [those] the U.S. spy satellites took in 1972 at the time of the first strategic arms accord," writes Peter Zimmerman, a remote-sensing and arms control expert, in a 1999 *Scientific American* article.

Indeed, when Ikonos released imagery of a top-secret North Korean missile base last January, the Federation of American Scientists (FAS), a U.S.-based nonprofit group, published a controversial analysis of the images contradicting U.S. military claims that the site is one of the most serious missile threats facing the United States. (The potential threat from this base is a key argument for proponents of a multibillion dollar missile shield, the construction of which would violate the anti-ballistic missile treaty.) Noting the absence of transportation links, paved roads, propellant storage, and staff housing, the FAS report found the facility "incapable of supporting the extensive test program that would be needed to fully develop a reliable missile system."

In addition, the private entrants in the satellite remote sensing business may spur the whole industry to become more accessible. Already, the new companies are beginning to seek partnerships with government imagery providers, so that customers are able to go to one place to buy a number of different types of images. For instance, Orbimage, which is scheduled to launch its first high resolution satellite this year, has made agreements to sell medium resolution images from the French SPOT series and radar images from the Canadian Radarsat. Such arrangements may make it easier for people to find and use imagery.

But private ownership of satellite data—making it available only at a price that not all beneficiaries can pay, or protecting it with copyright agreements—could also cause serious impediments to reversing ecological decline:

• One of the most important potential applications of remote sensing could be its use by non-governmental public interest groups, which provide a critical counterweight to the government and corporate sectors. NGOs that monitor arms control or humanitarian emergencies, for example, can use satellite data to pressure governments to live up to international nonproliferation agreements and foreign aid commitments. But a group that buys images from a private satellite company in order to publicize a humanitarian disaster or environmental threat could—depending on copyright laws and restrictions—find itself prohibited from posting the images on its website or distributing them to the media. While low-resolution imagery from government sources is easily shared, citizen groups and governments should quicky set a precedent for sharing information from high-resolution imagery.

• Only a few, very large agricultural operations can afford the high-tech equipment required to practice precision agriculture. Because farmers with small plots of land can simply walk into their fields for an assessment, the technology will remain most useful to larger operations.

FISHBONE FORESTS

These medium-resolution Landsat images show the legacy of the Brazilian government's Polonoreste project, which held out the offer of cheap land to bring farmers to this remote area of the Amazon. The centerpiece of this project, the Trans-Amazon Highway, can be seen following the Jiparaná River, running from the southeast to the northwest through the booming rural cities of Pimenta Bueno (center) and then Cacoal (top left). A tiny outpost in the early 1970s, Cacoal is now home to more than 50,000 people.

LANDSAT IMAGES COURTESY OF THE TROPICAL RAINFOREST INFORMATION CENTER OF THE BASIC SCIENCE AND REMOTE SENSING INITIATIVE AT MICHIGAN STATE UNIVERSITY, A NASA EARTH SCIENCE INFORMATION PARTNER. WWW.BSRSI.MSU.EDU

So although this tool could improve the way large-scale agriculture is practiced in the short term, it could also delay the long-term transition to more sustainable agricultural practices, which tend to require smaller-scale farms.

• Large companies looking for places to extract oil, minerals, or biological resources could purchase detailed satellite data and gain unfair advantage over cash-strapped developing nations in which the resources are located, whose governments cannot afford satellite imagery, GIS software, and technical support staff and systems needed to make and maintain such maps.

Finally, along with the question of who will own the technology, there is the related question of whether there is significant risk of its being badly misused. A few decades ago, the advent of commercially available high-resolution images would likely have been met with great alarm, as a manifestation of the "Big-Brother-is-watching-you" fear that pervaded the Cold War years. That fear may have receded, but what remains is a conundrum that has haunted every powerful new technology, from steel blades to genetic engineering. Could access to detailed images of their enemies cause belligerent nations to become more dangerous than they already are?

In any case, there's no turning back now. Many of the remote sensing satellites now scanning the Earth and scheduled for launch this year will be in orbit until long after the basic decisions affecting the planet's long-term environmental future—and perhaps the future of civilization—have been made. Ultimately, an educated global citizenry will be needed to make use of the flood of data being unleashed from both publicly and privately owned satellites. Policy analyst Ann Florini of the Carnegie Endowment writes: "With states, international organizations, and corporations all prodding one another to release ever more information, civil society can take that information, analyze and compile it, and disseminate it to networks of citizen groups and consumer organizations."

If the Internet has given the world's technological infrastructure a new nervous system, the Earth-observation satellites are giving it a new set of eyes. In precarious times, that could be useful. U.S. Vice President Al Gore, who understands the potential of remote sensing, has called for the completion of a "Digital Earth," a 1-meter resolution map of the world that would be widely accessible. According to Brian Soliday of Space Imaging, the Ikonos instrument alone might be able to assemble a cloud-free map of the world at 1-meter resolution within four to five years. That's about as long as it has taken to do the vaunted Human Genome map, and this map would be much bigger. Arguably, because it covers not only us humans but also the biological and climatic systems in which our genome evolved and must forever continue to depend, it could also be at least as valuable.

Molly O'Meara Sheehan is a research associate at the Worldwatch Institute.

Rondonia, Brazil, 1999

Highway BR 364, as it is called, has unzipped this part of Amazon, opening the vast forests to the telltale "fishbone pattern" of forest clearing. The roads that radiate out from the highway are flanked with once-forested land cleared for agriculture. Heavy rainfall leaches the soil of nutrients within a few seasons and crop yields quickly decline. The cycle of clearing starts anew when the degraded land is sold to cattle ranchers for pasture lands, which can be seen here as large rectangular clearings. The untouched tract of forest to the northwest is an indigenous preserve.

Article 34

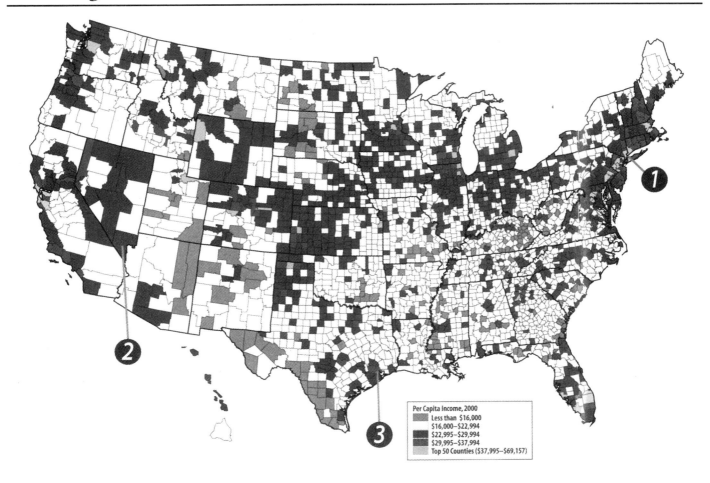

Counties With Cash

When you're hot, you're hot. When you're not—there's probably a reason.

BY JOHN FETTO

As April ushers in the tax season, it seemed like a good time to investigate where America's wealth is concentrated these days.

There are numerous ways to gauge wealth, from household income to median home value. But only one measure, per-capita income, indicates the amount of money available for each person to spend. Data provided by Washington, D.C.-based market research firm Woods & Poole Economics, Inc. was used to create our map that illustrates the distribution of per-capita income by county. The 50 counties with the highest per-capita income ... range from $37,995 per person in Union County, New Jersey, to $69,157 in New York, New York (1). The national average is $28,309.

The heaviest concentrations of wealth are found in and around New York City, San Francisco, and Washington, D.C. In fact, 20 of the top 50 counties are clustered around one of these three metro areas—ten in New York City, five in San Francisco, and five in D.C. As our map shows, there is plenty of wealth scattered throughout the country, mostly in major metropolitan areas.

But not all major metros have a county in the money: Las Vegas, Miami, Houston, and Salt Lake City are four that don't make the cut.

What gives? Factors such as the percent of the population employed by the service sector, percent of retirees, household size, and urban landscape all play a role, says Martin Holdrich, senior economist at Woods & Poole. The economy of Las Vegas (2), for example, is driven by the service sector, which pays

34. Counties With Cash

ROLLING IN IT

Counties with the highest per-capita income in 2000 and projected income for 2010, in 1999 dollars.

COUNTY	PER CAPITA INCOME 2000	2010
New York, NY	$69,157	$99,088
Marin, CA	$54,608	$85,243
Pitkin, CO	$54,076	$79,455
Fairfield, CT	$53,474	$80,234
Somerset, NJ	$51,605	$81,320
Alexandria (independent city), VA	$50,752	$78,255
Westchester, NY	$50,402	$74,206
Morris, NJ	$49,640	$76,613
Bergen, NJ	$48,137	$72,053
Arlington, VA	$47,252	$71,679
Montgomery, MD	$46,911	$69,648
Teton, WY	$45,758	$65,623
San Francisco, CA	$45,694	$68,942
Montgomery, PA	$45,553	$67,748
Fairfax + Fairfax City + Falls Church, VA	$45,493	$69,378
Lake, IL	$45,218	$69,154
Nassau, NY	$45,176	$68,399
Oakland, MI	$44,767	$69,393
Nantucket, MA	$44,534	$66,055
San Mateo, CA	$43,884	$65,378

Source: Woods & Poole Economics, Inc.

less than any other industry. "Unless, there is growth in the other sectors, Las Vegas counties will not be in the top 50," says Holdrich. Other cities affected by a high percentage of service workers are Reno, Honolulu, and Orlando.

Retirement counties also tend to rank lower on the wealth scale because typically, retiree incomes are low. "Retired persons do have income, but the flow of dollars going through their households annually is much lower than it was when they were working," says Holdrich. Phoenix, Tucson, Miami, and their surrounding counties are all affected in part by the retirement factor.

Another thing to consider in places like Phoenix, Tucson, and Miami is immigration. Immigrants tend to be younger and work in the service sector—a double whammy where income is concerned.

Immigrant households are also typically larger than those of nonimmigrants. "Income per capita is biased against places with large families," says Holdrich. That's also why you won't find Salt Lake City or many other Utah counties with large numbers of Mormon households near the top of the list, simply because they have larger-than-average households. In fact, we found that nine of the 20 lowest-ranking counties for per-capita income are among the top 20 counties with the most persons per household.

County size is important to consider when examining a map of income as well. Huge geographic areas, like Houston (3), incorporate vast disparities in income: from inner city to affluent suburbs, says Holdrich. Which is why you're less likely to see a county with high per-capita income in areas of the country with geographically large counties. Western counties' per-capita income tends to skew low because of their large size. Where counties are smaller, and wealth and poverty can be easily separated by a county line, per-capita income is up.

Just as there are metro areas that don't appear to have much wealth, there are non-metro areas that have loads of it, including seasonal escapes for the wealthy. Non-metro areas that score in the top 20 for per-capita income include Pitkin County, Colorado ($54,076), home to Aspen; Teton County, Wyoming ($45,758), Jackson Hole; and the island county of Nantucket, Massachusetts ($44,534).

Several upscale Florida communities appear near the top of the list as well. They are Sarasota County, Martin County, Collier County, and Indian River County.

Remember the old adage, it takes money to make money? Well, the richest counties are doing their best not to disprove that. In fact, very little will change with respect to wealth distribution in the next ten years.

In 2010, the only changes to our list of the ten wealthiest counties will be a little place-swapping here and there. Still, nobody will be packing up and moving off the list. The average per-capita income of the top ten is projected to increase by 51 percent between now and 2010. Now *that's* a raise.

Do We Still Need Skyscrapers?

The Industrial Revolution made skyscrapers possible.
The Digital Revolution makes them (almost) obsolete

by William J. Mitchell

Our distant forebears could create remarkably tall structures by exploiting the compressive strength of stone and brick, but the masonry piles they constructed in this way contained little usable interior space. At 146 meters (480 feet), the Great Pyramid of Cheops is a vivid expression of the ruler's power but inside it is mostly solid rock; the net-to-gross floor area is terrible. On a square base of 230 meters, it encloses the King's Chamber, which is just five meters across. The 52-meter spiraling brick minaret of the Great Mosque of Samarra does not have any interior at all. And the 107-meter stone spires of Chartres Cathedral, though structurally sophisticated, enclose nothing but narrow shafts of empty space and cramped access stairs.

The Industrial Revolution eventually provided ways to open up the interiors of tall towers and put large numbers of people inside. Nineteenth-century architects found that they could achieve greatly improved ratios of open floor area to solid construction by using steel and reinforced concrete framing and thin curtain walls. They could employ mechanical elevators to provide rapid vertical circulation. And they could integrate increasingly sophisticated mechanical systems to heat, ventilate and cool growing amounts of interior space. In the 1870s and 1880s visionary New York and Chicago architects and engineers brought these elements together to produce the modern skyscraper. Among the earliest full-fledged examples were the Equitable Building (1868–70), the Western Union Building (1872–75) and the Tribune Building (1873–75) in New York City, and Burnham & Root's great Montauk Building (1882) in Chicago.

These newfangled architectural contraptions found a ready market because they satisfied industrial capitalism's growing need to bring armies of office workers together at locations where they could conveniently interact with one another gain access to files and other work materials, and be supervised by their bosses. Furthermore, tall buildings fitted perfectly into the emerging pattern of the commuter city, with its high-density central business district, ring of low-density bedroom suburbs and radial transportation systems for the daily return journey. This centralization drove up property values in the urban core and created a strong economic motivation to jam as much floor area as possible onto every available lot. So as the 20th century unfolded, and cities such as New York and Chicago grew, downtown skylines sprouted higher while the suburbs spread wider.

But there were natural limits to this upward extension of skyscrapers, just as there are constraints on the sizes of living organisms. Floor and wind loads, people, water and supplies must ultimately be transferred to the ground, so the higher you go, the more of the floor area must be occupied by structural supports, elevators and ser-

Chrysler Building
Built 1930
Height 319 meters
New York

Tribune Building
Built 1875
Height 79 meters
New York

Western Union Building
Built 1875
Height 70 meters
New York

Equitable Building
Built 1870
Height 43 meters
New York

Chartres Cathedral
Built 13th century
Height 107 meters
France

Minaret of Samarra
Built 9th century
Height 52 meters
Iraq

Great Pyramid of Cheops
Built circa 2600 B.C.
Height 146 meters
Egypt

35. Do We Still Need Skyscrapers?

vice ducts. At some point, it becomes uneconomical to add additional floors; the diminishing increment of usable floor area does not justify the increasing increment of cost.

Urban planning and design considerations constrain height as well. Tall buildings have some unwelcome effects at ground level; they cast long shadows, blot out the sky and sometimes create dangerous and unpleasant blasts of wind. And they generate pedestrian and automobile traffic that strains the capacity of surrounding streets. To control these effects, planning authorities typically impose limits on height and on the ratio of floor area to ground area. More subtly, they may apply formulas relating allowable height and bulk to street dimensions—frequently yielding the stepped-back and tapering forms that so strongly characterize the Manhattan skyline.

The consequence of these various limits is that exceptionally tall buildings—those that really push the envelope—have always been expensive, rare and conspicuous. So organizations can effectively draw attention to themselves and express their power and prestige by finding ways to construct the loftiest skyscrapers in town, in the nation or maybe even in the world. They frequently find this worthwhile, even when it does not make much immediate practical sense.

There has, then, been an ongoing, century-long race for height. The Chrysler Building (319 meters) and the Empire State Building (381 meters) battled it out in New York in the late 1920s, adding radio antennas and even a dirigible mooring mast to gain the last few meters.

The contest heated up again in the 1960s and 1970s, with Lower Manhattan's World Trade Center twin towers (417 meters), Chicago's John Hancock tower (344 meters) and finally Chicago's gigantic Sears Tower (443 meters). More recently, Cesar Pelli's skybridge-linked Petronas Twin Towers (452 meters) in Kuala Lumpur have—for a while at least—taken the title of world's tallest building.

Along the way, there were some spectacular fantasy entrants as well. In 1900 Desiré Despradelle of the Massachusetts Institute of Technology proposed a 457-meter "Beacon of Progress" for the site of the Chicago World's Fair; like Malaysia's Petronas Towers of almost a century later, it was freighted with symbolism of a proud young nation's aspirations. Despradelle's enormous watercolor rendering hung for years in the M.I.T. design studio to inspire the students. Then, in 1956, Frank Lloyd Wright (not much more than five feet in his shoes and cape) topped it with a truly megalomaniac proposal for a 528-story, mile-high tower for the Chicago waterfront.

While this race has been running, though, the burgeoning Digital Revolution has been reducing the need to bring office workers together face-to-face, in expensive downtown locations. Efficient telecommunications have diminished the importance of centrality and correspondingly increased the attractiveness of less expensive suburban sites that are more convenient to the labor force. Digital storage and computer networks have increasingly supported decentralized remote access to data bases rather than reliance on cen-

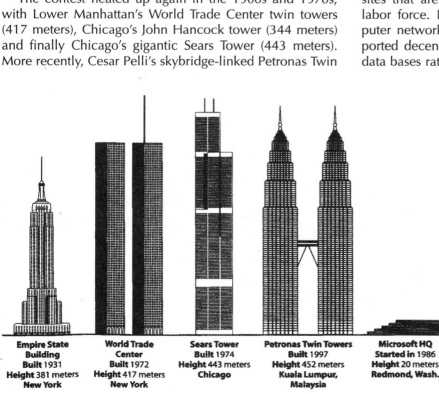
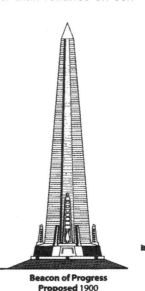

Empire State Building Built 1931 Height 381 meters New York

World Trade Center Built 1972 Height 417 meters New York

Sears Tower Built 1974 Height 443 meters Chicago

Petronas Twin Towers Built 1997 Height 452 meters Kuala Lumpur, Malaysia

Microsoft HQ Started in 1986 Height 20 meters Redmond, Wash.

Beacon of Progress Proposed 1900 Never built Height 457 meters Planned for Chicago

Mile High Tower Proposed 1956 Never built Height 1,609 meters Planned for Chicago

tralized paper files. And businesses are discovering that their marketing and public-relations purposes may now be better served by slick World Wide Web pages on the Internet and Superbowl advertising spots than by investments in monumental architecture on expensive urban sites.

We now find, more and more, that powerful corporations occupy relatively unobtrusive, low- or medium-rise suburban office campuses rather than flashy downtown towers. In Detroit, Ford and Chrysler spread themselves amid the greenery in this way—though General Motors has bucked the trend by moving into the lakeside Renaissance Center. Nike's campus in Beaverton, Ore., is pretty hard to find, but www.nike.com is not. Microsoft and Netscape battle it out from Redmond, Wash., and Mountain View, Calif., respectively, and—though their logos, the look and feel of their interfaces, and their Web pages are familiar worldwide—few of their millions of customers know or care what the headquarters buildings look like. And—a particularly telling straw in the wind—Sears has moved its Chicago workforce from the great Loop tower that bears its name to a campus in far-suburban Hoffman Estates.

Does this mean that skyscrapers are now dinosaurs? Have they finally had their day? Not quite, as a visit to the fancy bar high atop Hong Kong's prestigious Peninsula Hotel will confirm. Here the washroom urinals are set against the clear plate-glass windows so that powerful men can gaze down on the city while they relieve themselves. Obviously this gesture would not have such satisfying effect on the ground floor. In the 21st century, as in the time of Cheops, there will undoubtedly be taller and taller buildings, built at great effort and often without real economic justification, because the rich and powerful will still sometimes find satisfaction in traditional ways of demonstrating that they're on top of the heap.

WILLIAM J. MITCHELL is dean of the School of Architecture and Planning at the Massachusetts Institute of Technology.

BIO INVASION

No longer hindered by time and distance, disease can strike any species, anywhere

BY JANET GINSBURG

When reptile dealer Wayne Hill brought an ailing leopard tortoise into the veterinary clinic at the University of Florida at Gainesville in 1997, he was in for a big surprise. His pet, it turned out, had stowaways. Discreetly hidden beneath the animal's "armpits" were thumbnail-size African ticks, *Amblyomma marmoreum*. These critters can harbor a bacterium that causes heartwater, an animal disease that is endemic in sub-Saharan Africa but has spread to the Caribbean.

In the U.S., where cattle, sheep, deer, and elk have no immunity to the disease, heartwater could wipe out whole herds. Fortunately, the ticks were identified, says Michael J. Burridge, director of the Heartwater Research Project at UF. A quick inspection of Hill's facility revealed that it was crawling with the African ticks—and this was just one of a dozen infestations turned up by Burridge and his colleagues. The ticks proved to be heartwater-positive, raising alarms of a full-blown outbreak. And now worries about heartwater have spread to ranchers. "It could shut us down," says Jim Handley, executive vice-president of the Florida Cattlemen's Assn., who frets about a host of other new animal plagues as well. "West Nile encephalitis, screw worm—they all frighten us," he says "If those things are sneaking by, that's really scary."

And sneaking by they are. The tick that spreads heartwater is just the latest in a long list of foreign diseases that threaten ranch and farm economies throughout the world. These illnesses receive little media attention compared with exotic human afflictions such as Hanta virus or Ebola virus. But biologists say that alien animal and plant pests represent a much broader set of dangers than rare human illnesses do. And it's not either/or: In many cases, the animal plagues are closely associated with human illness as well.

Scientists and environmentalists have dubbed this phenomenon "bioinvasion." And the telltale signs of it are found all over the world. Mad cow disease, which decimated Britain's cattle industry, has now been identified in Belgium, France, Ireland, Switzerland, and Portugal. Even as Britain struggles with this blight, markets are slamming shut to its exports of hogs, which have been hit with an outbreak of classical swine fever. In Asia, Japan and Korea are battling outbreaks of foot-and-mouth disease (FMD) in cattle, which may have slipped in from China. In Mexico, just 200 miles from the Texas border, nearly 14 million chickens were slaughtered this spring because of a highly contagious virus called Exotic Newcastle Disease. A virus called Nepha has destroyed the Malaysian pork industry and killed 105 people. In North America, veterinarians are fighting a deadly parasitic disease called leishmaniasis. Rarely before found on these shores, it has sickened and killed hundreds of foxhounds in 21 states and Canada.

Experts blame the spread of these and other pests on an explosion in world trade, business travel, and tourism. Global trade policies aggravate the problem by putting strict limits on countries' abilities to ban animal trade. Meanwhile, in the U.S., years of flat budgets for border inspectors and disease researchers have left populations of animals—and humans—doubly exposed.

Such oversights can incur appalling costs. Even in the U.S., which has been spared the worst of the recent plagues, the price tag for battling agricultural blights ran

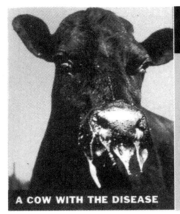

FOOT & MOUTH DISEASE
One of the most infectious diseases in the world, it is notoriously difficult to kill

ORIGINS
Recent outbreaks have occurred in Taiwan, Korea, Japan, and Brazil. Smuggled pigs from China or contaminated hay are suspected agents of infection. Agroterrorism may also be involved.

HOW IT SPREADS
The virus is spread by direct contact, ingestion, or inhalation.

AT RISK
Animals with cloven hooves, including horses, sheep, cows, and pigs.

DAMAGE
In 1997, 3.8 million pigs in Taiwan were slaughtered, costing billions. This year, 350,000 cattle in Korea and Japan have been killed. It is too soon to tell how many cattle may have to be killed in Argentina.

A COW WITH THE DISEASE

PHOTOGRAPH BY USDA/APHIS

to $9 billion last year, according to a Cornell University study. A serious outbreak of FMD would more than double that figure and ripple straight across the whole farm and food economy. "Our industry is at greater risk than ever of animal disease," says Terry L. Beals, executive director of the Texas Animal Health Commission. The implications are so dire that the National Intelligence Council—the research arm of the CIA—issued a report listing foreign animal diseases as a risk to national security. "We have been extremely fortunate so far," says Ernest W. Zirkle, president of the United States Animal Health Assn. (USAHA). "But everybody says [an outbreak] is a matter of when, and not if."

Public alarm over bioinvasion is most acute where the lines between animal and human risks are blurred. And that happens more frequently than most people realize. Hundreds of common diseases are zoonotic, meaning they affect animals and humans alike. Indeed, scientists estimate that as many as 70% of all pathogens are capable of jumping species.

Leishmaniasis and West Nile virus show why this is so frightening. In the U.S., the former has been detected only in hunting dogs. But in countries such as India, the parasite is endemic—and deadly—in humans as well. West Nile virus, which has killed a total of seven people around the New York metro region, may have entered the country on a smuggled bird or a jet-setting mosquito. It can also be fatal to chickens, pigs, and horses. And since it can be carried by ticks as well as by several species of mosquito, experts say the chances of eradication are slim. If the disease reaches Florida, for example, where the horse industry alone generates some 72,000 jobs, "the socio-economic damage would be immeasurable," says Leroy Coffman, the state's head veterinarian.

STRICKEN. Britain provides a chilling illustration of how economic costs and human suffering become intertwined in animal plagues. Herds of cows were probably first infected with bovine spongiform encephalopathy (BSE), or mad cow disease, in the early 1980s, when they were fed protein supplements that included the ground-up remains of sheep and cows. Some of those remains may have been infected with abnormal "prion" proteins, which proliferate in a live animal's brain, reducing it to sponge-like mush.

Britain has already spent an estimated $6.25 billion to clean up the mess—not including the jobs lost. Beef product exports are still down 99% from 1995, and the economic effects could easily stretch another 15 years. But the greater misery is that the illness seems to have jumped to humans. More than 80 people in Europe have been stricken with the new, human form of this always-fatal disease. As many as 136,000 may eventually become

BORDERS GROW MORE PERMEABLE
The U.S. plays host to millions of visitors annually. This year:

➤ More than 100 million people will travel to the U.S.

➤ 1.75 million illegal aliens will enter the U.S.

➤ 16.7 million farm animals, mostly livestock and poultry, will pass through inspection points

➤ 20 million wild animals will be brought in, many for the exotic pet trade

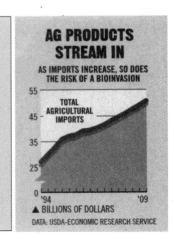

AG PRODUCTS STREAM IN
AS IMPORTS INCREASE, SO DOES THE RISK OF A BIOINVASION

TOTAL AGRICULTURAL IMPORTS

▲ BILLIONS OF DOLLARS
DATA: USDA-ECONOMIC RESEARCH SERVICE

36. Bio Invasion

MAD COW DISEASE — Called bovine spongiform encephalopathy, it has spread from England to several European countries

ORIGINS
BSE first appeared in Great Britain in the early 1980s.

HOW IT SPREADS
Animals develop the disease by eating feed contaminated with infectious proteins called prions. Slowly, malformed prions proliferate, turning the brain to mush.

AT RISK
Cows, sheep, deer, humans, and cats can develop forms of the disease.

DAMAGE
Since 1996, mad cow has cost Britain more than $6 billion. Over 4 million British cattle have been slaughtered, and more than 90 cases of the human version of the disease have been reported.

---

sick, according to an Oxford University study reported in the August issue of *Nature*.

North America is hardly insulated against such dangers. Animal researchers have long been tracking prion-linked diseases among wild populations of deer and elk. And in Vermont, the Agriculture Dept. recently ordered the destruction of three small flocks of imported sheep that may be carrying a new form of mad cow disease. There is evidence that before being brought into Vermont from Belgium, some of the sheep in the flock might have been exposed to mad cow through a feed supplement. Despite controversy over testing procedures that may have led to false positives, the USDA deemed the risk too great to let the sheep live.

The human errors that led to the BSE outbreak may be atypical in animal plagues. But the involvement of multiple species is hardly unique. The deadly HIV, which causes AIDS, may have originally leapt from monkeys to humans. And all human influenzas originate in birds and beasts, killing an estimated 40,000 a year.

But even where they do not affect human health, animal plagues can cripple whole industries. Japan and Korea are now struggling with FMD—their first outbreaks of the virus in more than 70 years. To date, Korea has had to slaughter 350,000 head of cattle at a cost of $90 million.

That loss pales, however, next to the 3.8 million pigs Taiwan was forced to destroy in a 1997 outbreak of FMD. The disease, thought to come from the mainland, could eventually cost Taiwan as much as $15 billion in lost exports of pork and other products. And the damage goes beyond the loss of revenue potential: After an outbreak, countries risk losing their place in global markets.

NIGHTMARE. Could a similar event take place in North America? Once again, it's a question of when, not if. Contamination spreads at the speed of transport. And the threat is vastly amplified by soaring agricultural trade, along with ever-greater numbers of world travelers. Since 1995, agricultural imports to the U.S. have risen 28%. And international travel to the U.S. is up 33%, to more than 100 million travelers annually, since 1988.

Most of those visitors make their trips in far less time than it takes diseases to incubate. And it's not just ailing passengers that worry health officials: It's what is in their luggage. Checking for contaminated products and insect stowaways is a logistical nightmare. By some estimates, worldwide customs officials inspect just 5% of all bags. What's more, about 30,000 people cross U.S. borders illegally every week. On top of that, exotic pets such as reptiles provide yet another channel for pathogens and parasites.

A virus such as FMD is brilliantly adapted to exploit these channels. Notoriously difficult to kill, it can survive for several weeks on clothes and for up to 24 hours in the human respiratory tract. "I can go from a farm in South America to one in Texas in a day," says Alfonso Torres, deputy administrator of the USDA's Veterinary Services Division. That's all the time it takes these days to start an outbreak.

Time and distance, in short, are no longer barriers for this disease. And the potential damages are astronomical. At risk are U.S. agricultural exports—currently worth $50 billion annually and expected to grow to $76 billion by 2009. A 1999 study by the University of California at Davis estimates that an FMD outbreak in the Golden State alone would result in losses of up to $13.5 billion. Many think that number is too low.

The scourge called classical swine fever shows how easily diseases travel. Epidemiologists believe that a 1997 outbreak in the Netherlands may have spread from Bosnia to Germany through contaminated pork brought in by returning German peacekeeping forces. Improperly treated food scraps from a military base probably spread the disease to local livestock. Infected pigs were then transported on trucks inadequately disinfected between deliveries. In short order, the pathogen was literally driven from Germany into the Netherlands, where the size and density of hog operations transformed the outbreak into a full-scale pandemic.

By 1998, 11 million pigs in the Netherlands had to be destroyed, about two-thirds of the national herd. Now, a similar catastrophe may be brewing in Britain. Biologists say the source is probably pork from Asia. In America, meanwhile, health authorities are watching these developments with alarm. "The Netherlands, Germany, and England all have superb veterinary systems, and they're getting hammered. So why on earth shouldn't it happen to the U.S.?" asks Keith Murray, director of the USDA's National Animal Disease Center in Ames, Iowa.

Beyond accidental transmission looms the very real danger of agroterrorism. Unlike human diseases, which

WEST NILE ENCEPHALITIS — A disease of humans and beasts, it has killed seven people in the New York area

ORIGINS
This virus, originally from Africa, mysteriously hopped to the U.S. in 1999. One possibility: An infected bird, mosquito, or person brought it here from the Middle East.

HOW IT SPREADS
Several different types of mosquitoes and ticks carry the virus.

AT RISK
Birds, horses, humans, and pigs can be infected with the virus.

DAMAGE
Hundreds of wild birds, nine horses, and seven people have died since the virus was discovered in the U.S. last year. Millions have been spent on surveillance and spraying.

---

generally require sophisticated manipulation in order to be "weaponized," many animal viruses such as FMD are ready to use as is. They're low-tech, low-cost high-impact, and very difficult to trace. "A person could smuggle in a virus, and they could just go to a feedlot," says Mowafak D. Salman, chair of USAHA's Foreign Animal Disease Committee. "If anything happened to our beef supplies, it would be more devastating [to society] than one or two deaths from West Nile fever."

Agroterrorism has a long history—even in the U.S. During World War I, German agents spread anthrax and glanders viruses in Maryland, Virginia, and New York in an attempt to kill horses and mules destined for Allied troops. By World War II, biowarfare programs had been set up in the U.S., Britain, Japan, Canada, and the Soviet Union. South Africa and Iraq have also had programs. While U.S. policy is to assume agricultural disease outbreaks are naturally occurring, says Floyd P. Horn, administrator of UDSA's Animal Research Service, "we really don't know." It's a point underscored by recent events in Brazil, where investigators say financial sabotage by rival ranchers may be behind a recent FMD outbreak in cattle.

RIGID POLICIES. Well-intentioned global trade policies may actually have weakened protections against bioinvasion. The WTO, for example, has sought to weed out fake health issues that nations sometimes use as ploys to shut out agricultural imports. But in its zeal, the WTO adopted agricultural trade policies that critics say are unrealistic and too rigidly scientific. Proof of a disease threat is often elusive. Free traders and WTO defenders have a different view. "Look at the facts on the ground," says Isi Siddique, senior trade adviser to Agriculture Secretary Daniel R. Glickman. "Have more animal diseases crept into the U.S. than in the years prior to WTO? I don't think so."

Trade policies aside, scientific risk assessments take time, money, and staff. Yet budgets for animal disease research in the U.S. have been stag-nant for at least a decade. In Ames, at one of the USDA's most critical lab complexes, the facilities urgently need updating and renovation. According to a recent department report, "virtually every critical system—ventilation, electrical, sewage treatment, biocontainment, incineration, and heating and cooling—is antiquated."

Conditions are not much better at Plum Island, N.Y., the only lab in the U.S. devoted to foreign animal disease research. It has neither the set-up nor staff to run heartwater tests, for example. So the Florida ticks have to be sent to a lab in Harare, Zimbabwe, and it takes five months to get results. "There are services we can no longer provide," complains the USDA's Torres.

Budgets for inspectors have also been pared to the bone. Veterinary Service (VS), the USDA division responsible for safeguarding livestock and poultry health, has just 98 full-time and 28 seasonal inspectors overseeing the import of nearly 17 million animals. And at the U.S. Fish & Wildlife Service (USFWS), there are only 91 field inspectors to process an estimated 20 million animals—200 million including fish. Forty-four positions for criminal investigators tracking animal smuggling are vacant due to lack of funding. "It's pathetic. We really do not have the support structure we need to protect animal agriculture in this country," says Elizabeth A. Lautner, a veterinarian with the National Pork Producers Council and vice-chair of the USDA's Secretary's Advisory Committee on Foreign Animal & Poultry Diseases.

While the U.S. has escaped large-scale animal plagues in recent years, several near-misses suggest that its luck finally may be running out. Since the 1950s, the U.S. has spent tens of millions of dollars eradicating screw worm—a horror story of a maggot that feeds and breeds on the open wounds of both man and beast—down to the southern Mexican border. But last March, a polo pony imported from Argentina, where the scourge is endemic, breezed through a USDA quarantine in Florida, only to have an infestation discovered by a private veterinarian less than 24 hours before the emergence of adult flies.

Exotic Newcastle Disease (END) also looks like a narrow miss. The poultry plague in Mexico was contained. But that doesn't mean there aren't consequences in the U.S. According to a USDA field report, the Mexican operations of U.S.-based Tyson Foods Inc. have borne the brunt of the pest, with more than 80% of the losses. Ty-

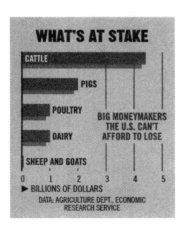

WHAT'S AT STAKE

BIG MONEYMAKERS THE U.S. CAN'T AFFORD TO LOSE

DATA: AGRICULTURE DEPT., ECONOMIC RESEARCH SERVICE

EXOTIC NEWCASTLE DISEASE
The blight this year dealt a huge blow to Tyson Foods in Mexico.

ORIGINS
An outbreak occurred this spring in Mexico. Origins unknown.

HOW IT SPREADS
The virus is primarily spread by bird-to-bird contact. It can also be spread via contaminated feed, water, or clothing, especially shoes.

AT RISK
Poultry and parrots can be infected with the virus.

DAMAGE
This year, 13.6 million birds were destroyed, at a loss of approximately $14 million. Additional cleanup and replacement expenses boost the cost to about $25 million.

---

son won't reveal specific costs, but poultry experts estimate the value of lost birds to be at least $11 million. Cleanup, restocking, and lost production time add to the tab. According to the company's Securities & Exchange Commission filings, Tyson's international division posted a $25 million decline in third-quarter profits this year.

END hasn't made its way north—yet. But agriculture experts vividly recall the last outbreak, in the early 1970s, when 12 million birds were destroyed and 46,000 square miles of California and Arizona were placed under quarantine. "You let Newcastle get into the U.S., and people won't be so concerned about the cost of a gallon of gas,"
says James Grimm of the Texas Poultry Federation. "They won't get any chicken tenders and chicken breasts." The END threat is so serious that in 1996, the mere rumor of an outbreak in Missouri was enough for China to cut off its poultry trade with the state for nearly two months. A full-scale outbreak would "end the export market," says Charles Beard, vice-president for research at the U.S. Poultry & Egg Assn.

WALL STREET WORRIES. So what can be done? Global trade and travel are facts of modern life. But there are signs that Washington and Wall Street now recognize the problem. The threat of a foreign animal disease outbreak could cause investors to flee farm and food stocks. "All it would take is one incident, one company to miss a quarter," says Credit Suisse First Boston analyst David C. Nelson.

On the legislative front, both the House and Senate agriculture sub-committees have recommended allocating funds to redesign the USDA lab complex in Ames. Plum Island is also slated for an overhaul—but political battles may hold up plans to build new, Bio-level 4 facilities to handle agroterrorism research.

On the regulatory front, the USDA has now banned the import of African tortoises known to carry heartwater ticks and restricted intrastate trade. In the meantime, USDA staff are racing to develop a national preparedness program for animal health emergencies. These steps don't begin to answer the kinds of questions that plague the USDA's Torres. "How much risk is acceptable?" he asks. "Is it one incident every 200 years? 25 years? That's the big question, and we don't have a good answer." You can bet scientists and legislators will be working on the problem though. The future of U.S. agriculture depends upon it.

With Ellen Licking in New York

A PLETHORA OF PATHOGENS

These deadly diseases can devastate economies.

CLASSICAL SWINE FEVER Highly contagious disease, which decimated pig herds in The Netherlands in 1997 and 1998. New outbreak in England.

NIPAH Newly identified virus causes brain inflammation in pigs and people. Since 1998, 1 million pigs have been destroyed in Malaysia and Singapore.

CHICKEN INFLUENZA 5 million chickens killed in a 1984 Pennsylvania outbreak, costing $63 million. In 1997 the flu jumps to humans, killing six in Hong Kong.

HEARTWATER DISEASE Tick-borne disease causes high fever and respiratory failure in cattle and horses. This infection is deadly unless discovered very early.

SCREW WORM Maggots buried in living tissue cause inflammation, even death. The U.S. has spent millions to eradicate it, but sightings in Texas and Florida indicate it is sneaking in.

Unit 5

Unit Selections

37. **Before the Next Doubling,** Jennifer D. Mitchell
38. **A Turning-Point for AIDS?** *The Economist*
39. **The End of Cheap Oil,** Colin J. Campbell and Jean H. Laherrere
40. **Gray Dawn: The Global Aging Crisis,** Peter G. Peterson
41. **A Rare and Precious Resource,** Houria Tazi Sedeq
42. **Rethinking Flood Prediction: Does the Traditional Approach Need to Change?** William D. Gosnold Jr., Julie A. LeFever, Paul E. Todhunter, and Leon F. Osborne Jr.
43. **Risking the Future of Our Cities,** Thomas Hackett
44. **The End of Urban Man? Care to Bet?** *The Economist*
45. **China's Changing Land: Population, Food Demand and Land Use in China,** Gerhard K. Heilig
46. **Helping the World's Poorest,** Jeffrey Sachs

Key Points to Consider

❖ How are you personally affected by the population explosion?

❖ Give examples of how economic development adversely affects the environment. How can such adverse effects be prevented?

❖ How do you feel about the occurrence of starvation in developing world regions?

❖ What might it be like to be a refugee?

❖ In what forms is colonialism present today?

❖ How is Earth a system?

❖ For how long are world systems sustainable?

❖ What is your scenario of the world in the year 2010?

 **Links** www.dushkin.com/online/

25. **African Studies WWW (U.Penn)**
    http://www.sas.upenn.edu/African_Studies/AS.html
26. **Human Rights and Humanitarian Assistance**
    http://info.pitt.edu/~ian/resource/human.htm
27. **Hypertext and Ethnography**
    http://www.umanitoba.ca/faculties/arts/anthropology/tutor/aaa_presentation.new.html
28. **Research and Reference (Library of Congress)**
    http://lcweb.loc.gov/rr/
29. **Space Research Institute**
    http://arc.iki.rssi.ru/Welcome.html

These sites are annotated on pages 6 and 7.

Population, Resources, and Socioeconomic Development

The final unit of this anthology includes discussions of several important problems facing humankind. Geographers are keenly aware of regional and global difficulties. It is hoped that their work with researchers from other academic disciplines and representatives of business and government will help bring about solutions to these serious problems.

Probably no single phenomenon has received as much attention in recent years as the so-called population explosion. World population continues to increase at unacceptably high rates. The problem is most severe in the less developed countries, where in some cases, populations are doubling in less than 20 years.

The human population of the world passed the 6 billion mark in 1999. It is anticipated that population increase will continue well into the twenty-first century, despite a slowing in the rate of population growth globally since the 1960s. The first article in this section deals with issues of population growth. The second reviews the devastation of AIDS in Africa. Declining petroleum reserves are discussed in a *Scientific American* article. "Gray Dawn" reviews the increase in the number of elderly globally. Then, freshwater is considered as a scarce resource in "A Rare and Precious Resource." The next article considers changes in flood prediction methods. Next, two articles deal with aspects of urbanization. Gerhard Heilig outlines changes in land use in China as that country shifts its economic focus. The last article, "Helping the World's Poorest," makes a plea for developed world assistance to the poorest countries.

Before the Next Doubling

Nearly 6 billion people now inhabit the Earth—almost twice as many as in 1960. At some point over the course of the next century, the world's population could double again. But we don't have anything like a century to prevent that next doubling; we probably have less than a decade.

by Jennifer D. Mitchell

In 1971, when Bangladesh won independence from Pakistan, the two countries embarked on a kind of unintentional demographic experiment. The separation had produced two very similar populations: both contained some 66 million people and both were growing at about 3 percent a year. Both were overwhelmingly poor, rural, and Muslim. Both populations had similar views on the "ideal" family size (around four children); in both cases, that ideal was roughly two children smaller than the actual average family. And in keeping with the Islamic tendency to encourage large families, both generally disapproved of family planning.

But there was one critical difference. The Pakistani government, distracted by leadership crises and committed to conventional ideals of economic growth, wavered over the importance of family planning. The Bangladeshi government did not: as early as 1976, population growth had been declared the country's number one problem, and a national network was established to educate people about family planning and supply them with contraceptives. As a result, the proportion of couples using contraceptives rose from around 6 percent in 1976 to about 50 percent today, and fertility rates have dropped from well over six children per woman to just over three. Today, some 120 million people people live in Bangladesh, while 140 million live in Pakistan—a difference of 20 million.

Bangladesh still faces enormous population pressures—by 2050, its population will probably have increased by nearly 100 million. But even so, that 20 million person "savings" is a colossal achievement, especially given local conditions. Bangladeshi officials had no hope of producing the classic "demographic transition," in which improvements in education, health care, and general living standards tend to push down the birth rate. Bangladesh was—and is—one of the poorest and most densely populated countries on earth. About the size of England and Wales, Bangladesh has twice as many people. Its per capita GDP is barely over $200. It has one doctor for every 12,500 people and nearly three-quarters of its adult population are illiterate. The national diet would be considered inadequate in any industrial country, and even at current levels of population growth, Bangladesh may be forced to rely increasingly on food imports.

All of these burdens would be substantially heavier than they already are, had it not been for the family planning program. To appreciate the Bangladeshi achievement, it's only necessary to look at Pakistan: those "additional" 20 million Pakistanis require at least 2.5 million more houses, about 4 million more tons of grain each year, millions more jobs, and significantly greater investments in health care—or a significantly greater burden of disease. Of the two nations, Pakistan has the more robust economy—its per capita GDP is twice that of Bangladesh. But the Pakistani economy is still primarily agricultural, and the size of the average farm is shrinking, in part because of the expanding population. Already, one fourth of the country's farms are under 1 hectare, the standard minimum size for economic viability, and Pakistan is looking increasingly towards the international grain markets to feed its people. In 1997, despite its third consecutive year of near-record harvests, Pakistan attempted to double its wheat imports but was not able to do so because it had exhausted its line of credit.

And Pakistan's extra burden will be compounded in the next generation. Pakistani women still bear an average of well over five children, so at the current birth rate, the 10 million or so extra couples would produce at least 50 million children. And these in turn could bear nearly 125 million children of their own. At its current fertility rate, Pakistan's population will double in just 24 years—that's more than twice as fast as Bangladesh's population is growing. H. E. Syeda Abida Hussain, Pakistan's Minister of Population Welfare, explains the problem bluntly: "If we achieve success in lowering our population growth substantially, Pakistan has a future. But if, God forbid, we should not—no future."

The Three Dimensions of the Population Explosion

Some version of Mrs. Abida's statement might apply to the world as a whole. About 5.9 billion people currently inhabit the Earth. By the middle of the next century, according to U.N. projections, the population will probably reach 9.4 billion—and all of the net increase is likely to occur in the developing world. (The total population of the industrial countries is expected to decline slightly over the next 50 years.) Nearly 60 percent of the increase will occur in Asia, which will grow from 3.4 billion people in 1995 to more than 5.4 billion in 2050. China's population will swell from 1.2 billion to 1.5 billion, while India's is projected to soar from 930 million to 1.53 billion. In the Middle East and North Africa, the population will probably more than double, and in sub-Saharan Africa, it will triple. By 2050, Nigeria alone is expected to have 339 million people—more than the entire continent of Africa had 35 years ago.

Despite the different demographic projections, no country will be immune to the effects of population growth. Of course, the countries with the highest growth rates are likely to feel the greatest immediate burdens—on their educational and public health systems, for instance, and on their forests, soils, and water as the struggle to grow more food intensifies. Already some 100 countries must rely on grain imports to some degree, and 1.3 billion of the world's people are living on the equivalent of $1 a day or less.

But the effects will ripple out from these "front-line" countries to encompass the world as a whole. Take the water predicament in the Middle East as an example. According to Tony Allan, a water expert at the University of London, the Middle East "ran out of water" in 1972, when its population stood at 122 million. At that point, Allan argues, the region had begun to draw more water out of its aquifers and rivers than the rains were replenishing. Yet today, the region's population is twice what it was in 1972 and still growing. To some degree, water management now determines political destiny. In Egypt, for example, President Hosni Mubarak has announced a $2 billion diversion project designed to pump water from the Nile River into an area that is now desert. The project—Mubarak calls it a "necessity imposed by population"—is designed to resettle some 3 million people outside the Nile flood plain, which is home to more than 90 percent of the country's population.

Elsewhere in the region, water demands are exacerbating international tensions; Jordan, Israel, and Syria, for instance, engage in uneasy competition for the waters of the Jordan River basin. Jordan's King Hussein once said that water was the only issue that could lead him to declare war on Israel. Of course, the United States and the western European countries are deeply involved in the region's antagonisms and have invested heavily in its fragile states. The western nations have no realistic hope of escaping involvement in future conflicts.

Yet the future need not be so grim. The experiences of countries like Bangladesh suggest that it is possible to build population policies that are a match for the threat. The first step is to understand the causes of population growth. John Bongaarts, vice president of the Population Council, a non-profit research group in New York City, has identified three basic factors. (See figure on the next page.)

Unmet demand for family planning. In the developing world, at least 120 million married women—and a large but undefined number of unmarried women—want more control over their pregnancies, but cannot get family planning services. This unmet demand will cause about one-third of the projected population growth in developing countries over the next 50 years, or an increase of about 1.2 billion people.

Desire for the large families. Another 20 percent of the projected growth over the next 50 years, or an increase of about 660 million people, will be caused by couples who may have access to family planning services, but who choose to have more than two children. (Roughly two children per family is the "replacement rate," at which a population could be expected to stabilize over the long term.)

Population momentum. By far the largest component of population growth is the least commonly understood. Nearly one-half of the increase projected for the next 50 years will occur simply because the next reproductive generation—the group of people currently entering puberty or younger—is so much larger than the current reproductive generation. Over the next 25 years, some 3 billion people—a number equal to the entire world population in 1960—will enter their reproductive years, but only about 1.8 billion will leave that phase of life. Assuming that the couples in this reproductive bulge begin to have children at a fairly early age, which is the global norm, the global population would still expand by 1.7 billion, even if all of those couples had only two children—the longterm replacement rate.

Meeting the Demand

Over the past three decades, the global percentage of couples using some form of family planning has increased dramatically—from less than 10 to more than 50 percent. But due to the growing population, the absolute number of women not using family planning is greater today than it was 30 years ago. Many of these women fall into that first category above—they want the services but for one reason or another, they cannot get them.

Sometimes the obstacle is a matter of policy: many governments ban or restrict valuable methods of contraception. In Japan, for instance, regulations discourage the use of birth control pills in favor of condoms, as a public health measure against sexually transmitted diseases. A study conducted in 1989 found that some 60 countries required a husband's permission before a woman can be sterilized; several required a husband's consent for all forms of birth control.

Elsewhere, the problems may be more logistical than legal. Many developing countries lack clinics and pharmacies in rural areas. In some rural areas of sub-Saharan Africa, it takes an average of two hours to reach the nearest contraceptive provider. And often contraceptives are too expensive for most people. Sometimes the products or services are of such poor quality that they are not simply ineffective, but dangerous. A woman who has been injured by a badly made or poorly inserted IUD may well be put off by contraception entirely.

In many countries, the best methods are simply unavailable. Sterilization is often the only available nontraditional option, or the only one that has gained wide acceptance. Globally, the procedure accounts for about 40 percent of contraceptive use and in some countries the fraction is much higher: in the Dominican Republic and India, for example, it stands at 69 percent. But women don't generally resort to sterilization until well into their childbearing years, and in some countries, the procedure isn't permitted until a woman reaches a certain age or bears a certain number of children. Sterilization is therefore no substitute for effective temporary methods like condoms, the pill, or IUDs.

There are often obstacles in the home as well. Women may be prevented from seeking family planning services by disapproving husbands or in-laws. In Pakistan, for example, 43 percent of husbands object to family planning. Frequently,

Population of Developing Countries, 1950–95, with Projected Growth to 2050

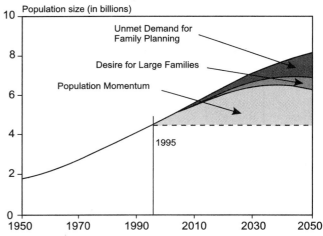

Source: U.N., *World Population Prospects: The 1996 Revision* (New York: October 1998); and John Bongaarts, "Population Policy Options in the Developing World," *Science,* 11 February 1994.

such objections reflect a general social disapproval inculcated by religious or other deeply-rooted cultural values. And in many places, there is a crippling burden of ignorance: women simply may not know what family planning services are available or how to obtain them.

Yet there are many proven opportunities for progress, even in conditions that would appear to offer little room for it. In Bangladesh, for instance, contraception was never explicitly illegal, but many households follow the Muslim custom of *purdah,* which largely secludes women in their communities.

Since it's very difficult for such women to get to family planning clinics, the government brought family planning to them: some 30,000 female field workers go door-to-door to explain contraceptive methods and distribute supplies. Several other countries have adopted Bangladesh's

approach. Ghana, for instance, has a similar system, in which field workers fan out from community centers. And even Pakistan now deploys 12,000 village-based workers, in an attempt to reform its family planning program, which still reaches only a quarter of the population.

Reducing the price of contraceptives can also trigger a substantial increase in use. In poor countries, contraceptives can be an extremely price-sensitive commodity even when they are very cheap. Bangladesh found this out the hard way in 1990, when officials increased contraceptive prices an average of 60 percent. (Under the increases, for example, the cheapest condoms cost about 1.25 U.S. cents per dozen.) Despite regular annual sales increases up to that point, the market slumped immediately: in 1991, condom sales fell by 29 percent and sales of the pill by 12 percent. The next year, prices were rolled back; sales rebounded and have grown steadily since then.

Additional research and development can help broaden the range of contraceptive options. Not all methods work for all couples, and the lack of a suitable method may block a substantial amount of demand. Some women, for instance, have side effects to the pill; others may not be able to use IUDs because of reproductive tract infections. The wider the range of available methods, the better the chance that a couple will use one of them.

Planning the Small Family

Simply providing family planning services to people who already want them won't be enough to arrest the population juggernaut. In many countries, large families are still the ideal. In Senegal, Cameroon, and Niger, for example, the average woman still wants six or seven children. A few countries have tried to legislate such desires away. In India, for example, the Ministry of Health and Family Welfare is interested in promoting a policy that would bar people who have more than two children from political careers, or deny them promotion if they work within the civil service bureaucracy. And China's well-known policy allows only one child per family.

But coercion is not only morally questionable—it's likely to be ineffective because of the backlash it invites. A better starting point for policy would be to try to understand why couples want large families in the first place. In many developing countries, having lots of children still seems perfectly rational: children are a source of security in old age and may be a vital part of the family economy. Even when they're very young, children's labor can make them an asset rather than a drain on family income. And in countries with high child mortality rates, many births may be viewed as necessary to compensate for the possible deaths (of course, the cumulative statistical effect of such a reaction is to *over*-compensate).

Religious or other cultural values may contribute to the big family ideal. In Pakistan, for instance, where 97 percent of the population is Muslim, a recent survey of married women found that almost 60 percent of them believed that the number of children they have is "up to God." Preference for sons is another widespread factor in the big family psychology: many large families have come about from a perceived need to bear at least one son. In India, for instance, many Hindus believe that they need a son to perform their last rites, or their souls will not be released from the cycle of births and rebirths. Lack of a son can mean abandonment in this life too. Many husbands desert wives who do not bear sons. Or if a husband dies, a son is often the key to a woman's security: 60 percent of Indian women over 60 are widows, and widows tend to rely on their sons for support. In some castes, a widow has no other option since social mores forbid her from returning to her birth village or joining a daughter's family. Understandably, the fear of abandonment prompts many Indian women to continue having children until they have a son. It is estimated that if son preference were eliminated in India, the fertility rate would decline by 8 percent from its current level of 3.5 children per woman.

Yet even deeply rooted beliefs are subject to reinterpretation. In Iran, another Muslim society, fertility rates have dropped from seven children per family to just over four in less than three decades. The trend is due in some measure to a change of heart among the government's religious authorities, who had become increasingly concerned about the likely effects of a population that was growing at more than 3 percent per year. In 1994, at the International Conference on Population and Development (ICPD) held in Cairo, the Iranian delegation released a "National Report on Population" which argued that according to the "quotations from prophet Mohammad ... and verses of [the] holy Quran, what is standing at the top priority for the Muslims' community is the social welfare of Muslims." Family planning, therefore, "not only is not prohibited but is emphasized by religion."

Promotional campaigns can also change people's assumptions and behavior, if the campaigns fit into the local social context. Perhaps the most successful effort of this kind is in Thailand, where Mechai Viravidaiya, the founder of the Thai Population and Community Development Association, started a program that uses witty songs, demonstrations, and ads to encourage the use of contraceptives. The program has helped foster widespread awareness of family planning throughout Thai society. Teachers use population-related examples in their math classes; cab drivers even pass out condoms. Such efforts have paid off: in less than three decades, contraceptive use among married couples has risen from 8 to 75 percent and population growth has slowed from over 3 percent to about 1 percent—the same rate as in the United States.

Better media coverage may be another option. In Bangladesh, a recent study found that while local journalists recognize the importance of family planning, they do not understand population issues well enough to cover them effectively and objectively. The study, a collaboration between the University Research Corporation of Bangladesh and Johns Hopkins University in the United States, recommended five ways to improve coverage: develop easy-to-use information for journalists (press releases, wall charts, research summaries), offer training and workshops, present awards for population journalism, create a forum for communication between journalists and family planning professionals, and establish a population resource center or data bank.

Often, however, the demand for large families is so tightly linked to social conditions that the conditions themselves must be viewed as part of the problem. Of course, those conditions vary greatly from one society to the next, but there are some common points of leverage:

Reducing child mortality helps give parents more confidence in the future of the children they already have. Among the most effective ways of reducing mortality are child immunization programs, and the promotion of "birth spacing"—lengthening the time between births. (Children born less than a year and a half apart are twice as likely to die as those born two or more years apart.)

Improving the economic situation of women provides them with alternatives to child-bearing. In some countries, officials could reconsider policies or customs that limit women's job opportunities or other economic rights, such as the right to inherit property. Encouraging "micro-leaders" such as Bangladesh's Grameen Bank can also be an effective tactic. In Bangladesh, the Bank has made loans to well over a million villagers—mostly impoverished women—to help them start or expand small businesses.

Improving education tends to delay the average age of marriage and to further the two goals just mentioned. Compulsory school attendance for children undercuts the economic incentive for larger families by reducing the opportunities for child labor. And in just about every society, higher levels of education correlate strongly with small families.

Momentum: The Biggest Threat of All

The most important factor in population growth is the hardest to counter—and to understand. Population momentum can be easy to overlook because it isn't directly captured by the statistics that attract the most attention. The global growth rate, after all, is dropping: in the mid-1960s, it amounted to about a 2.2 percent annual increase; today the figure is 1.4 percent. The fertility rate is dropping too: in 1950, women bore an average of five children each; now they bear roughly three. But despite these continued declines, the absolute number of births won't taper off any time soon. According to U.S. Census Bureau estimates, some 130 million births will still occur annually for the next 25 years, because of the sheer number of women coming into their child-bearing years.

The effects of momentum can be seen readily in a country like Bangladesh, where more than 42 percent of the population is under 15 years old—a typical proportion for many poor countries. Some 82 percent of the population growth projected for Bangladesh over the next half century will be caused by momentum. In other words, even if from now on, every Bangladeshi couple were to have only two children, the country's population would still grow by 80 million by 2050 simply because the next reproductive generation is so enormous.

The key to reducing momentum is to delay as many births as possible. To understand why delay works, it's helpful to think of momentum as a kind of human accounting problem in which a large number of births in the near term won't be balanced by a corresponding number of deaths over the same period of time. One side of the population ledger will contain those 130 million annual births (not all of which are due to momentum, of course), while the other side will contain only about 50 million annual deaths. So to put the matter in a morbid light, the longer a substantial number of those births can be delayed, the longer the death side of the balance sheet will be when the births eventually occur. In developing countries, according to the Population Council's Bongaarts, an average 2.5-year delay in the age when a woman bears her first child would reduce population growth by over 10 percent.

One way to delay childbearing is to postpone the age of marriage. In Bangladesh, for instance, the median age of first marriage among women rose from 14.4 in 1951 to 18 in 1989, and the age at first birth followed suit. Simply raising the legal age of marriage may be a useful tactic in countries that permit marriage among the very young. Educational improvements, as already mentioned, tend to do the same thing. A survey of 23 developing countries found that the median age of marriage for women with secondary education exceeded that of women with no formal education by four years.

Another fundamental strategy for encouraging later childbirth is to help women break out of the "sterilization syndrome" by providing and promoting high-quality, temporary contraceptives. Sterilization might appear to be the ideal form of contraception because it's permanent. But precisely because it is permanent, women considering sterilization tend to have their children early, and then resort to it. A family planning program that relies heavily on sterilization may therefore be working at cross purposes with itself: when offered as a primary form of contraception, sterilization tends to promote early childbirth.

What Happened to the Cairo Pledges?

At the 1994 Cairo Conference, some 180 nations agreed on a 20-year reproductive health package to slow population growth. The agreement called for a progressive rise in annual funding over the life of the package; according to U.N. estimates, the annual price tag would come to about $17 billion by 2000 and $21.7 billion by 2015. Developing countries agreed to pay for two thirds of the program, while the developed countries were to pay for the rest. On a global scale, the package was fairly modest: the annual funding amounts to less than two weeks' worth of global military expenditures.

Today, developing country spending is largely on track with the Cairo agreement, but the developed countries are not keeping their part of the bargain. According to a recent study by the U.N. Population Fund (UNFPA), all forms of developed country assistance (direct foreign aid, loans from multilateral agencies, foundation grants, and so on) amounted to only $2 billion in 1995. That was a 24 percent increase over the previous year, but preliminary estimates indicate that support declined some 18 percent in 1996 and last year's funding levels were probably even lower than that.

The United States, the largest international donor to population programs, is not only failing to meet its Cairo commitments, but is toying with a policy that would undermine international family planning efforts as a whole. Many members of the U.S. Congress are seeking reimposition of the "Mexico City Policy" first enunciated by President Ronald Reagan at the 1984 U.N. population conference in Mexico City, and repealed by the Clinton administration in 1993. Essentially, a resurrected Mexico City Policy would extend the current U.S. ban on funding abortion services to a ban on funding any organization that:

- funds abortions directly, or
- has a partnership arrangement with an organization that funds abortions, or
- provides legal services that may facilitate abortions, or
- engages in any advocacy for the provision of abortions, or
- participates in any policy discussions about abortion, either in a domestic or international forum.

The ban would be triggered even if the relevant activities were paid for entirely with non-U.S. funds. Because of its draconian limits even on speech, the policy has been dubbed the "Global Gag Rule" by its critics, who fear that it could stifle, not just abortion services, but many family planning operations involved only incidentally with abortion. Although Mexico City proponents have not managed to enlist enough support to reinstate the policy, they have succeeded in reducing U.S. family planning aid from $547 million in 1995 to $385 million in 1997. They have also imposed an unprecedented set of restrictions that meter out the money at the rate of 8 percent of the annual budget per month—a tactic that *Washington Post* reporter Judy Mann calls "administrative strangulation."

If the current underfunding of the Cairo program persists, according to the UNFPA study, 96 million fewer couples will use modern contraceptives in 2000 than if commitments had been met. One-third to one-half of these couples will resort to less effective traditional birth control methods; the rest will not use any contraceptives at all. The result will be an additional 122 million unintended pregnancies. Over half of those pregnancies will end in births, and about 40 percent will end in abortions. (The funding shortfall is expected to produce 16 million more abortions in 2000 alone.) The unwanted pregnancies will kill about 65,000 women by 2000, and injure another 844,000.

Population funding is always vulnerable to the illusion that the falling growth rate means the problem is going away. Worldwide, the annual population increase had dropped from a high of 87 million in 1988 to 80 million today. But dismissing the problem with that statistic is like comforting someone stuck on a railway crossing with the news that an oncoming train has slowed from 87 to 80 kilometers an hour, while its weight has increased. It will now take 12.5 years instead of 11.5 years to add the next billion people to the world. But that billion will surely arrive—and so will at least one more billion. Will still more billions follow? That, in large measure, depends on what policymakers do now. Funding alone will not ensure that population stabilizes, but lack of funding will ensure that it does not.

The Next Doubling

In the wake of the Cairo conference, most population programs are broadening their focus to include improvements in education, women's health, and women's social status among their many goals. These goals are worthy in their own right and they will ultimately be necessary for bringing population under control. But global population growth has gathered so much momentum that it could simply overwhelm a development agenda. Many countries now have little choice but to tackle their population problem in as direct a fashion as possible—even if that means temporarily ignoring other social problems. Population growth is now a global social emergency. Even as officials in both developed and developing countries open up their program agendas, it is critical that they not neglect their single most effective tool for dealing with that emergency: direct expenditures on family planning.

The funding that is likely to be the most useful will be constant, rather than sporadic. A fluctuating level of commitment, like sporadic condom use, can

end up missing its objective entirely. And wherever it's feasible, funding should be designed to develop self-sufficiency—as, for instance, with UNFPA's $1 million grant to Cuba, to build a factory for making birth control pills. The factory, which has the capacity to turn out 500 million tablets annually, might eventually even provide the country with a new export product. Self-sufficiency is likely to grow increasingly important as the fertility rate continues to decline. As Tom Merrick, senior population advisor at the World Bank explains, "while the need for contraceptives will not go away when the total fertility rate reaches two—the donors will."

Even in narrow, conventional economic terms, family planning offers one of the best development investments available. A study in Bangladesh showed that for each birth prevented, the government spends $62 and saves $615 on social services expenditures—nearly a tenfold return. The study estimated that the Bangladesh program prevents 890,000 births a year, for a net annual savings of $547 million. And that figure does not include savings resulting from lessened pressure on natural resources.

Over the past 40 years, the world's population has doubled. At some point in the latter half of the next century, today's population of 5.9 billion could double again. But because of the size of the next reproductive generation, we probably have only a relatively few years to stop that next doubling. To prevent all of the damage—ecological, economic, and social—that the next doubling is likely to cause, we must begin planning the global family with the same kind of urgency that we bring to matters of trade, say, or military security. Whether we realize it or not, our attempts to stabilize population—or our failure to act—will likely have consequences that far outweigh the implications of the military or commercial crisis of the moment. Slowing population growth is one of the greatest gifts we can offer future generations.

Jennifer D. Mitchell is a staff researcher at the Worldwatch Institute.

SCIENCE AND TECHNOLOGY

A turning-point for AIDS?

The impact of the global AIDS epidemic has been catastrophic, but many of the remedies are obvious. It is now a question of actually doing something

DURBAN

WHEN Thabo Mbeki, South Africa's president, opened the world AIDS conference in Durban on July 9th, he was widely expected to admit that he had made a mistake. Mr Mbeki has been flirting with the ideas of a small but vociferous group of scientists who, flying in the face of all the evidence, maintain that AIDS is not caused by the human immunodeficiency virus (HIV). His speech at the opening ceremony would have been the ideal opportunity for a graceful climbdown. Instead, he blustered and prevaricated, pretending that there was a real division of opinion among scientists about the matter, and arguing that the commission that he has appointed to look into this non-existent division would resolve it.

AIDS is the most political disease around. People talk a lot about AIDS "exceptionalism", and in many ways it is exceptional. For a start, it is difficult to think of another disease that would have brought the host country's head of state out of his office to open a conference, confused though his ideas may be. It is also exceptional, in modern times, in the attitudes of the healthy towards the infected. Illness usually provokes sympathy. But in many parts of the world those who have HIV are treated rather as lepers were in biblical days. Indeed, Gugu Dlamini, a community activist in KwaZulu Natal, the South African province in which Durban lies, was stoned to death by her neighbours when she revealed that she had the virus. In many parts of the world, as Kevin De Cock, of America's Centres for Disease Control (CDC), pointed out to the conference, attitudes to AIDS can be summed up in four words: silence, stigma, discrimination and denial. Mr Mbeki himself is at least guilty of denial.

Nevertheless, the fact that this, the 13th such AIDS conference, was held in Africa shows that some progress is being made. The previous conference, in Geneva in 1998, claimed to be "bridging the gap" between the treatment of the disease in the rich and poor worlds. It did no such thing. This one set as its goal to "break the silence". It may have succeeded. What needed to be shouted from the rooftops was that, contrary to some popular views, AIDS is not primarily a disease of gay western men or of intravenous drug injectors. It is a disease of ordinary people leading ordinary lives, except that most of them happen to live in a continent, Africa, that the rich countries of the world find it easy to ignore.

In some places the problem is so bad that it is hard to know where to begin. According to United Nations estimates, 25m of the 34m infected people in the world live in Africa. In absolute terms, South Africa has the most cases (4m, or about 20% of the adult population), but several of its neighbours have even worse ratios. In Botswana, for example, 36% of the adult population is now infected with HIV. Barring some currently unimaginable treatment—unimaginable in both efficacy and cost—almost all of these people will die as a result.

The hydra-headed monster

And mere numbers are not the only issue. People talk, rhetorically, of waging war on diseases. In the case of AIDS, the rhetoric could be inverted, for the effects of the illness on human populations are similar to those of war. Most infectious diseases tend to kill infants and the old. AIDS, like war, kills those in the prime of life. Indeed, in one way it is worse than war. When armies fight, it is predominantly young men who are killed. AIDS kills young women, too.

The result is social dislocation on a grand scale. As the diagram on the next page shows, the age-distribution of Botswana's population will change from the "pyramid" that is typical of countries with rapidly growing populations, to a "chimney-shaped" graph from which the young have been lopped out. Ten years from now, according to figures released at the conference by USAID, the American government's agency for international development, the life expectancy of somebody born in Botswana will have fallen to 29. In 20 years' time, the old will outnumber the middle-aged. Nor are things much better in other countries in southern Africa. In Zimbabwe and Namibia, two of Botswana's neighbours, life expectancy in 2010 will be 33. In South Africa it will be 35.

The destruction of young adults means that AIDS is creating orphans on an unprecedented scale. There are 11.2m of them, of whom 10.7m live in Africa. On top of that, vast numbers of children are infected as they are born. These are the exception to the usual rule that infants do not get the disease. Children are rarely infected in the womb, but they may acquire the virus from their mothers' vaginal fluids when they are born, or from breast milk. More than 5m children are reckoned to have been infected in this way. Almost 4m of them are already dead.

It sounds hopeless. And yet it isn't. Two African countries, Uganda and Senegal, seem to have worked out how to cope with the disease. Their contrasting experiences serve both as a warning and as a lesson to other countries in the world, particularly those in Asia that now have low infection rates and may be feeling complacently smug about them. The warning: act early, or you will be sorry. The lesson: it is, even so, never too late to act. Senegal began its anti-AIDS programme in 1986, before the virus had got a proper grip. It has managed to keep its infection rate below 2%. Uganda began its programme in the early 1990s, when 14% of the adult population was already infected. Now that figure is down to 8% and falling. In these two countries, the epidemic seems to have been stopped in its tracks.

As Roy Anderson, a noted epidemiologist from Oxford University, pointed out to the conference, stopping an epidemic requires one thing: that the average number of people infected by somebody who already has the disease be less than one. For a sexually transmitted disease, this average has three components: the "transmissibility" of the disease, the average rate that an infected person acquires new and uninfected partners, and the average length of time for which somebody is infectious.

Cutting off the hydra's heads

The easiest of these to tackle has been transmissibility. Surprisingly, perhaps, AIDS is not all that easily transmissible compared with other diseases. But there are three ways—one certain, one as yet a pious hope, and one the subject of some controversy—to reduce the rate of transmission between adults still further. The first is to use condoms. The second is to develop a microbicide that will kill the virus in the vagina. And the third is to treat other sexually transmitted diseases.

Both Senegal and Uganda have been strong on the use of condoms. In Senegal, for example, the annual number of condoms used rose from 800,000 in 1988 to 9m in 1997. Nevertheless, it still takes a lot of encouragement to persuade people to use them. Partly, this is a question of discounting the future. For decades African lives have been shorter, on average, than those in the rest of the world. With AIDS, they are getting shorter still. A Botswanan who faces the prospect of death before his 30th birthday is likely to be more reckless than an American who can look forward to well over twice that lifespan; a short life might as well be a merry one.

There is also the question of who wears the condom. Until recently, there was no choice. Only male condoms were available. And women in many parts of Africa are in a weak negotiating position when it comes to in-

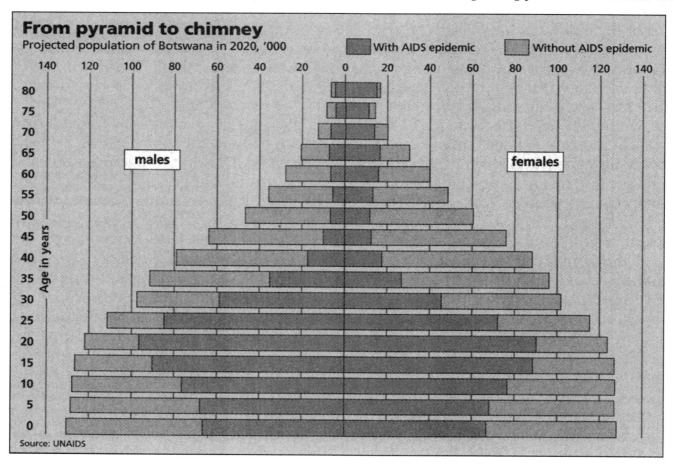

38. Turning-Point for AIDS?

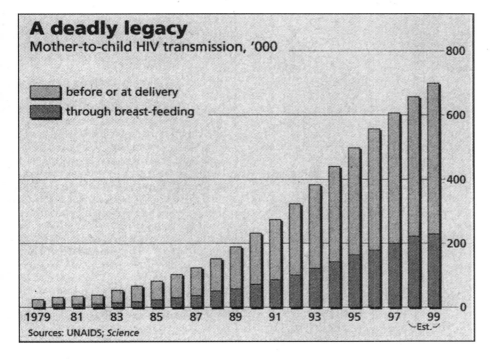

A deadly legacy
Mother-to-child HIV transmission, '000
- before or at delivery
- through breast-feeding

Sources: UNAIDS; *Science*

sisting that a man put one on. The best way out of this is to alter the balance of power. That, in general, means more and better education, particularly for girls. This, too, has been an important component of the Senegalese and Ugandan anti-AIDS programmes. A stop-gap, though, is the female condom, a larger version of the device that fits inside the vagina, which is proving surprisingly popular among groups such as Nairobi prostitutes. But an even less intrusive—and to a man invisible—form of protection would be a vaginal microbicide that kills the virus before it can cross the vaginal wall.

Here, however, the news is bad. Much hope had been pinned on a substance called nonoxynol-9 (the spermicide used to coat condoms that are intended to prevent pregnancy rather than disease). Unfortunately, the results of a major United Nations trial announced at the conference have confirmed the suspicion that nonoxynol-9 does not work against HIV. So researchers have gone back to the drawing-board and are searching for suitable (and suitably cheap) substances among the cast-offs from antiviral drugs used to treat AIDS in rich countries.

More equivocal is the value of treating other sexually transmitted diseases as a way of preventing the transmission of HIV. Clearly, such treatment is a good thing in its own right. But a study carried out a few years ago at Mwanza, Tanzania, suggested that it also stymies HIV. That would not be surprising, since the vaginal lesions that other venereal diseases produce should make excellent entry points for the virus. Yet a more recent study at Rakai, Uganda, suggests that other venereal diseases make no difference; the matter is now a subject of much debate.

Sex is not the only way that HIV is transmitted. Infected mothers can give it to their children. But here, too, transmissibility can be reduced dramatically.

A first way of doing this is to test pregnant women to see if they have the virus. If they do, they are unlikely to pass it on to the fetus in the womb, but they are quite likely to do so in the act of giving birth. According to figures presented to the conference by Ruth Nduati, of the University of Nairobi, up to 40% of children born to untreated infected women catch AIDS this way. But that number can be reduced drastically—to around 20%—by giving infected pregnant women a short course of an antiviral drug just before they give birth.

Until recently, the preferred drug was AZT. Many African governments balked at using this because, although it is cheap by western standards, it can stretch African health budgets to breaking point. However, recent studies carried out in Kenya and South Africa have shown that an even cheaper drug called nevirapine will do just as well. A course of this costs $4, still a fair whack for an impoverished country, but worth it both for the life of a child and for the cost-saving of not having to treat that child's subsequent illness.

Once safely born, the child of an infected mother is still not out of the woods. This is because it can be infected via its mother's milk. Oddly, this is a more intractable problem than transmission at birth. Nobody knows (because nobody has tried to find out) if carrying on with AZT or nevirapine would keep a mother's milk virus-free. But the cost would be prohibitive anyway. The only alternative is not to breast-feed.

That may sound easy, but it is not. First, the formula milk that could substitute for breast milk costs money. Second, unless clean water is available to mix with it, the result is likely to be a diarrhoeal disease that may kill the child anyway. And third, by failing to breast-feed, a mother in many parts of Africa is in effect announcing that she has the virus, and thus exposing herself to both stigma and discrimination. Not breast-feeding is, nevertheless, an effective addition to pre-natal antiviral drugs. According to Dr Nduati, combining both methods can bring the infection rate below 8%.

The second of Dr Anderson's criteria, the rate of acquisition of new and uninfected partners, is critical to the speed with which AIDS spreads, but is also far harder to tackle. The reasons why AIDS has spread faster in some places than in others are extremely complicated. But one important factor is so-called disassortative mating.

As far as is known, all AIDS epidemics start with the spread of the disease in one or more small, high-risk groups. These groups include prostitutes and their clients, male homosexuals and injecting drug users. The rate at which an epidemic spreads to lower-risk groups depends a great deal on whether different groups mate mainly among themselves (assortative mating) or whether they mate a lot with other people (disassortative mating). The more disassortative mating there is, the faster the virus will spread.

Sub-Saharan Africa and the Caribbean (the second-worst affected part of the world) have particularly high levels of disassortative mating between young girls and older men. And in an area where AIDS is already highly prevalent, older men are a high-risk group; they are far more likely to have picked up the virus than younger ones. This helps to explain why the rate of infection is higher in young African women than it is in young African men.

191

Inter-generational churning may thus, according to Dr Anderson's models, go a long way towards explaining why Africa and the Caribbean have the highest levels of HIV infection in the world. And it suggests that, as with condom use, a critical part of any anti-AIDS campaign should be to give women more power. In many cases, young women are coerced or bribed into relationships with older men. This would diminish if girls were better educated—not least because they would then find it easier to earn a living.

The partial explanation for Africa's plight that disassortative mating provides should not, however, bring false comfort in other areas. Dr Anderson's models suggest that lower levels of disassortative mating cannot stop an epidemic, they merely postpone it. Those countries, such as Ukraine, where HIV is spreading rapidly through a high-risk group (in Ukraine's case, injecting drug users), need to act now, even if the necessary action, such as handing out clean needles, is politically distasteful. Countries such as India, where lower-risk groups are starting to show up in the statistics, and where the prevalence rates in some states are already above 2%, needed to act yesterday, and to aim their message more widely. The example of Senegal (and, indeed, the strongly worded, morally neutral advertising campaigns conducted in many western countries in the 1980s), shows the value of early action as surely as do Dr Anderson's models.

To tackle the third element of those models—the length of time that somebody is infectious—really requires a vaccine. Drugs can reduce it to some extent, by bringing people's viral load down to the point where they will not pass on the disease. But effective therapies are currently expensive and, despite the widespread demands at the conference for special arrangements that would lower their price in poor countries, are unlikely to become cheap enough for routine use there for some time. On top of that, if drugs are used carelessly, resistant strains of the virus can emerge, rendering the therapies useless. A study by the CDC, published to coincide with the conference, showed resistant strains in the blood of three-quarters of the participants in a United Nations AIDS drug-access programme in Uganda.

The search for a vaccine

Vaccines are not immune to the emergence of resistant strains. But they are one-shot treatments and so are not subject to the whims of patient compliance with complex drug regimes. Non-compliance is the main cause of the emergence of resistant strains, since the erratic consumption of a particular drug allows populations of resistant viruses to evolve and build up.

In total, 21 clinical trials of vaccines are happening around the world, but only five are taking place in poor countries, and only two are so-called phase 3 trials that show whether a vaccine will work effectively in the real world. Preliminary results from these two trials, which are being conducted by an American company called VaxGen, are expected next year. They are eagerly awaited, for even a partially effective vaccine could have a significant impact on the virus's spread. A calculation by America's National Institutes of Health shows that, over the course of a decade, a 60%-effective vaccine introduced now would stop nearly twice as many infections as a 90%-effective one introduced five years hence.

Even then, there is the question of cost. This is being addressed by the International AIDS Vaccine Initiative (IAVI), a New York-based charity. IAVI, according to its boss Seth Berkley, acts like a venture-capital firm. At the moment, that capital amounts to about $100m, gathered from various governments and foundations. IAVI provides small firms with seed money to develop new products, but instead of demanding a share of the equity in return, it requires that the eventual product, if any, should be sold at a low profit margin—about 10%. If a sponsored firm breaks this arrangement, IAVI can give the relevant patents to anybody it chooses. At the moment, IAVI has four such partnerships, and it chose the conference to announce that one—a collaboration with the Universities of Oxford and Nairobi—has just received regulatory approval and will start trials in September.

None of these things alone will be enough to stop the epidemic in its tracks, but in combination they may succeed. And one last lesson from Dr Anderson's equations is not to give up just because a policy does not seem to be working. Those equations predict that applying a lot of effort to an established epidemic will have little initial effect. Then, suddenly, infection rates will drop fast. The message is: "hang in there". AIDS may be exceptional, but it is not that exceptional. Good science and sensible public policy can defeat it. There is at least a glimmer of hope.

The End of Cheap Oil

Global production of conventional oil will begin to decline sooner than most people think, probably within 10 years

by Colin J. Campbell and Jean H. Laherrère

In 1973 and 1979 a pair of sudden price increases rudely awakened the industrial world to its dependence on cheap crude oil. Prices first tripled in response to an Arab embargo and then nearly doubled again when Iran dethroned its Shah, sending the major economies sputtering into recession. Many analysts warned that these crises proved that the world would soon run out of oil. Yet they were wrong.

Their dire predictions were emotional and political reactions; even at the time, oil experts knew that they had no scientific basis. Just a few years earlier oil explorers had discovered enormous new oil provinces on the north slope of Alaska and below the North Sea off the coast of Europe. By 1973 the world had consumed, according to many experts' best estimates, only about one eighth of its endowment of readily accessible crude oil (so-called conventional oil). The five Middle Eastern members of the Organization of Petroleum Exporting Countries (OPEC) were able to hike prices not because oil was growing scarce but because they had managed to corner 36 percent of the market. Later, when demand sagged, and the flow of fresh Alaskan and North Sea oil weakened OPEC's economic stranglehold, prices collapsed.

The next oil crunch will not be so temporary. Our analysis of the discovery and production of oil fields around the world suggests that within the next decade, the supply of conventional oil will be unable to keep up with demand. This conclusion contradicts the picture one gets from oil industry reports, which boasted of 1,020 billion barrels of oil (Gbo) in "proved" reserves at the start of 1998. Dividing that figure by the current production rate of about 23.6 Gbo a year might suggest that crude oil could remain plentiful and cheap for 43 more years—probably longer, because official charts show reserves growing.

Unfortunately, this appraisal makes three critical errors. First, it relies on distorted estimates of reserves. A second mistake is to pretend that production will remain constant. Third and most important, conventional wisdom erroneously assumes that the last bucket of oil can be pumped from the ground just as quickly as the barrels of oil gushing from wells today. In fact, the rate at which any well—or any country—can produce oil always rises to a maximum and then, when about half the oil is gone, begins falling gradually back to zero.

From an economic perspective, when the world runs completely out of oil is thus not directly relevant: what matters is when production begins to taper off. Beyond that point, prices will rise unless demand declines commensurately. Using several different techniques to estimate the current reserves of conventional oil and the amount still left to be discovered, we conclude that the decline will begin before 2010.

Digging for the True Numbers

We have spent most of our careers exploring for oil, studying reserve figures and estimating the amount of oil left to discover, first while employed at major oil companies and later as independent consultants. Over the years, we have come to appreciate that the relevant statistics are far more complicated than they first appear.

Consider, for example, three vital numbers needed to project future oil production. The first is the tally of how much oil has been extracted to date, a figure known as cumulative production. The second is an estimate of reserves, the amount that companies can pump out of known oil fields before having to abandon them. Finally, one must have an educated guess at the quantity of conventional oil that remains to be discovered and exploited. Together they add up to ultimate recovery, the total number of barrels that will have been extracted when production ceases many decades from now.

The obvious way to gather these numbers is to look them up in any of several publications. That approach works well enough for cumulative production statistics because companies meter the oil as it flows from their wells. The record of production is not perfect (for example, the two billion barrels of Kuwaiti oil wastefully burned by Iraq in 1991 is usually not included in official statistics), but errors are relatively easy to spot and rectify. Most experts agree that the industry had removed just over 800 Gbo from the earth at the end of 1997.

Getting good estimates of reserves is much harder, however. Almost all the publicly available statistics are taken from surveys conducted by the *Oil and Gas Journal* and *World Oil*. Each year these two trade journals query oil firms and governments around the world. They then publish whatever production and reserve numbers they receive but are not able to verify them.

The results, which are often accepted uncritically, contain systematic errors. For one, many of the reported figures are unrealistic. Estimating reserves is an inexact science to begin with, so petroleum engineers assign a probability to their assessments. For example, if, as geologists estimate, there is a 90 percent chance that the Oseberg field in Norway contains 700 million barrels of recoverable oil but only a 10 percent chance that it will yield 2,500 million more barrels, then the lower figure should be cited as the so-called P90 estimate (P90 for "probability 90 percent") and the higher as the P10 reserves.

In practice, companies and countries are often deliberately vague about the likelihood of the reserves they report, preferring instead to publicize whichever figure, within a P10 to P90 range, best suits them. Exaggerated estimates can, for instance, raise the price of an oil company's stock.

The members of OPEC have faced an even greater temptation to inflate their reports because the higher their reserves, the more oil they are allowed to export. National companies, which have exclusive oil rights in the main OPEC countries, need not (and do not) release detailed statistics on each field that could be used to verify the country's total reserves. There is thus good reason to suspect that when, during the late

5 ❖ POPULATION, RESOURCES, AND SOCIOECONOMIC DEVELOPMENT

COURTESY OF THE SOCIETY OF EXPLORATION GEOPHYSICISTS

FLOW OF OIL starts to fall from any large region when about half the crude is gone. Adding the output of fields of various sizes and ages (*bottom curves at right*) usually yields a bell-shaped production curve for the region as a whole. M. King Hubbert (*left*), a geologist with Shell Oil, exploited this fact in 1956 to predict correctly that oil from the lower 48 American states would peak around 1969.

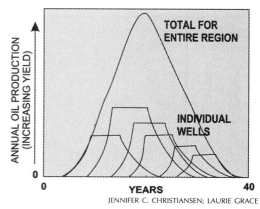

JENNIFER C. CHRISTIANSEN; LAURIE GRACE

1980s, six of the 11 OPEC nations increased their reserve figures by colossal amounts, ranging from 42 to 197 percent, they did so only to boost their export quotas.

Previous OPEC estimates, inherited from private companies before governments took them over, had probably been conservative, P90 numbers. So some upward revision was warranted. But no major new discoveries or technological breakthroughs justified the addition of a staggering 287 Gbo. That increase is more than all the oil ever discovered in the U.S.—plus 40 percent. Non-OPEC countries, of course, are not above fudging their numbers either: 59 nations stated in 1997 that their reserves were unchanged from 1996. Because reserves naturally drop as old fields are drained and jump when new fields are discovered, perfectly stable numbers year after year are implausible.

Unproved Reserves

Another source of systematic error in the commonly accepted statistics is that the definition of reserves varies widely from region to region. In the U.S., the Securities and Exchange Commission allows companies to call reserves "proved" only if the oil lies near a producing well and there is "reasonable certainty" that it can be recovered profitably at current oil prices, using existing technology. So a proved reserve estimate in the U.S. is roughly equal to a P90 estimate.

Regulators in most other countries do not enforce particular oil-reserve definitions. For many years, the former Soviet countries have routinely released wildly optimistic figures—essentially P10 reserves. Yet analysts have often misinterpreted these as estimates of "proved" reserves. *World Oil* reckoned reserves in the former Soviet Union amounted to 190 Gbo in 1996, whereas the *Oil and Gas Journal* put the number at 57 Gbo. This large discrepancy shows just how elastic these numbers can be.

Using only P90 estimates is not the answer, because adding what is 90 percent likely for each field, as is done in the U.S., does not in fact yield what is 90 percent likely for a country or the entire planet. On the contrary, summing many P90 reserve estimates always understates the amount of proved oil in a region. The only correct way to total up reserve numbers is to add the mean, or average, estimates of oil in each field. In practice, the median estimate, often called "proved and probable," or P50 reserves, is more widely used and is good enough. The P50 value is the number of barrels of oil that are as likely as not to come out of a well during its lifetime, assuming prices remain within a limited range. Errors in P50 estimates tend to cancel one another out.

We were able to work around many of the problems plaguing estimates of conventional reserves by using a large body of statistics maintained by Petroconsultants in Geneva. This information, assembled over 40 years from myriad sources, covers some 18,000 oil fields worldwide. It, too, contains some dubious reports, but we did our best to correct these sporadic errors.

According to our calculations, the world had at the end of 1996 approximately 850 Gbo of conventional oil in P50 reserves—substantially less than the 1,019 Gbo reported in the *Oil and Gas Journal* and the 1,160 Gbo estimated by *World Oil*. The difference is actually greater than it appears because our value represents the amount most likely to come out of known oil fields, whereas the larger number is supposedly a cautious estimate of proved reserves.

For the purposes of calculating when oil production will crest, even more critical than the size of the world's reserves is the size of ultimate recovery—all the cheap oil there is to be had. In order to estimate that, we need to know whether, and how fast, reserves are moving up or down. It is here that the official statistics become dangerously misleading.

Diminishing Returns

According to most accounts, world oil reserves have marched steadily upward over the past 20 years. Extending that apparent trend into the future, one could easily conclude, as the U.S. Energy Information Administration has, that oil production will continue to rise unhindered for decades to come, increasing almost two thirds by 2020.

Such growth is an illusion. About 80 percent of the oil produced today flows

EARTH'S CONVENTIONAL CRUDE OIL is almost half gone. Reserves (defined here as the amount as likely as not to come out of known fields) and future discoveries together will provide little more than what has already been burned.

UNDISCOVERED: 150 BILLION BARRELS   RESERVES: 850 BILLION BARRELS

194

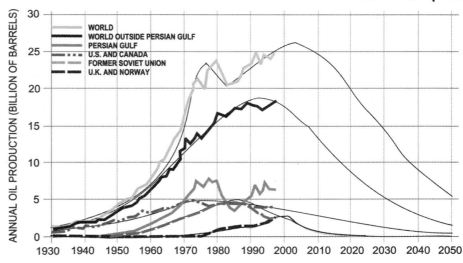

GLOBAL PRODUCTION OF OIL, both conventional and unconventional, recovered after falling in 1973 and 1979. But a more permanent decline is less than 10 years away, according to the authors' model, based in part on multiple Hubbert curves (*thin lines*). U.S. and Canadian oil topped out in 1972; production in the former Soviet Union has fallen 45 percent since 1987. A crest in the oil produced outside the Persian Gulf region now appears imminent.

from fields that were found before 1973, and the great majority of them are declining. In the 1990s oil companies have discovered an average of seven Gbo a year; last year they drained more than three times as much. Yet official figures indicated that proved reserves did not fall by 16 Gbo, as one would expect—rather they expanded by 11 Gbo. One reason is that several dozen governments opted not to report declines in their reserves, perhaps to enhance their political cachet and their ability to obtain loans. A more important cause of the expansion lies in revisions: oil companies replaced earlier estimates of the reserves left in many fields with higher numbers. For most purposes, such amendments are harmless, but they seriously distort forecasts extrapolated from published reports.

To judge accurately how much oil explorers will uncover in the future, one has to backdate every revision to the year in which the field was first discovered—not to the year in which a company or country corrected an earlier estimate. Doing so reveals that global discovery peaked in the early 1960s and has been falling steadily ever since. By extending the trend to zero, we can make a good guess at how much oil the industry will ultimately find.

We have used other methods to estimate the ultimate recovery of conventional oil for each country [*see box*, "Earth's Conventional Crude Oil"] and we calculate that the oil industry will be able to recover only about another 1,000 billion barrels of conventional oil. This number, though great, is little more than the 800 billion barrels that have already been extracted.

It is important to realize that spending more money on oil exploration will not change this situation. After the price of crude hit all-time highs in the early 1980s, explorers developed new technology for finding and recovering oil, and they scoured the world for new fields. They found few: the discovery rate continued its decline uninterrupted. There is only so much crude oil in the world, and the industry has found about 90 percent of it.

Predicting the Inevitable

Predicting when oil production will stop rising is relatively straightforward once one has a good estimate of how much oil there is left to produce. We simply apply a refinement of a technique first published in 1956 by M. King Hubbert. Hubbert observed that in any large region, unrestrained extraction of a finite resource rises along a bell-shaped curve that peaks when about half the resource is gone. To demonstrate his theory, Hubbert fitted a bell curve to production statistics and projected that crude oil production in the lower 48 U.S. states would rise for 13 more years, then crest in 1969, give or take a year. He was right: production peaked in 1970 and has continued to follow Hubbert curves with only minor deviations. The flow of oil from several other regions, such as the former Soviet Union and the collection of all oil producers outside the Middle East, also follows Hubbert curves quite faithfully.

The global picture is more complicated, because the Middle East members of OPEC deliberately reined back their oil exports in the 1970s, while other nations continued producing at full capacity. Our analysis reveals that a number of the largest producers, including Norway and the U.K., will reach their peaks around the turn of the millennium unless they sharply curtail production. By 2002 or so the world will rely on Middle East nations, particularly five near the Persian Gulf (Iran, Iraq, Kuwait, Saudi Arabia and the United Arab Emirates), to fill in the gap between dwindling supply and growing demand. But once approximately 900 Gbo have been consumed, production must soon begin to fall. Barring a global recession, it seems most likely that world production of conventional oil will peak during the first decade of the 21st century.

Perhaps surprisingly, that prediction does not shift much even if our estimates are a few hundred billion barrels high or low. Craig Bond Hatfield of the University of Toledo, for example, has conducted his own analysis based on a 1991 estimate by the U.S. Geological Survey of 1,550 Gbo remaining—55 percent higher than our figure. Yet he similarly concludes that the world will hit maximum oil production within the next 15 years. John D. Edwards of the University of Colorado publish-

How Much Oil Is Left to Find?

We combined several techniques to conclude that about 1,000 billion barrels of conventional oil remain to be produced. First, we extrapolated published production figures for older oil fields that have begun to decline. The Thistle field off the coast of Britain, for example, will yield about 420 million barrels (*a*). Second, we plotted the amount of oil discovered so far in some regions against the cumulative number of exploratory wells drilled there. Because larger fields tend to be found first—they are simply too large to miss—the curve rises rapidly and then flattens, eventually reaching a theoretical maximum: for Africa, 192 Gbo. But the time and cost of exploration impose a more practical limit of perhaps 165 Gbo (*b*). Third, we analyzed the distribution of oil-field sizes in the Gulf of Mexico and other provinces. Ranked according to size and then graphed on a logarithmic scale, the fields tend to fall along a parabola that grows predictably over time. (*c*). (Interestingly, galaxies, urban populations and other natural agglomerations also seem to fall along such parabolas.) Finally, we checked our estimates by matching our projections for oil production in large areas, such as the world outside the Persian Gulf region, to the rise and fall of oil discovery in those places decades earlier (*d*).

—C.J.C. and J.H.L.

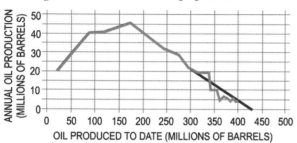

We can predict the amount of remaining oil from the decline of aging fields...

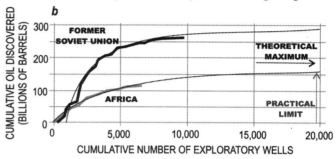

...from the diminishing returns on exploration in larger regions...

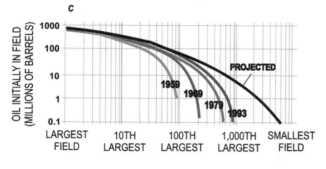

...by extrapolating the size of new fields into the future...

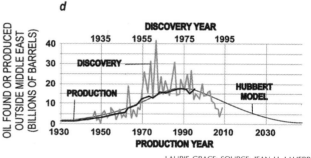

...and by matching production to earlier discovery trends.

LAURIE GRACE; SOURCE: JEAN H. LAHERRÈRE

---

ed last August one of the most optimistic recent estimates of oil remaining: 2,036 Gbo. (Edwards concedes that the industry has only a 5 percent chance of attaining that very high goal.) Even so, his calculations suggest that conventional oil will top out in 2020.

Smoothing the Peak

Factors other than major economic changes could speed or delay the point at which oil production begins to decline. Three in particular have often led economists and academic geologists to dismiss concerns about future oil production with naive optimism.

First, some argue, huge deposits of oil may lie undetected in far-off corners of the globe. In fact, that is very unlikely. Exploration has pushed the frontiers back so far that only extremely deep water and polar regions remain to be fully tested, and even their prospects are now reasonably well understood. Theoretical advances in geochemistry and geophysics have made it possible to map productive and prospective fields with impressive accuracy. As a result, large tracts can be condemned as barren. Much of the deepwater realm, for example, has been shown to be absolutely nonprospective for geologic reasons.

What about the much touted Caspian Sea deposits? Our models project that oil production from that region will grow until around 2010. We agree with analysts at the USGS World Oil Assessment program and elsewhere who rank the total resources there as roughly equivalent to those of the North Sea—that is, perhaps 50 Gbo but certainly not several hundreds of billions as sometimes reported in the media.

A second common rejoinder is that new technologies have steadily increased the fraction of oil that can be recovered from fields in a basin—the so-called recovery factor. In the 1960s oil companies assumed as a rule of thumb that only 30 percent of the oil in a field was typically recoverable; now they bank on an average of 40 or 50 percent. That progress will continue and will extend global reserves for many years to come, the argument runs.

Of course, advanced technologies will buy a bit more time before production starts to fall [see "Oil Production in the 21st Century," by Roger N. Anderson*]. But most of the apparent improvement in recovery factors is an artifact of reporting. As oil fields grow old, their owners often deploy newer technology to slow their decline. The falloff also allows engineers to gauge the size of the field more accurately and to correct previous underestimation—in particular P90 estimates that by definition were 90 percent likely to be exceeded.

Another reason not to pin too much hope on better recovery is that oil companies routinely count on technological pro-

39. End of Cheap Oil

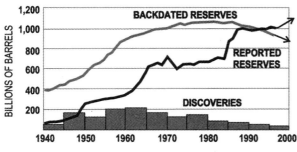

GROWTH IN OIL RESERVES since 1980 is an illusion caused by belated corrections to oil-field estimates. Back-dating the revisions to the year in which the fields were discovered reveals that reserves have been falling because of a steady decline in newfound oil (*bottom bars*).

gress when they compute their reserve estimates. In truth, advanced technologies can offer little help in draining the largest basins of oil, those onshore in the Middle East where the oil needs no assistance to gush from the ground.

Last, economists like to point out that the world contains enormous caches of unconventional oil that can substitute for crude oil as soon as the price rises high enough to make them profitable. There is no question that the resources are ample: the Orinoco oil belt in Venezuela has been assessed to contain a staggering 1.2 trillion barrels of the sludge known as heavy oil. Tar sands and shale deposits in Canada and the former Soviet Union may contain the equivalent of more than 300 billion barrels of oil [see "Mining for Oil," by Richard L. George*]. Theoretically, these unconventional oil reserves could quench the world's thirst for liquid fuels as conventional oil passes its prime. But the industry will be hard-pressed for the time and money needed to ramp up production of unconventional oil quickly enough.

Such substitutes for crude oil might also exact a high environmental price. Tar sands typically emerge from strip mines. Extracting oil from these sands and shales creates air pollution. The Orinoco sludge contains heavy metals and sulfur that must be removed. So governments may restrict these industries from growing as fast as they could. In view of these potential obstacles, our skeptical estimate is that only 700 Gbo will be produced from unconventional reserves over the next 60 years.

On the Down Side

Meanwhile global demand for oil is currently rising at more than 2 percent a year. Since 1985, energy use is up about 30 percent in Latin America, 40 percent in Africa and 50 percent in Asia. The Energy Information Administration forecasts that worldwide demand for oil will increase 60 percent (to about 40 Gbo a year) by 2020.

The switch from growth to decline in oil production will thus almost certainly create economic and political tension. Unless alternatives to crude oil quickly prove themselves, the market share of the OPEC states in the Middle East will rise rapidly. Within two years, these nations' share of the global oil business will pass 30 percent, nearing the level reached during the oil-price shocks of the 1970s. By 2010 their share will quite probably hit 50 percent.

The world could thus see radical increases in oil prices. That alone might be sufficient to curb demand, flattening production for perhaps 10 years. (Demand fell more than 10 percent after the 1979 shock and took 17 years to recover.) But by 2010 or so, many Middle Eastern nations will themselves be past the midpoint. World production will then have to fall.

With sufficient preparation, however, the transition to the post-oil economy need not be traumatic. If advanced methods of producing liquid fuels from natural gas can be made profitable and scaled up quickly, gas could become the next source of transportation fuel [see "Liquid Fuels from Natural Gas," by Safaa A. Fouda*]. Safer nuclear power, cheaper renewable energy, and oil conservation programs could all help postpone the inevitable decline of conventional oil.

Countries should begin planning and investing now. In November a panel of energy experts appointed by President Bill Clinton strongly urged the administration to increase funding for energy research by $1 billion over the next five years. That is a small step in the right direction, one that must be followed by giant leaps from the private sector.

The world is not running out of oil—at least not yet. What our society does face, and soon, is the end of the abundant and cheap oil on which all industrial nations depend.

The Authors

COLIN J. CAMPBELL and JEAN H. LAHERRÈRE have each worked in the oil industry for more than 40 years. After completing his Ph.D. in geology at the University of Oxford, Campbell worked for Texaco as an exploration geologist and then at Amoco as chief geologist for Ecuador. His decade-long study of global oil-production trends has led to two books and numerous papers. Laherrère's early work on seismic refraction surveys contributed to the discovery of Africa's largest oil field. At Total, a French oil company, he supervised exploration techniques worldwide. Both Campbell and Laherrère are currently associated with Petroconsultants in Geneva.

Further Reading

UPDATED HUBBERT CURVES ANALYZE WORLD OIL SUPPLY. L. F. Ivanhoe in *World Oil*, Vol. 217, No. 11, pages 91–94; November 1996.

THE COMING OIL CRISIS. Colin J. Campbell. Multi-Science Publishing and Petroconsultants, Brentwood, England, 1997.

OIL BACK ON THE GLOBAL AGENDA. Craig Bond Hatfield in *Nature*, Vol. 387, page 121; May 8, 1997.

---

*Editor's note: All of the articles mentioned in this article can be found in Scientific American, March 1998.

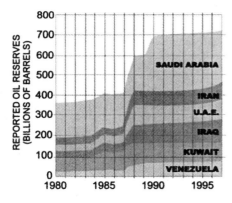

SUSPICIOUS JUMP in reserves reported by six OPEC members added 300 billion barrels of oil to official reserve tallies yet followed no major discovery of new fields.

Gray Dawn:
The Global Aging Crisis

Peter G. Peterson

DAUNTING DEMOGRAPHICS

THE LIST of major global hazards in the next century has grown long and familiar. It includes the proliferation of nuclear, biological, and chemical weapons, other types of high-tech terrorism, deadly super-viruses, extreme climate change, the financial, economic, and political aftershocks of globalization, and the violent ethnic explosions waiting to be detonated in today's unsteady new democracies. Yet there is a less-understood challenge—the graying of the developed world's population—that may actually do more to reshape our collective future than any of the above.

Over the next several decades, countries in the developed world will experience an unprecedented growth in the number of their elderly and an unprecedented decline in the number of their youth. The timing and magnitude of this demographic transformation have already been determined. Next century's elderly have already been born and can be counted—and their cost to retirement benefit systems can be projected.

Unlike with global warming, there can be little debate over whether or when global aging will manifest itself. And unlike with other challenges, even the struggle to preserve and strengthen unsteady new democracies, the costs of global aging will be far beyond the means of even the world's wealthiest nations—unless retirement benefit systems are radically reformed. Failure to do so, to prepare early and boldly enough, will spark economic crises that will dwarf the recent meltdowns in Asia and Russia.

40. Gray Dawn

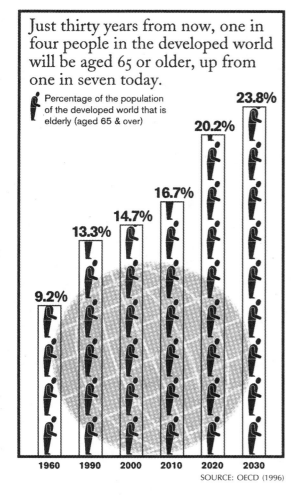

Just thirty years from now, one in four people in the developed world will be aged 65 or older, up from one in seven today.

Percentage of the population of the developed world that is elderly (aged 65 & over)

1960: 9.2%
1990: 13.3%
2000: 14.7%
2010: 16.7%
2020: 20.2%
2030: 23.8%

SOURCE: OECD (1996)

How we confront global aging will have vast economic consequences costing quadrillions of dollars over the next century. Indeed, it will greatly influence how we manage, and can afford to manage, the other major challenges that will face us in the future.

For this and other reasons, global aging will become not just the transcendent economic issue of the 21st century, but the transcendent political issue as well. It will dominate and daunt the public-policy agendas of developed countries and force the renegotiation of their social contracts. It will also reshape foreign policy strategies and the geopolitical order.

The United States has a massive challenge ahead of it. The broad outlines can already be seen in the emerging debate over Social Security and Medicare reform. But ominous as the fiscal stakes are in the United States, they loom even larger in Japan and Europe, where populations are aging even faster, birthrates are lower, the influx of young immigrants from developing countries is smaller, public pension benefits are more generous, and private pension systems are weaker.

Aging has become a truly global challenge, and must therefore be given high priority on the global policy agenda. A gray dawn fast approaches. It is time to take an unflinching look at the shape of things to come.

The Floridization of the developed world. Been to Florida lately? You may not have realized it, but the vast concentration of seniors there—nearly 19 percent of the population—represents humanity's future. Today's Florida is a demographic benchmark that every developed nation will soon pass. Italy will hit the mark as early as 2003, followed by Japan in 2005 and Germany in 2006. France and Britain will pass present-day Florida around 2016; the United States and Canada in 2021 and 2023.

Societies much older than any we have ever known. Global life expectancy has grown more in the last fifty years than over the previous five thousand. Until the Industrial Revolution, people aged 65 and over never amounted to more than 2 or 3 percent of the population. In today's developed world, they amount to 14 percent. By the year 2030, they will reach 25 percent and be closing in on 30 in some countries.

An unprecedented economic burden on working-age people. Early in the next century, working-age populations in most developed countries will shrink. Between 2000 and 2010, Japan, for example, will suffer a 25 percent drop in the number of workers under age 30. Today the ratio of working taxpayers to nonworking pensioners in the developed world is around 3:1. By 2030, absent reform, this ratio will fall to 1.5:1, and in some countries, such as Germany and Italy, it will drop all the way down to 1:1 or even lower. While the longevity revolution represents a miraculous triumph of modern medicine and the extra years of life will surely be treasured by the elderly and their families, pension plans and other retirement benefit programs were not designed to provide these billions of extra years of payouts.

The aging of the aged: the number of "old old" will grow much faster than the number of "young old." The United Nations projects that by 2050, the number of people aged 65 to 84 worldwide will grow from 400 million to 1.3 billion (a threefold increase), while the number of people aged 85 and over will grow from 26 million to 175 million (a sixfold increase)—and the number aged 100 and over from 135,000 to 2.2 million (a sixteenfold increase). The "old old" consume far more health care than the "young old"—about two to three times as much. For nursing-home care, the ratio is roughly 20:1. Yet little of this cost is figured in the official projections of future public expenditures.

Falling birthrates will intensify the global aging trend. As life spans increase, fewer babies are being born. As recently as the late 1960s, the worldwide total fertility rate (that is, the average number of lifetime births per woman) stood at about 5.0, well within the historical range. Then came a behavioral revolution, driven by growing affluence, urbanization, feminism, rising female participation in the workforce, new birth control technologies, and legalized abortion. The result: an unprecedented and unexpected decline in the global fertility rate to about 2.7—a drop fast approaching the replacement rate of 2.1 (the rate required merely to maintain a constant population). In the developed world alone, the average fertility rate has plummeted to 1.6. Since 1995, Japan has had fewer births annually than in any year since 1899. In Germany, where the rate has fallen to 1.3, fewer babies are born each year than in Nepal, which has a population only one-quarter as large.

A shrinking population in an aging developed world. Unless their fertility rates rebound, the total populations of western Europe and Japan will shrink to about one-half of their current size before the end of the next century. In 1950, 7 of the 12 most populous nations were in the developed world: the United States, Russia, Japan, Germany, France, Italy, and the United Kingdom. The United Nations projects that by 2050, only the United States will remain on the list. Nigeria, Pakistan, Ethiopia, Congo, Mexico, and the Philippines will replace the others. But since developing countries are also experiencing a drop in fertility, many are now actually aging faster than the typical developed country. In France, for example, it took over a century for the elderly to grow from 7 to 14 percent of the population. South Korea, Taiwan, Singapore, and China are projected to traverse that distance in only 25 years.

From worker shortage to rising immigration pressure. Perhaps the most predictable consequence of the gap in fertility and population growth rates between developed and developing countries will be the rising demand for immigrant workers in older and wealthier societies facing labor shortages. Immigrants are typically young and tend to bring with them the family practices of their native culture—including higher fertility rates. In many European countries, non-European foreigners already make up roughly 10 percent of the population. This includes 10 million to 13 million Muslims, nearly all of whom are working-age or younger. In Germany, foreigners will make up 30 percent of the total population by 2030, and over half the population of major cities like Munich and Frankfurt. Global aging and attendant labor shortages will therefore ensure that immigration remains a major issue in developed countries for decades to come. Culture wars could erupt over the bal-

199

kanization of language and religion; electorates could divide along ethnic lines; and émigré leaders could sway foreign policy.

GRAYING MEANS PAYING

OFFICIAL PROJECTIONS suggest that within 30 years, developed countries will have to spend at least an extra 9 to 16 percent of GDP simply to meet their old-age benefit promises. The unfunded liabilities for pensions (that is, benefits already earned by today's workers for which nothing has been saved) are already almost $35 trillion. Add in health care, and the total jumps to at least twice as much. At minimum, the global aging issue thus represents, to paraphrase the old quiz show, a $64 trillion question hanging over the developed world's future.

To pay for promised benefits through increased taxation is unfeasible. Doing so would raise the total tax burden by an unthinkable 25 to 40 percent of every worker's taxable wages—in countries where payroll tax rates sometimes already exceed 40 percent. To finance the costs of these benefits by borrowing would be just as disastrous. Governments would run unprecedented deficits that would quickly consume the savings of the developed world.

And the $64 trillion estimate is probably low. It likely underestimates future growth in longevity and health care costs and ignores the negative effects on the economy of more borrowing, higher interest rates, more taxes, less savings, and lower rates of productivity and wage growth.

There are only a handful of exceptions to these nightmarish forecasts. In Australia, total public retirement costs as a share of GDP are expected to rise only slightly, and they may even decline in Britain and Ireland. This fiscal good fortune is not due to any special demographic trend, but to timely policy reforms—including tight limits on public health spending, modest pension benefit formulas, and new personally owned savings programs that allow future public benefits to shrink as a share of average wages. This approach may yet be emulated elsewhere.

Failure to respond to the aging challenge will destabilize the global economy, straining financial and political institutions around the world. Consider Japan, which today runs a large current account surplus making up well over half the capital exports of all the surplus nations combined. Then imagine a scenario in which Japan leaves its retirement programs and fiscal policies on autopilot. Thirty years from now, under this scenario, Japan will be importing massive amounts of capital to prevent its domestic economy from collapsing under the weight of benefit outlays. This will require a huge reversal in global capital flows. To get some idea of the potential volatility, note that over the next decade, Japan's annual pension deficit is projected to grow to roughly 3 times the size of its recent and massive capital exports to the United States; by 2030, the annual deficit is expected to be 15 times as large. Such reversals will cause wildly fluctuating interest and exchange rates, which may in turn short-circuit financial institutions and trigger a serious market crash.

As they age, some nations will do little to change course, while others may succeed in boosting their national savings rate, at least temporarily, through a combination of fiscal restraint and household thrift. Yet this too could result in a volatile disequilibrium in supply and demand for global capital. Such imbalance could wreak havoc with international institutions such as the European Union.

In recent years, the EU has focused on monetary union, launched a single currency (the euro), promoted cross-border labor mobility, and struggled to harmonize fiscal, monetary, and trade policies. European leaders expect to have their hands full smoothing out differences between members of the Economic and Monetary Union (EMU)—from the timing of their business cycles to the diversity of their credit institutions and political cultures. For this reason, they established official public debt and deficit criteria (three percent of GDP for EMU membership) in order to discourage maverick nations from placing undue economic burdens on fellow members. But the EU has yet to face up to the biggest challenge to its future viability: the likelihood of varying national responses to the fiscal pressures of demographic aging. Indeed, the EU does not even include unfunded pension liabilities in the official EMU debt and deficit criteria—which is like measuring icebergs without looking beneath the water line.

When these liabilities come due and move from "off the books" to "on the books," the EU will, under current constraints, be required to penalize EMU members that exceed the three percent deficit cap. As a recent IMF report concludes, "over time it will become increasingly difficult for most countries to meet the deficit ceiling without comprehensive social security reform." The EU could, of course, retain members by raising the deficit limit. But once the floodgates are opened, national differences in fiscal policy may mean that EMU members rack up deficits at different rates. The European Central Bank, the euro, and a half-century of progress toward European unity could be lost as a result.

The total projected cost of the age wave is so staggering that we might reasonably

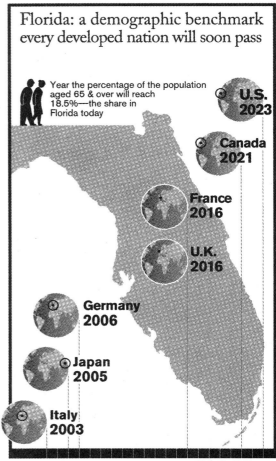

SOURCE: OECD (1996)
SOURCE: OECD (1996); author's calculations

conclude it could never be paid. After all, these numbers are projections, not predictions. They tell us what is likely to happen if current policy remains unchanged, not whether it is likely or even possible for this condition to hold. In all probability, economies would implode and governments would collapse before the projections ever materialize. But this is exactly why we must focus on these projections, for they call attention to the paramount question: Will we change course sooner, when we still have time to control our destiny and reach a more sustainable path? Or later, after unsustainable economic damage and political and social trauma cause a wrenching upheaval?

A GRAYING NEW WORLD ORDER

WHILE THE fiscal and economic consequences of global aging deserve serious discussion, other important consequences must also be examined. At the top of the list is the impact of the age wave on foreign policy and international security.

Will the developed world be able to maintain its security commitments? One need not be a Nobel laureate in economics to understand that a country's GDP growth is the

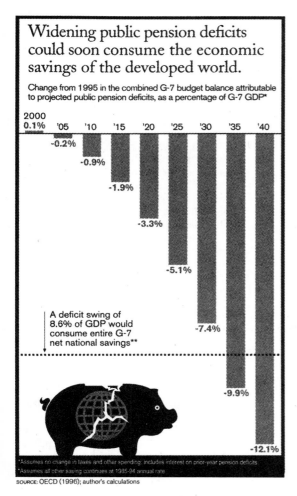

Widening public pension deficits could soon consume the economic savings of the developed world.

product of workforce and productivity growth. If workforces shrink rapidly, GDP may drop as well, since labor productivity may not rise fast enough to compensate for the loss of workers. At least some developed countries are therefore likely to experience a long-term decline in total production of goods and services—that is, in real GDP.

Economists correctly focus on the developed world's GDP per capita, which can rise even as its workforce and total GDP shrink. But anything with a fixed cost becomes a national challenge when that cost has to be spread over a smaller population and funded out of shrinking revenues. National defense is the classic example. The West already faces grave threats from rogue states armed with biological and chemical arsenals, terrorists capable of hacking into vulnerable computer systems, and proliferating nuclear weapons. None of these external dangers will shrink to accommodate our declining workforce or GDP.

Leading developed countries will no doubt need to spend as much or more on defense and international investments as they do today. But the age wave will put immense pressure on governments to cut back. Falling birthrates, together with a rising demand for young workers, will also inevitably mean smaller armies. And how many parents will allow their only child to go off to war?

With fewer soldiers, total capability can be maintained only by large increases in technology and weaponry. But boosting military productivity creates a Catch-22. For how will governments get the budget resources to pay for high-tech weaponry if the senior-weighted electorate demands more money for high-tech medicine? Even if military capital is successfully substituted for military labor, the deployment options may be dangerously limited. Developed nations facing a threat may feel they have only two extreme (but relatively inexpensive) choices: a low-level response (antiterrorist strikes and cruise-missile diplomacy) or a high-level response (an all-out attack with strategic weapons).

Will Young/Old become the next North/South fault line? Historically, the richest industrial powers have been growing, capital-exporting, philanthropic giants that project their power and mores around the world. The richest industrial powers of the future may be none of these things. Instead, they may be demographically imploding, capital-importing, fiscally starving neutrals who twist and turn to avoid expensive international entanglements. A quarter-century from now, will the divide between today's "rich" and "poor" nations be better described as a divide between growth and decline, surplus and deficit, expansion and retreat, future and past? By the mid-2020s, will the contrast between North and South be better described as a contrast between Young and Old?

If today's largest low-income societies, especially China, set up fully funded retirement systems to prepare for their own future aging, they may well produce ever larger capital surpluses. As a result, today's great powers could someday depend on these surpluses to keep themselves financially afloat. But how should we expect these new suppliers of capital to use their newly acquired leverage? Will they turn the tables in international diplomacy? Will the Chinese, for example, someday demand that the United States shore up its Medicare system the way Americans once demanded that China reform its human rights policies as a condition for foreign assistance?

As Samuel Huntington recently put it, "the juxtaposition of a rapidly growing people of one culture and a slowly growing or stagnant people of another culture generates pressure for economic and/or political adjustments in both societies." Countries where populations are still exploding rank high on any list of potential trouble spots, whereas the countries most likely to lose population—and to see a weakening of their commitment to expensive defense and global security programs—are the staunchest friends of liberal democracy.

In many parts of the developing world, the total fertility rate remains very high (7.3 in the Gaza Strip versus 2.7 in Israel), most people are very young (49 percent under age 15 in Uganda), and the population is growing very rapidly (doubling every 26 years in Iran). These areas also tend to be the poorest, most rapidly urbanizing, most institutionally unstable—and most likely to fall under the sway of rogue leadership. They are the same societies that spawned most of the military strongmen and terrorists who have bedeviled the United States and Europe in recent decades. The Pentagon's long-term planners predict that outbreaks of regional anarchy will occur more frequently early in the next century. To pinpoint when and where, they track what they call "youth bulges" in the world's poorest urban centers.

Is demography destiny, after all? Is the rapidly aging developed world fated to decline? Must it cede leadership to younger and faster-growing societies? For the answer to be no, the developed world must redefine that role around a new mission. And what better way to do so than to show the younger, yet more tradition-bound, societies—which will soon age in their turn—how a world dominated by the old can still accommodate the young.

WHOSE WATCH IS IT, ANYWAY?

FROM PRIVATE discussions with leaders of major economies, I can attest that they are well briefed on the stunning demographic trends that lie ahead. But so far they have responded with paralysis rather than action. Hardly any country is doing what it should to prepare. Margaret Thatcher confesses that she repeatedly tried to raise the aging issue at G-7 summit meetings. Yet her fellow leaders stalled. "Of course aging is a profound challenge," they replied, "but it doesn't hit until early in the next century—after my watch."

Americans often fault their leaders for not acknowledging long-term problems and for not facing up to silent and slow-motion challenges. But denial is not a peculiarly American syndrome. In 1995, Silvio Berlusconi's *Forza Italia* government was buffeted by a number of political storms, all of which it weathered—except for pension reform, which shattered the coalition. That same year, the Dutch parliament was forced to repeal a recent cut in retirement benefits after a strong Pension Party, backed by the elderly, emerged from nowhere to punish the reformers. In 1996, the French government's modest proposal to trim pensions triggered strikes and even riots. A year later the Socialists overturned the ruling government at the polls.

Each country's response, or nonresponse, is colored by its political and cultural institutions. In Europe, where the welfare state is more expansive, voters can hardly imagine that the promises made by previous generations of politicians can no longer be kept. They therefore support leaders, unions, and party coalitions that make generous unfunded pensions the very cornerstone of social democracy. In the United States, the problem has less to do with welfare-state dependence than the uniquely American notion that every citizen has personally earned and is therefore entitled to whatever benefits government happens to have promised.

How governments ultimately prepare for global aging will also depend on how global aging itself reshapes politics. Already some of the largest and most strident interest groups in the United States are those that claim to speak for senior citizens, such as the American Association of Retired Persons, with its 33 million members, 1,700 paid employees, ten times that many trained volunteers, and an annual budget of $5.5 billion.

Senior power is rising in Europe, where it manifests itself less through independent senior organizations than in labor unions and (often union-affiliated) political parties that formally adopt pro-retiree platforms. Could age-based political parties be the wave of the future? In Russia, although the Communist resurgence is usually ascribed to nationalism and nostalgia, a demographic bias is at work as well. The Communists have repositioned themselves as the party of retirees, who are aggrieved by how runaway inflation has slashed the real value of their pensions. In the 1995 Duma elections, over half of those aged 55 and older voted Communist, versus only ten percent of those under age 40.

Commenting on how the old seem to trump the young at every turn, Lee Kuan Yew once proposed that each taxpaying worker be given two votes to balance the lobbying clout of each retired elder. No nation, not even Singapore, is likely to enact Lee's suggestion. But the question must be asked: With ever more electoral power flowing into the hands of elders, what can motivate political leaders to act on behalf of the long-term future of the young?

A handful of basic strategies, all of them difficult, might enable countries to overcome the economic and political challenges of an aging society: extending work lives and postponing retirement; enlarging the workforce through immigration and increased labor force participation; encouraging higher fertility and investing more in the education and productivity of future workers; strengthening intergenerational bonds of responsibility within families; and targeting government-paid benefits to those most in need while encouraging and even requiring workers to save for their own retirements. All of these strategies unfortunately touch raw nerves—by amending existing social contracts, by violating cultural expectations, or by offending entrenched ideologies.

TOWARD A SUMMIT ON GLOBAL AGING

ALL COUNTRIES would be well served by collective deliberation over the choices that lie ahead. For that reason I propose a Summit on Global Aging. Few venues are as well covered by the media as a global summit. Leaders have been willing to convene summits to discuss global warming. Why not global aging, which will hit us sooner and with greater certainty? By calling attention to what is at stake, a global aging summit could shift the public discussion into fast forward. That alone would be a major contribution. The summit process would also help provide an international framework for voter education, collective burden-sharing, and global leadership. Once national constituencies begin to grasp the magnitude of the global aging challenge, they will be more inclined to take reform seriously. Once governments get into the habit of cooperating on what in fact is a global challenge, individual leaders will not need to incur the economic and political risks of acting alone.

This summit should launch a new multilateral initiative to lend the global aging agenda a visible institutional presence: an Agency on Global Aging. Such an agency would examine how developed countries should reform their retirement systems and how developing countries should properly set them up in the first place. Perhaps the most basic question is how to weigh the interests and well-being of one generation against the next. Then there is the issue of defining the safety-net standard of social adequacy. Is there a minimum level of retirement income that should be the right of every citizen? To what extent should retirement security be left to people's own resources? When should government pick up the pieces, and how can it do so without discouraging responsible behavior? Should government compel people in advance to make better life choices, say, by enacting a mandatory savings program?

Another critical task is to integrate research about the age wave's timing, magnitude, and location. Fiscal projections should be based on assumptions that are both globally consistent and—when it comes to longevity, fertility, and health care costs—more realistic than those now in use. Still to be determined: Which countries will be hit earliest and hardest? What might happen to interest rates, exchange rates, and cross-border capital flows under various political and fiscal scenarios?

But this is not all the proposed agency could do. It could continue to build global awareness, publish a high-visibility annual report that would update these calculations, and ensure that the various regular multilateral summits (from the G-7 to ASEAN and APEC) keep global aging high on their discussion agendas. It could give coherent voice to the need for timely policy reform around the world, hold up as models whatever major steps have been taken to reduce unfunded liabilities, help design funded benefit programs, and promote generational equity. On these and many other issues, nations have much to learn from each other, just as those who favor mandatory funded pension plans are already benefiting from the examples of Chile, Britain, Austria, and Singapore.

Global aging could trigger a crisis that engulfs the world economy. This crisis may even threaten democracy itself. By making tough choices now world leaders would demonstrate that they genuinely care about the future, that they understand this unique opportunity for young and old nations to work together, and that they comprehend the price of freedom. The gray dawn approaches. We must establish new ways of thinking and new institutions to help us prepare for a much older world.

PETER G. PETERSON is the author of *Gray Dawn: How the Coming Age Wave Will Transform America—and the World.* He is Chairman of The Blackstone Group, a private investment bank, Chairman of The Institute for International Economics, Deputy Chairman of The Federal Reserve Bank of New York, Co-founder and President of The Concord Coalition, and Chairman of The Council on Foreign Relations.

A rare and precious resource

Fresh water is a scarce commodity. Since it's impossible to increase supply, demand and waste must be reduced. But how?

Houria Tazi Sadeq*

Water is a bond between human beings and nature. It is ever-present in our daily lives and in our imaginations. Since the beginning of time, it has shaped extraordinary social institutions, and access to it has provoked many conflicts.

But most of the world's people, who have never gone short of water, take its availability for granted. Industrialists, farmers and ordinary consumers blithely go on wasting it. These days, though, supplies are diminishing while demand is soaring. Everyone knows that the time has come for attitudes to change.

Few people are aware of the true extent of fresh water scarcity. Many are fooled by the huge expanses of blue that feature on maps

Sharper vision

Desalinization, state of the art irrigation systems, techniques to harvest fog—technological solutions like these are widely hailed as the answer to water scarcity. But in searching for the "miracle" solution, hydrologists and policy-makers often lose sight of the question: how can we use and safeguard this vital resource? UNESCO's International Hydrological Programme (IHP) takes an interdisciplinary approach to this question. On the one hand, IHP brings together scientists from 150 countries to develop global and regional assessments of water supplies and, for example, inventories of groundwater contamination. At the same time, the programme focuses on the cultural and socio-economic factors involved in effective policy-making. For example, groundwater supplies in Gaza (Palestinian Authority) are coming under serious strain, partly because of new business investment in the area. IHP has a two-pronged approach. First, train and help local hydrologists accurately assess the supplies. Second, work with government officials to set up a licensing system for pumping groundwater.

By joining forces with the World Water Council, an international think-tank on hydrological issues, IHP is now hosting one of the most ambitious projects in the field: World Water Vision. Hundreds of thousands of hydrologists, policy-makers, farmers, business leaders and ordinary citizens will take part in public consultations to develop regional scenarios as to how key issues like contamination will evolve in the next 25 years.

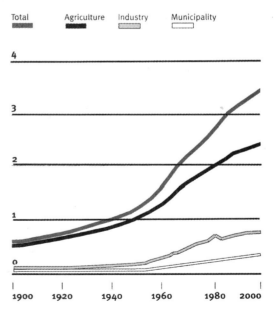

Global water withdrawals (1900-2000) in thousands of km³ per year

*Moroccan jurist, president of the Maghreb–Machrek Water Union, vice-president of the International Water Secretariat

of the world. They do not know that 97.5 per cent of the planet's water is salty—and that most of the world's fresh water—the remaining 2.5 per cent—is unusable: 70 per cent of it is frozen in the icecaps of Antarctica and Greenland and almost all the rest exists in the form of soil humidity or in water tables which are too deep to be tapped. In all, barely one per cent of fresh water—0.007 per cent of all the water in the world, is easily accessible.

Over the past century, population growth and human activity have caused this precious resource to dwindle. Between 1900 and 1995, world demand for water increased more than sixfold—compared with a threefold increase in world population. The ratio between the stock of fresh water and world population seems to show that in overall terms there is enough water to go round. But in the most vulnerable regions, an estimated 460 million people (8 per cent of the world's population) are short of water, and another quarter of the planet's inhabitants are heading for the same fate. Experts say that if nothing is done, two-thirds of humanity will suffer from a moderate to severe lack of water by the year 2025.

Inequalities in the availability of water—sometimes even within a single country—are reflected in huge differences in consumption levels. A person living in rural Madagascar uses 10 litres a day, the minimum for survival, while a French person uses 150 litres and an American as many as 425.

Scarcity is just one part of the problem. Water quality is also declining alarmingly. In some areas, contamination levels are so high that water can no longer be used even for industrial purposes. There are many reasons for this—untreated sewage, chemical waste, fuel leakages, dumped garbage, contamination of soil by chemicals used by farmers. The worldwide extent of such pollution is hard to assess because data are lacking for several countries. But some figures give an idea of the problem. It is thought for example that 90 per cent of waste water in developing countries is released without any kind of treatment.

Things are especially bad in cities, where water demand is exploding. For the first time in human history, there will soon be more people living in cities than in the countryside and so water consumption will continue to increase. Soaring urbanization will sharpen the rivalry between the different kinds of water users.

Curbing the explosion in demand

Today, farming uses 69 per cent of the water consumed in the world, industry 23 per cent and households 8 per cent. In developing countries, agriculture uses as much as 80 per cent. The needs of city-dwellers, industry and tourists are expected to increase rapidly, at least as much as the need to produce more farm products to feed the planet. The problem of increasing water supply has long been seen as a technical one, calling for technical solutions such as building more dams and desalination plants. Wild ideas like towing chunks of icebergs from the poles have even been mooted.

But today, technical solutions are reaching their limits. Economic and socio-ecological arguments are levelled against building new dams, for example: dams are costing more and more because the best sites have already been used, and they take millions of people out of their environment and upset ecosystems. As a result, twice as many dams were built on average between 1951 and 1977 than during the past decade, according to the US environmental research body Worldwatch Institute.

Hydrologists and engineers have less and less room for manoeuvre, but a new consensus with new actors is taking shape. Since supply can no longer be expanded—or only at prohibitive cost for many countries—the explosion in demand must be curbed along with wasteful practices. An estimated 60 per cent of the water used in irrigation is lost through inefficient systems, for example.

Economists have plunged into the debate on water and made quite a few waves. To obtain "rational use" of water, i.e. avoiding waste and maintaining quality, they say consumers must be made to pay for it. Out of the question, reply those in favour of free water, which some cultures regard as "a gift from heaven." And what about the poor, ask the champions of human rights and the right to water? Other important and prickly questions being asked by decisionmakers are how to

The water from the fountain glides, flows and dreams as, almost dumb, it licks the mossy stone.

Antonio Machado (1875–1939), Spain

calculate the "real price" of water and who should organize its sale.

The state as mediator

The principle of free water is being challenged. For many people, water has become a commodity to be bought and sold. But management of this shared resource cannot be left exclusively to market forces. Many elements of civil society—NGOs, researchers, community groups—are campaigning for the cultural and social aspects of water management to be taken into account.

Even the World Bank, the main advocate of water privatization, is cautious on this point. It recognizes the value of the partnerships between the public and private sectors which have sprung up in recent years. Only the state seems to be in a position to ensure that practices are fair and to mediate between the parties involved—consumer groups, private firms and public bodies. At any rate, water regulation and management systems need to be based on other than purely financial criteria. If they aren't, hundreds of millions of people will have no access to it.

- A person can survive for about a month without food, but only about a week without water.
- About 70 per cent of human skin consists of water.
- Women and children in most developing regions travel an average of 10 to 15 kilometres each day to get water.
- Some 34,000 people die a day from water-related diseases like diarrhoea and parasitic worms. This is the equivalent to casualties from 100 jumbo jets crashing every day!
- A person needs five litres of water a day for drinking and cooking and another 25 litres for personal hygiene.
- The average Canadian family uses 350 litres of water a day. In Africa, the average is 20 litres and in Europe, 165 litres.
- A dairy cow needs to drink about four litres of water a day to produce one litre of milk.
- A tomato is about 95 per cent water.
- About 9,400 litres of water are used to make four car tires.
- About 1.4 billion litres of water are needed to produce a day's supply of the world's newsprint.

Sources: International Development Initiative of McGill University, Canada; Saint Paul Water Utility, Minnesota, USA

Lack of access to safe water and basic sanitation, by region, 1990-1996 (percent)

Region	People without access to safe water	People without access to basic sanitation
Arab States	21	30
Sub-Saharan Africa	48	55
South-East Asia and the Pacific	35	45
Latin America and the Caribbean	23	29
East Asia	32	73
East Asia (excluding China)	13	—
South Asia	18	64
Developing countries	29	58
Least developed countries	43	64

Source: *Human Development Report 1998*, New York, UNDP

Periods of complete renewal of the earth's water resources

Kinds of water	Period of renewal
Biological water	several hours
Atmospheric water	8 days
Water in river channels	16 days
Soil moisture	1 year
Water in swamps	5 years
Water storages in lakes	17 years
Groundwater	1 400 years
Mountain glaciers	1 600 years
World ocean	2 500 years
Polar ice floes	9 700 years

Source: *World Water Balance and Water Resources of the Earth*, Gidrometeoizdat, Leningrad, 1974 (in Russian)

5 ❖ POPULATION, RESOURCES, AND SOCIOECONOMIC DEVELOPMENT

A thirsty planet

We now have less than half the amount of water available per capita than we did 50 years ago. In 1950, world reserves, (after accounting for agricultural, industrial and domestic uses) amounted to 16.8 thousand cubic metres per person. Today, global reserves have dropped to 7.3 thousand cubic metres and are expected to fall to 4.8 thousand in just 25 years.

Scientists have developed many ways of measuring supplies and evaluating water scarcity. In the maps at right, "catastrophic" levels mean that reserves are unlikely to sustain a population in the event of a crisis like drought. Low supplies refer to levels which put in danger industrial development or ability to feed a population.

Just 50 years ago, not a country in the world faced catastrophic water supply levels. Today, about 35 per cent of the population lives under these conditions. By 2025, about two-thirds will have to cope with low if not catastrophic reserves. In contrast, "water rich" regions and countries—such as northern Europe, Canada, almost everywhere in South America, Central Africa, the Far East and Oceania—will continue to enjoy ample reserves.

The sharp declines reflect the soaring water demands of growing populations, agricultural needs and industrialization. In addition, nature has been far from even-handed. More than 40 per cent of the water in rivers, reservoirs and lakes is concentrated in just six countries: Brazil, Russia, Canada, the United States, China and India. Meanwhile just two per cent of river, reservoir and lake water is found in about 40 per cent of the world's land mass.

As a result, in 2025 Europe and the United States will have half the per capita reserves they did in 1950, while Asia and Latin America will have just a quarter of what they previously enjoyed. But the real drama is likely to hit Africa and the Middle East, where available supplies by 2025 may be only an eighth of what they were in 1950.

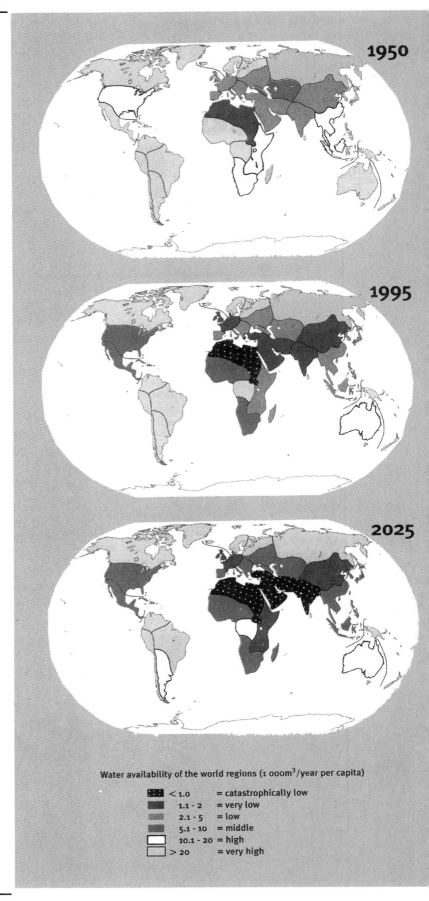

Water availability of the world regions (1 000m³/year per capita)

- < 1.0 = catastrophically low
- 1.1 - 2 = very low
- 2.1 - 5 = low
- 5.1 - 10 = middle
- 10.1 - 20 = high
- > 20 = very high

Rethinking Flood Prediction: Does the Traditional Approach Need to Change?

William D. Gosnold Jr., Julie A. LeFever, Paul E. Todhunter and Leon F. Osborne Jr.

At 5:00 the morning of April 18, 1997, we awoke to the sound of a police officer yelling through a bullhorn, telling us to abandon our homes in Grand Forks, N.D., and East Grand Forks, Minn. A few hours before dawn that morning, the Red River of the North, laden with ice and melted snow, surged over the top of a temporary sandbag dike at the north end of the Lincoln Park Golf Course in Grand Forks.

Although the tops of the dikes protecting both cities sat 61 centimeters above the predicted crest for the river's annual spring flood, the river heeded neither the prediction nor the dikes. Floodwaters quickly flowed north and west down streets and in storm sewers, through residential neighborhoods and into the heart of downtown. At 3:30 p.m., the dike protecting East Grand Forks from the Red Lake River failed, and East Grand Forks was quickly inundated by several meters of water.

By that evening, 75 percent of the population of Grand Forks and the entire population of East Grand Forks had been evacuated. The water level continued to rise an additional 72 centimeters over four days and spread horizontally about five kilometers east and west from the north-flowing Red River. A large section of downtown Grand Forks was destroyed by water and fire and 24,031 homes were either partially or completely destroyed by flood waters. Only 27 homes in East Grand Forks, Minn., survived unscathed and the total economic loss in Grand Forks and East Grand Forks was approximately $3.6 billion, according to the U.S. Department of Commerce.

As geoscientists, we were awed by this display of nature's might, but we were also dismayed by the failure of our science to foresee and prevent this disaster. As we slowly recovered from its effects, we each began to seek answers to questions we had about this particular flood and flood science in general. Through mutual contact at various forums we began to share ideas and information and eventually came to the understanding we present here.

We conclude that the process of predicting floods is not simply one of engineering hydrology, but should include changes in climate and land use as major factors. Flood prediction must go beyond the traditional

> **We face a dilemma in that our flood mitigation program is based on floods that do not cause much damage and ignores the floods (paleofloods and other extreme floods) that cause the vast majority of damage.**

statistical, engineering approach and adopt an earth-system science approach. Developing new guidelines for flood prediction will require substantial new research to determine applicable factors and their effects.

Flood recurrence intervals

The cornerstone of flood protection planning is knowledge of flood recurrence intervals. Calculations of these intervals determine legal definitions, engineering decisions, and civic zoning and building guidelines. They are the basis for the Federal Emergency Management Agency's National Flood Insurance Program.

In the strictest sense, flood recurrence interval is the reciprocal of exceedance probability, i.e., the probability that flood flow will exceed a specific stream discharge at a location where peak stream flow has been continuously recorded. Although recurrence interval calculations can be expressed in terms of either stream discharge or river stage (river height) values, they are usually interpreted in terms of river stage so that areas of potential inundation in the floodplain can be easily mapped. Specific elevation levels in flood plains are thus designated as five-year, 20-year, 100-year, 200-year, etc., flood levels. Most people perceive flood probability in absolute terms: that a flood will reach the 100-year flood level only once per 100 years, rather than that the probability of reaching the 100-year level in any year is one chance in 100.

The federal guidelines for estimating flood recurrence, *Guidelines for Determining Flood Flow Frequency*, known as *Bulletin 17–B*, have evolved through three revisions since they were first developed in 1967. The initial guidelines established the type of statistical distribution for estimating flood flows, the log Pearson type III distribution, and subsequent revisions refined the techniques and increased the statistical confidence limits for the projections. These guidelines are the standards used by all federal and most state agencies responsible for flood protection and flood recovery.

The 1997 flood of the Red River of the North was the largest flood recorded by the USGS gage in East Grand Forks, Minn., since its 1882 installation. We wanted to know the 1997 flood's recurrence interval at Grand Forks. This interest in recurrence intervals was widespread and soon a number of estimates began appearing in the news media. Various media reports ranged from 150 years to 400 years. At that time, we weren't familiar with the intricacies of recurrence interval estimates so we relied on those who are—specifically, colleagues in the field of hydrology and the U.S. Army Corps of Engineers. But their estimates added to our confusion. They ranged from 117 years to 1,010 years.

Our initial research showed us we were not alone in wondering if a statistical analysis by itself is sufficient to predict floods. In a 1994 paper, Victor Baker of the University of Arizona argued that the flaw in statistical approaches is that they do not deal with real flood phenomena but with idealized parameters that have flood-like properties. He further argued that natural floods are far more diverse and complex than the simplifications used in statistical analysis and that such analysis tends to ignore the rarest and largest floods—which are, in fact, the main focus for hazard mitigation. This point is particularly key because the vast majority of flood damages are associated with only a small number of floods. For example, a recent North Dakota Geological Survey report revealed that about 75 percent of all historical flood damage in Grand Forks were due to one event: the 1997 flood.

Baker's point is that it is precisely these rare and huge events that do not fit the statistical approach, and they are the very events we should be examining because they cause the most damage. The log Pearson type III distribution works well for 25-year events used for calculating road design, for example. But such floods cause little damage.

We face a dilemma in that our flood mitigation program is based on floods that do not cause much damage and ignores the floods (paleofloods and other extreme floods) that cause the majority of damage.

Rethinking assumptions

We sought to reproduce the calculations of recurrence intervals to understand why they vary among different sources. This required that we analyze the methods used in *Bulletin 17–B*. In doing so, we discovered several flaws in the methods' underlying assumptions—primarily that "in hydrologic analysis it is conventional to assume flood flows are not affected by climatic trends or cycles." Under this presumption, *Bulletin 17–B* also states that "an array of annual maximum peak flow rates may be considered a sample of random and independent events" and that "measurement errors are usually random." We now know that climate varies on scales of decades to centuries as well as on longer time scales. Interestingly, recognizing climate change as an important factor would not have been likely in 1981, when *Bulletin 17–B* was last revised.

42. Rethinking Flood Prediction

A number of researchers have reported that a small change in climate can lead to a large change in flood magnitudes. For example, Jim Knox of the University of Wisconsin has found that a change of only 1 to 2 degrees Celsius in mean annual temperature and 10 to 20 percent in mean annual precipitation can produce an order of magnitude change in recurrence probability.

A correlation between climate change and flood magnitudes in the Red River drainage basin would be significant because large climate changes have occurred in the basin during the past century. Average annual ground-surface temperature has increased 2.5 degrees Celsius and the average annual air temperature has increased 2 degrees Celsius. Precipitation and temperature data for the Northern Great Plains region, which includes the Red River drainage basin, show that the climate has varied from cool and wet between 1870 and 1910, to warm and dry from 1920 to 1940, to warm and wet since 1950. Warming and precipitation reached new levels in the 1990s and both are still increasing.

In a 1988 book on flood geomorphology, Katherine Hirschboeck of the University of Arizona discussed the effects of climate change on flood recurrence estimates and recommended a reevaluation of assumptions about climate and flood hydrology. She specifically suggested using flood data from distinct climate patterns rather than mixing different populations of data that represent different climate patterns.

The problems inherent in mixing populations of data is a critical point. By using the log Pearson type III distribution for discharge data, one assumes that a log transformation of the entire data set produces a normal distribution which can then be analyzed using established statistical methods. But mixing dry-period data with wet-period data uses two separate populations and invalidates the assumption of a homogenous data set. If more than one population of data is present, the distribution is not normal and the statistical methods for normal distributions do not apply.

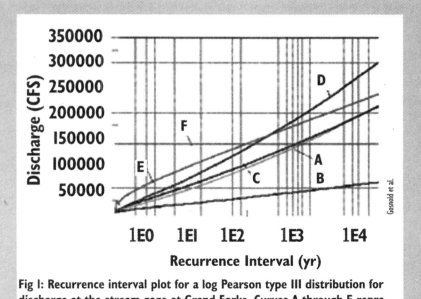

Fig I: Recurrence interval plot for a log Pearson type III distribution for discharge at the stream gage at Grand Forks. Curves A through E represent the time periods 1941-1980, 1921-1940, 1981-1998, 1882-1920, 1993-1998.

We found an example that illustrates this point in a 1984 U.S. Geological Survey paper by Jeffrey Miller and Dale Frink analyzing changes in flood response of the Red River of the North. Miller and Frink calculated flood recurrence curves at Grand Forks for four consecutive, 25-year periods beginning in 1882. Because the climate was different for each 25-year period, the curves yielded distinctly different recurrence intervals. For example, recurrence intervals calculated for a discharge of 80,000 cubic feet per second (cfs) with data from the two relatively wet periods (1882–1904 and 1955–1979) were 12 and 11 years, respectively. Recurrence intervals of the 80,000 cfs discharge with data from the relatively dry periods (1930–1954 and 1905–1929) were 50 years and greater than 100 years.

We made two similar analyses of recurrence intervals, but divided the data by characteristic climate rather than by fixed time intervals. In the first analysis, we calculated recurrence intervals using the methodology of Bulletin 17–B. Curves D and E in Fig. 1 (this page), based on data from 1882–1920 and 1993–1998, indicate that a flood discharge of 136,900 cfs (line F)—the discharge of the 1997 flood—has a recurrence interval of 125 and 90 years, respectively. Curves A and C (1941–1980 and 1981–1998) indicate a recurrence interval of about 600 years for the same discharge, while curve B (1921–1940) indicates that such a discharge would be unlikely at any time scale. Using the complete data set (curve not shown), without including historical floods that occurred prior to the 1882 installation of the gage, yields a recurrence interval of 400 years.

In the second analysis (see Fig. 2 next page), we used most of the same climate periods to determine the recurrence of a lesser flood, such as 20,000 cfs, which corresponds to a gage height of 10 meters (measured above river bottom). Between 1882 and 1920 the climate was cool and wet and 12 floods exceeding 10 meters occurred. From 1921 until 1940, the climate was warm and dry, and there were no floods exceeding 10 meters. From 1941 until 1980 the climate became wetter and warmer and 20 floods reaching 10 meters occurred. Since 1980 the climate has become wetter and warmer than any preceding period and the area has experienced 12 floods hitting 10 meters.

used to calculate flood recurrence intervals. Between 1900 and 1967, 3,176 miles of legal ditches had been excavated in the Red River basin for draining crop land. By 1997, the number had increased to 28,000 miles.

Integrating past, present and future

Victor Baker suggests that predicting future flooding trends is beyond our ability, but that understanding the past can provide a guide for preparedness. We might ask: If floods of a particular level have occurred with a known frequency, can we suggest that they will continue to do so? The findings of Hirschboeck and Knox suggest that the answer to this question can be affirmative only if the climate during the period of observation (the past) matches the climate of the period in question (the future). Small changes in climate can cause large changes in flood magnitude. We are presently experiencing a wetter and warmer climate that is not sampled by historical gage data, but that may have been sampled by the geologic record.

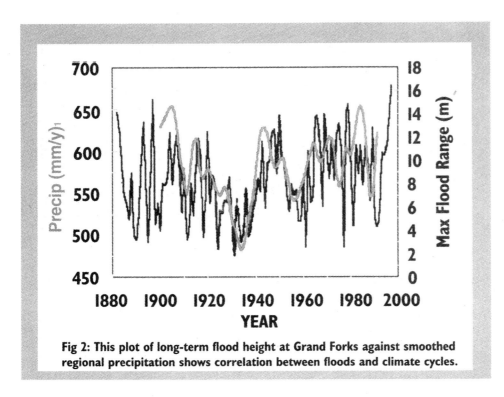

Fig 2: This plot of long-term flood height at Grand Forks against smoothed regional precipitation shows correlation between floods and climate cycles.

It is clear to us that warmer and wetter conditions increase the magnitude of flows at all recurrence intervals. Consequently, we suggest that using data from the most recent, analogous climate type provides a more reliable estimate of recurrence interval than does the traditional approach of using the entire historical record. In the Northern Great Plains region, using the entire historical record results in mixing data from the dry Dust Bowl years of the 1930s with data from the current wet period. We find that this approach can drastically underestimate flood recurrence and flood danger.

Land-use changes

Changes that have occurred in land use and watershed management also change the character of the data

These ditches now drain 4,285,454 square miles within the Red River basin.

The effects of such land-use changes are significant. In her 1998 master's thesis at Bemidji State University in Bemidji, Minn., Linda Kingery found significant differences in discharge associated with land-use changes in the Red River Basin. As with climate change, using flood data from time periods having different drainage patterns created by changes in land use and watershed management mixes data from separate rainfall-runoff regimes and violates the assumption of independence of data. Again we have come to the conclusion that only data from the time period that includes the present watershed structure should be used to estimate recurrence intervals.

However, quantitative correlation of past climates with past flood magnitudes may not be possible. Although paleoflood data can show that large floods occurred during wetter climate periods, precise estimates of paleoflood discharge would be questionable. Also, the effects of land-use change would likely disallow any quantitative relationship.

Although a quantitative relationship between paleoclimates and paleofloods may not be attainable, we believe that a qualitative relationship can be recognized and would be useful. An example of a qualitative approach is found in the paleoclimatic record of Devils Lake in eastern North Dakota. The 8,100-square-kilometer basin is closed and contains a chain of more than 20 interconnected lakes that have experienced a 14-meter rise in water level since 1941. Geologic investigations show that in the past 5,000 years the basin has undergone eight extreme wet-dry cycles and has overflowed into a tributary of the Red River during three of the wet cycles, most recently about 1,200 years ago. The water table in the Devils Lake basin is perched on several hundred meters of impermeable shale.

Thus, if the closed basin fills or dries, the cause is a prolonged episode of precipitation or drought. Research by North Dakota Geological Survey geologists shows that these climate-driven drying and filling cycles last for hundreds of years and can change lake levels by up to 17 meters. Both wet and dry extremes indicated by the geologic record exceed the limits observed in the stream-gage record. If the climate continues to become wetter as is predicted by some climate model simulations, the correlation between climate and flood magnitude implies that high-discharge floods may become more frequent. In this case, the catastrophic flood of 1997 in Grand Forks may be a precursor of even greater floods.

How is this useful? Current plans for protecting Grand Forks and East Grand Forks are based on the historical gage record and, in particular, on the largest recorded flood, i.e., the 1997 flood. The link between climate and flood magnitude and the prediction of a wetter climate in the future demand that the heights of planned dikes be reinvestigated. Our key question is: Can the traditional approach to flood prediction be modified to consider climate and land-use change?

We have no certain solutions to the problem of flood prediction. However, we suggest that new research efforts to correlate floods with climate and land-use change can lead to better planning and preparation for future floods.

Gosnold is a professor of geophysics in the Department of Geology and Geological Engineering at the University of North Dakota.

LeFever earned her master's degree in 1982 from California State University, Northridge, and has worked for the North Dakota Geological Survey since 1980.

Todhunter is a professor of geography at the University of North Dakota.

Osborne is a professor of atmospheric science and director of the Regional Weather Information Center at the University of North Dakota.

Risking the Future of Our Cities

Poorly planned growth has left many neighborhoods behind.

Can the pattern be changed?

BY THOMAS HACKETT

A new sign has been put up on the way into Park Forest, Ill. Meant to renew business interests and rekindle hope, it points optimistically to the village's downtown. But that downtown—never more than a shopping mall to begin with—is a vast parking lot surrounding empty retail spaces.

Park Forest was carved into cornfields 30 miles south of Chicago in the boom years that followed World War II. It was to be haven of belonging and bustling civic involvement for the young families swept up in the "great outward movement from the inner city," William H. Whyte Jr. wrote in his 1956 classic "The Organization Man." Forty-three years later, the arrow on the new sign seems to point backward, to a promise that has been lost for many aging suburbs around the country.

Something is happening to the "inner-ring" suburbs—like those south of Chicago, east of Cleveland, north of Detroit, southeast of Los Angeles—that clashes with all the clichés of untroubled blandness. For many major metropolitan areas, the struggle to create good jobs, provide affordable housing and meet the social needs that arise with concentrated poverty is no longer just for inner-city neighborhoods to contend with. Older suburbs, too, are rapidly losing ground as a kind of centrifugal force—fed by decades of government policies favoring the construction of highways and new housing—throws jobs and businesses and prospering families even farther from urban centers.

Despite a robust national economy, urban researchers and community organizers say that all too often what is left behind is a minority population that is poorer, more isolated and less equipped to improve its lot than at any time in the last 25 years, when the country began to shift from an industrial to a service economy.

Indeed, when Whyte said that Park Forest was a harbinger "of the way it's going to be" for many American communities at the end of the century, he was more prescient than anyone supposed. With inner-city neighborhoods and aging suburbs floundering across the country, urban planners are looking for ways to develop regional responses to the uneven, runaway growth of many metropolitan areas. With support from the Ford Foundation, they also hope to counter the impression that certain communities are disposable, not worth saving.

"We're the only nation on earth that does this to our cities," says Melvin L. Oliver, the Foundation's vice president for Asset Building and Community Development. "A lot of these suburbs are more than 40 years old, and they're showing signs of deterioration. Rather than tend to the problems, it seems easier to just leave them. But we're the only country without a concerted policy to help these areas change their form and function over time."

Many planners now stress the point that no community in a metropolitan area exists in isolation. The particular difficulties of a Park Forest are also symptomatic of a more general disorder confronting that entire region. And the only way to effectively address those difficulties, they argue, is to face the area's larger problems.

"These trends have been going on for a long time," says MarySue Barrett, president of Chicago's Metropolitan Planning Council, a nonpartisan organization that advocates for more balanced development. "But the pace is picking up, and that's what's so troubling. If we don't like what we see from past development patterns"—endless strip developments, tortuous commutes, the disappearance of forest and fields—"we're really not going to like where we're heading."

The concerns go beyond esthetics. Some urban planners, real estate developers and transportation experts argue that with an economy requiring a less-concentrated workforce, the market is simply following suit by developing new communities centered around office parks in outlying areas. Sprawl, when it results in cheaper housing and better schools and supports the largest middle class in the world, isn't an enemy, they say.

The problem is that new growth has come at the expense of established communities. Nationally, most loans from the Small Business Administration go to more prosperous exurban counties, according to a recent study by the Woodstock Institute. In the Chicago area, the Northeastern Illinois Planning Commission reports that from 1970 to 1990 the outer-fringe suburbs grew by nearly a million people, while the city and inner suburbs diminished in population by nearly as many. Of course, there isn't an exact correlation here. The commission doesn't conclude that the same people pioneering the outskirts all came from the city and older suburbs; indeed, city neighborhoods are constantly being rediscovered by new residents. But overall, Chicago and other metropolitan areas have been spreading themselves thin. While the population of greater Chicago grew by only 4 percent in that 20-year period, it spread across 35 percent more land, the commission reports, with large-lot homes and strip developments swallowing up nearly a quarter of the farmland in northeastern Illinois.

A subtle effect of the population decentralization is that now, more than ever, living on the wrong side of a region can be as much of a marginalizing factor as race and class. Instead of being near the center of economic activity—once a defining virtue of the American bedroom community—inner-ring suburbs are now experiencing much of the same isolation that rural communities have always known. "If that's where you can afford housing," Barrett says, "the harsh reality is, the jobs just aren't there."

As regional planning proponents see more communities like Park Forest being slighted by policy decisions that encourage sprawl (in particular, the investment in outlying highways and zoning in new developments that require large lots and large homes that working families usually cannot afford), new metropolitan coalitions are beginning to form that see the link between land-use planning and poverty and between livability and job development.

Although the agenda "isn't bumper-sticker clear yet," as Barrett notes, there is a growing consensus that no community can solve its problems by itself. For one, many of the quality-of-life issues that enter the discussion—air and water quality, resource management, transportation—are concerns that transcend municipal boundaries.

The work has to be undertaken regionwide, with new forms of regional cooperation, argues Myron Orfield, director of the Metropolitan Area Research Corporation, which maps demographic trends. It has to involve tax reformers, transportation planners, environmentalists, members of minority groups, developers, middle-class suburbanites and advocates for the poor. It has to go beyond decades of parochial thinking in which race and class divisions have conspired against the development of a regional perspective. And it must happen soon, he adds, while a healthy economy has opened a window of opportunity.

"What's happening in places like Chicago's south suburbs—the urban stresses and strains of places like Chicago Heights, the racial separation and economic decline of places like Ford Heights, the middle-class flight of Park Forest—are the same patterns that tore apart the inner-city neighborhoods of Chicago in the 1940's, 50's, and 60's," Orfield says. "It may be happening with more violence and dramatic impact in the Chicago region, but I think we're seeing the future of many of the country's suburbs."

It's not a pretty picture. Take a drive down Route 30 south of Chicago. Look at the Dixie Square Mall in Harvey, which 30 years ago supplanted the first regional shopping mall in the country, in Park Forest, but has more or less been abandoned for the last 20. Tour Chicago Heights with Mayor Angelo Ciambrone and let him describe all the bakeries and delicatessens and other small family businesses that used to line Main Street, buildings that were demolished when yet another mall went up down the road, leaving garbage-filled empty lots behind.

It is easy to see what Vice President Al Gore meant when he spoke last year at the Brookings Institution of an "ugliness" across the country "that leaves us with a quiet sense of sadness." Though few people would disagree with the general observations of Gore's "quality of life" speech, the problem for almost every metropolitan region has been organizing vague and sometimes divergent concerns around a unifying theme.

"This is a new message," says Bruce Katz, director of the Center on Urban and Metropolitan Policy at the Brookings Institution. "At this point, we're almost in a reeducation stage, an issue-identification phase. There needs to be a fundamental sinking in of how profound the trends are."

A few places have done this. For years, the Minneapolis and St. Paul area had been relatively resistant to the nation's growing urban poverty. But in the 1980's, the poverty rate began to rise noticeably in the Twin Cities, tripling the number of schoolchildren who qualified for free or reduced-price school lunches. As Orfield's maps revealed marked patterns of disparity—showing just how quickly the decline was occurring and how concentrated it was in the central-city neighborhoods and older suburbs—those communities, which had often been at odds with each other, united to reform regional governance and pass landmark state land use, tax-base sharing and fair-housing legislation.

> **These towns are casualties of planning and policy decisions, and not simply vague demographic trends.**

Similarly, city and suburban leaders in Portland, Ore., saw that a plan that curbed outward development would draw development and investment back toward the city's central core, preserving the surrounding landscape. The city passed a "smart growth" plan that created boundaries in development, and reduced dependence on cars with a light rail system and more pedestrian-friendly city streets. Far from being hindered by the boundaries, Portland's total employment grew by 30 percent in the last decade, according to the state's employment department.

In Milwaukee, instead of building more highways that divide and devitalize established neighborhoods, the city is demolishing the Park East Freeway at a cost of $20 million. Three-term Mayor John O. Norquist argued successfully that such an action is the best way to restore life and commerce to the city.

But it is the rough-and-tumble, racially polarized and economically segregated big cities like Cleveland, Detroit, Washington, D.C., Atlanta and Los Angeles that represent the real test of regionalism. Greater Chicago, because it is fractured into hundreds of municipalities and units of government and because its social divisions are so deeply entrenched, may represent the biggest challenge of all.

To help bring their work out of the realm of policy discussions and into the world of political action, Katz and Orfield have worked with church-affiliated community development groups that cross traditional racial and economic borders in the more combative big cities. "Government bureaucracy doesn't have a good track record," Katz explains. "It's community and church organizations that are going to do a lot of the legwork. And we are trying to provide a clear understanding of these very complex forces so that community leaders can put their work—whether it's affordable housing or revitalizing commercial districts—into the larger context of demographic and market trends and the policy that affects those trends."

One of the most active of these organizations is Mary Gonzales's Metropolitan Alliance of Congregations in Chicago. Gonzales, 58, is a community organizer whose career has largely been in the trenches of inner-city politics. But in the last few years, she has been one of the few organizers to unite urban and suburban interests. While inner-city neighborhoods may make more noise politically, it is the suburbs—in particular, the poorer suburbs, often identified as the swing districts in American politics—that have the votes and decide elections.

Having had partial success in pushing for fairer school funding on the state level, Gonzales recently brought a group of city and suburban church leaders together with representatives of Chicago's six largest banks, asking the banks to pledge some $90 million in affordable mortgages for underserved communities like Harvey and Chicago Heights. A contentious issue was whether the banks would let church leaders control the lending process. It was nuts-and-bolts community-development work, and there was no question that Gonzales was the galvanizing force in the room.

William Whyte wrote that in the years after World War II, people and small businesses moved to Park Forest and the other suburbs "for quite rational, and eminently sensible, reasons." Presumably, they began moving out of those communities for similarly sensible reasons. But in the process, the commercial base of those communities deteriorated, revenues diminished, schools declined and property taxes rose.

A conclusion drawn by Orfield, Katz and others is that the problems these postwar suburbs now face were never their own doing. These towns are casualties of planning and policy decisions, and not simply vague demographic trends. Yet, if people like Gonzales can bring together disconnected constituencies to reconsider those decisions and redirect those forces, if neighboring towns can arrive at a unified vision for a region's future, if reinvestment in places like Park Forest can be made as attractive as chewing up a green field, then major metropolitan areas will start to "become as oriented inward as they are outward," Katz believes.

When that happens, it won't be for sentimental reasons that people and businesses move back to places like Park Forest. They will move back because it makes sense.

Thomas Hackett is a freelance writer based in New York.

The end of urban man? Care to bet?

Humanity for millennia has chosen to live in cities. They serve its needs

"MAN", wrote Aristotle in roughly 330BC, "is by nature a city beast." So, he could have added, are rats.

Not that he would have, because his "city" was the little city-state, the *polis*, of ancient Greece, usually fed by its own hinterland, filthy, mostly ill-built and unpaved, but (except for its slaves) a coherent, durable social unit. His adjective *politikon* really meant, for him, what it suggests to us: he was on about politics, not town-planning.

Yet today the comparison with rats looks apt. The Athens of Aristotle's day was a monster, by that day's standards, with 150,000 people, hinterland included; and it had been bigger. But even at its peak it was a village to today's Athens of 3m people, traffic jams and fumes. And that in turn is nothing to some modern cities: "metro" Mexico city—the conurbation—holds around 18m people.

And today's cities are ballooning. Bombay in 1960 was a jam-packed city of 4m people. A peninsula, a bit like Manhattan, surely it could take no more? It did: today's metro Mumbai, bursting over its landward boundary, holds 18m. Rich cities have barely one person per room; Jakarta has three, Lagos six.

Other ills are just as visible. Most rich cities have cleared the smoke of the 19th century: London breathed its last pea-soup fog in 1954, the velvety soot on Manchester's buildings has been scoured away. Not so in poor countries. Mexico city's air is famously filthy, as is that of many Indian, Chinese, even Russian and East European cities.

Then there is the car. It pollutes, witness the smogs of Los Angeles or Chile's Santiago. It strangles: by 1940 New York had 1m motor vehicles, in Manhattan averaging less than 5mph (8kph). Poor cities can now beat that: add the car and Cairo's 10m people to streets fit for a tenth as many, and the car often goes slower than the people. Not to forget sewage and waste disposal, nor the inner-city decay as the richer rats flee to the suburbs. And so the litany goes on. The city is surely doomed, as Lewis Mumford, an American sociologist, foretold 60 years ago, and many others since.

But hold on. Sixty years? That's a long time to spend dying, and in fact most cities have spent it doing the opposite. Not even rich countries—which could afford it—have seen mass desertion of their cities. For every Detroit left (temporarily, at that) to rot, there are umpteen city centres still full of willing citizens; look at the rents. In poor countries, 40 years ago, it was easy to foresee riot and bloodshed, as grim, already bursting, cities swelled. It has seldom happened. Is Lagos a coherent, durable social unit? Is Calcutta? Maybe not. But where is the revolution? Overcrowded rats turn and rend each other. Humans might—but they haven't.

The usefulness of cities

To know why not, ask why cities grew. Europe's, over the centuries, reply: they served ends that could not have been served otherwise. In 999, as when Aristotle wrote, man was not, in practice, a city beast: most people lived in the countryside. Yet not by choice. Today, almost half of humanity lives in towns. And it does so because it wants to; or, at least, chooses the lesser of two evils.

Man as always lived in groups: it makes life safer and easier (and more fun). Some of Europe's towns grew round the stronghold of a local lord. Most, including these, developed as buying and selling did; trade needs a market, and markets need a certain mass. They are also self-reinforcing; a good one wins more business, a bad one, with few buyers, or the wrong goods or prices, dies.

All successful towns met this economic need—or others. The English "villein" of 1100 could hope to escape the tie to his land and lord if he decamped to town. For any peasant anywhere, town was the one place where he might make his fortune, like Dick Whittington, a real person who did "turn again", became lord mayor of London and died in 1423. When harvests failed, the nearby town offered hope of survival.

> Hark, hark, the dogs do bark,
> the beggars are come to town

is not just a nursery rhyme, it is social history. Centuries later, country people all over Europe flocked to towns for work in the new factories. The same motivations drive the ex-peasants of poor countries to town today.

Above all, though, the towns served their own citizens. Provided they were

5 ❖ POPULATION, RESOURCES, AND SOCIOECONOMIC DEVELOPMENT

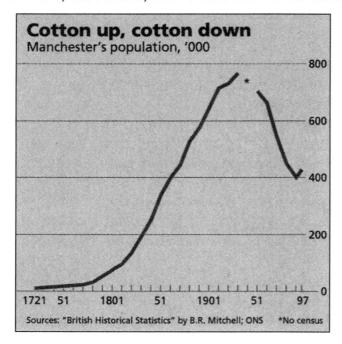

in turn served by them. Geography—rich soil, a safe harbour or navigable river, ample fresh water, easy defence, later minerals like coal—was the start of many a town. But men made the difference. Let the river silt up, as Bruges did around 1500, and decline was near.

Bruges—whose Buers family, on the Buerseplaats, gave us the bourse, the merchants' exchange—was in fact already declining. It had been a huge cloth market. Foreign traders crowded it, ships came from the Mediterranean as early as 1300. Later the dukes of Burgundy set up there. Here were courtly splendour, wealth, richly endowed religious houses—and somehow enterprise got lost. Antwerp took the trade. Not just was it nearer the growing supply of cheap English cloth, but it chose to compete: Bruges tried to keep the cheap cloth out, Antwerp bought it unfinished, and finished it for resale.

Rather the same happened in England's textile industry. Wool came from flocks in the Cotswolds, west of Oxford, or East Anglia (whose vast churches witness still to the wealth it brought). Yet it was Yorkshire whose weavers, later, created new cities; because it had enterprise and its rivals had less. Likewise it was a climate of enterprise—and existing mechanical skills—not of the damp air dear to old geography books, that made Manchester the cotton capital of the world.

Mischance could defeat the best-laid plans. Antwerp bought its cloth, and Portuguese spices, with German silver; then 16th century Spaniards found South American silver, and the spice/silver trade moved to Lisbon. Lübeck lived by trans-shipping Baltic goods to cross Denmark—till traders learned to sail round that country. Yet Florence showed how a city could live on its wits. Its hinterland is not rich; it became so. It is not a port; yet its merchants built a Europe-wide trade. Its river Arno gave it plenty of water, essential for making cloth, but no more than some other cities; yet it beat most as a maker and market of cloth. Above all, it virtually invented banking, and for 250 years led Europe at it. That was no accident. It was the result of a community, men learning from and competing with each other, just like London's Elizabethan theatre or Detroit's car makers; in sum, a city at work.

Venice built its medieval greatness on trade. Rome sold the faith, drawing in contributions to the papacy, and pilgrims in sometimes embarrassing but always enriching numbers. Since 1850, Paris—though also a great industrial centre, housing, not least, Gustave Eiffel's ironworks—has sold pleasure, just as Las Vegas sells gambling.

All cities, of course, sold and sell sex. This was a huge trade in the 19th century, though seldom as glamourously practised as by the *grandes horizontales* of 19th-century Paris or 16th- or 17th-century Rome and Venice. Amsterdam has records of a licensed, well-organised red-light district 350 years ago.

The sex trade, then and now, is typically an urban one, and it witnesses to something else that cities sell: a degree both of tolerance and of anonymity. Tolerance was limited, as many a pogrom in European cities testifies of the past, or sectarian riots in South Asian or Indonesian cities do today. Yet the Jewish community in its 16th-century ghetto, chosen or imposed, was far safer (and big enough to provide rabbis and synagogues) than would the lonely rural Jew have been, or was the lonely Albanian or Serb in 1999 Kosovo. For all its faults, the city serves.

It has served education well, if only because children had to be assembled in one place, and older students to be housed. Not until the 19th-century English boarding-school was any secondary school set up except in a town, and few then: Westminster school in central London and Galatasaray in Istanbul are far more typical. No university, till very lately, was plonked down in virgin countryside. Cities did not always like the results. Medieval Paris often saw conflict between lay and clerical authority over miscreant students. Oxford and Cambridge are still to outgrow the mutual dislike of town and gown. But the urban university, in a town and eager to be of it, is the pattern.

Cities served politics too. A few housed central government (hence the rise of Berlin and later, in part, thanks to public procurement, of its great electrical industry); many challenged it. City councils often disputed royal, noble or churchly power. They were rarely democratic. Many cities had a ruling elite, which even the new rich found hard to join: Venice's patricians admitted no fresh blood from 1381 to 1646. And their politics could be rough; in Italian cities like Florence, often bloody. Yet cities were a countervailing force, notably in the Low Countries, 40–50% urbanised by 1600, where they coexisted uneasily with the local lord; and, till about 1550, in Germany, recognising imperial authority, but in practice governing themselves.

Many a ruler had reason to rue the power of cities, or their mobs. In Castile in 1520–21, commoners claimed rights still denied in Spain 450 years later. In England's 1640s civil war, London was a pillar of Parliament against the king. In the great killing of Protestants in 1572, the Paris mob outran its rulers; in 1789, far more radical than the countryside, it overthrew them; it was crushed only by ruthless force in 1848 and 1871. Russia's Bolshevik revolution was in fact a coup backed by the urban working class. For good or ill, the city was always an agent of change.

The price

The human price of city growth has been huge, of course. Witness Odon de Deuil, on Constantinople in the 1140s:

> Filthy, stinking... perpetual darkness... richer than anywhere else, and wickeder.

Or Alexis de Tocqueville in 1835 on Manchester, the first industrial city, suddenly swollen by its cotton mills, a city of "half-daylight", as he put it, smothered by black smoke:

> Heaps of dung, building rubble... one-storey houses whose ill-fitting planks and broken windows suggest a last refuge between poverty and death... yet below some a row of cellars, 12 to 15 human beings crowded into each repulsive hole... [Yet] from this foul drain, the greatest stream of human industry flows out to fertilise the whole world, from this filthy sewer pure gold flows.

Could things have been otherwise? For all their filth, medieval cities were not chaos. They housed real communities, each trade run by its own guild, sometimes concentrated in its own district. They were small, to today's eyes. Constantinople had perhaps 750,000 people around 1000, but Florence not quite 100,000 in 1300 (and half as many in 1400, after the Black Death). They grew slowly. Some were planned: Salisbury was laid out as a new town before 1250; a pope was driving straight streets through Rome 400 years before Haussmann did it in 1860s Paris.

The explosion began in the 18th century. Berlin's inhabitants quadrupled from 1700 to 1800, if only to 170,000 (with 950,000, London was Europe's largest city by far in 1800). Paris grew 3½-fold, to 2m, from 1800 to 1870; Chicago 12-fold, to 3½m, from 1870 to 1930. Planners did their best, with some success in the New World: Philadelphia was designed, on an elegant grid, in 1681–82, Washington, on a grid, plus diagonal avenues, in 1791–92. Space shaped architecture. Cheap land gave sites for the detached houses of 19th-century Chicago, just as dear land was producing the tall, multi-family, bourgeois courtyard blocks of Paris. Many cities quite early had rules on heights or plot sizes: hence the narrow one-family houses of Amsterdam, or, later, the backward-stretching tenements of Berlin or New York. But sheer numbers and market forces often defeated good intentions. The half-acre plots from which early Philadelphians were to feed themselves were built over; Frank Lloyd Wright's one-acre dreams of the 1930s never left the drawing-board. The "garden city" was a fine idea; with rare exceptions, it just hasn't happened.

In contrast, gold did indeed flow from putrid Manchester (until its cotton industry shrank after 1950, and the city with it). And note: while the gold remains, the filth has gone (if only to the exploding cities of Asia). One could blench 30 years ago at Europe's megalopolis of 56m people, stretching from the Ruhr up to the Amsterdam-Rotterdam "Randstad", down to Charleroi and Lille, and over to south-east England. It now has 59m, richer, people—but far cleaner cities.

Yet if pollution, traffic and the suburban shopping mall cannot kill the city, will tele-working and the net? Will downtowns like Houston's be abandoned to decay, their office towers unpeopled as the pyramids, leaving suburbia to rule? Futurologists love to tell us so. Let them tell the birds.

China's changing land

Population, food demand and land use in China

Gerhard K. Heilig

China's land-use changes directly affect the country's capacity to generate sufficient food supplies. Losses of arable land due to natural disasters, agricultural restructuring, and infrastructure expansion might lead to food production deficits in the future.

China's food prospects, however, are also of geostrategic and geopolitical relevance to the West. Food deficits in China might destabilize the country and jeopardize the process of economic and political reform. Despite China's strong commitment to self-sufficiency, the country may become a major importer of (feed) grain. This is of great economic interest to large grain exporters such as the United States, France, or Australia.

There is also the danger that climate change might affect China's vegetation cover—especially the large grasslands, which are important for the country's livestock production. An increased frequency of catastrophic weather events could trigger massive floods or extended droughts that would affect major agricultural areas in various parts of China. These trends pose serious risks for the country's future food security.

What are the major trends in China's land-use changes?

There are few places in the world where people have changed the land so intensively, and over such a long period, as in China. The Loess Plateau of Northern China, for instance, was completely deforested in pre-industrial times. The Chinese started systematic land reclamation and irrigation schemes, converting large areas of natural land into rice paddies, as far back as the early Han Dynasty, in the fourth and third centuries BC. This process, which was scientifically planned and coordinated by subsequent dynastic bureaucracies, reached its first climax in the 11th and 12th centuries. Another period of massive land modification followed in the second half of the 18th and first half of the 19th century.

The LUC project, however, deals with a more recent phase—approximately since the foundation of the People's Republic of China in 1949—and in particular the period since economic reforms began in 1978. Although there has been widespread speculation in the scientific literature about losses of cultivated land in China due to pollution and urban expansion, very little hard data was available when the LUC project started. Through its collaboration with Chinese partners and other sources, LUC has received highly detailed land-use and land-cover data for China. This information includes statistical data, mapped information and remote sensing data. Table 1 presents some results from LUC's analyses of the most recent trends, based on new surveys from the Chinese State Land Administration:

- Conversion of cropland into horticulture was the most important factor in land-use change in China in recent years. Between 1988 and 1995 farmers converted some 1.2 million hectares (ha) of cropland into horticulture. This conversion is a positive trend, indicating a growing market orientation of Chinese agriculture.
- Manifold construction activities (roads, settlements, industry and mining) diminished China's cultivated land by 980,000 ha between 1988 and 1995.
- The third most important type of land-use change was reforestation, diverting almost 970,000 ha of previously cultivated land.
- Between 1988 and 1995, China also lost some 850,000 ha due to natural disasters–mainly flooding and droughts.

These data clearly indicate that anthropogenic factors are mainly responsible for recent land-use changes in China. The growing demand for meat, fish, fruit, and vegetables drives much of the agricultural restructuring, such as conversion to horticultural land and fishponds. Environmental programs are promoting reforestation, and the expansion of infrastructure is a function of rapid economic development and urbanization.

What are important drivers of land-use change in China?

Proximate determinants, such as those discussed above, are just the last step in a chain of causation that triggers land-use change in China. There are, however, three fundamental factors behind these trends:

- Population growth. Most recent projections from the United Nations Population Division (see Figure 1) assume that China's population will increase to some 1.49 billion people by 2025 and then slowly decline to 1.48 by 2050.

45. China's Changing Land

Table 1: Increase, decrease and net-change of cultivated land in China by region, 1988 – 1995 (in hectares).

	Increase				Decrease Conversion						Net-Change
	Reclamation	Drainage	Re-use of abandoned land	Conv. from agricultural land	Construct. (1)	to horticulture	to forest land	to grassland	to fishponds	Disasters	
North	289.733	18.233	106,028	41,080	−229.825	−298.595	−111.895	−41.255	−7.979	−105,535	−340.020
Northeast	396.867	21.991	39.957	25.852	−109,662	−87.417	−156.501	−90.631	−9.426	−250.871	−219.842
East	43.250	21.147	34.284	44.628	−242.824	−178.677	−24.765	−1.416	−45.624	−29.328	−379.324
Central	79.943	7.665	13.159	26.529	−91.802	−101.858	−111.782	−2.689	−61.519	−41.456	−283.771
South	320.001	22.686	18.870	96.681	−117.023	−116.966	−81.465	−10.361	−87.096	−56.534	−11.206
Southwest	358.704	8.687	40.260	90.450	−92.583	−114.235	−161.770	−93.864	−7.902	−132.915	−105.169
Plateau	28.044	980	3.135	3.935	−7.927	−163	−363	−11.874	−1	−486	15.281
Northwest	681.902	19.798	93.428	84.270	−88.588	−325.316	−321.214	−296.605	−6.322	−239.252	−397.900
TOTAL	2.198.444	121.177	349.121	413.465	−980.235	−1.223.229	−969.756	−548.694	−225.868	−856.377	−1.721.951

Source: State Land Administration, Statistical Information on the Land of China in 1995. Beijing, 1996. And equivalent reports for 1988 to 1994.

Note: (1) "Construction" includes all kinds of infrastructure, industrial areas and residential areas. In the original data tables this category includes all contruction by state-owned units (cities, towns, mining and factories, railways, highways, water reservoirs, public buildings) and constructions by rural communities (rural roads, township and village enterprises, rural water reservoirs, offices, education and sanitation, rural private resident housing).

- Income growth. By all measures, China had spectacular economic growth rates in recent years. The number of people in poverty declined by some 200 million.
- Urbanization. Although China still has a large agricultural population, most experts predict a rapid increase in the number and size of towns and cities in the future. This urbanization is associated with changes in labor-force participation and lifestyle.

Population growth increases the demand for food and thus leads to intensification of agriculture and expansion of cultivated land. However, as the LUC analysis shows, there is a high error range in projecting China's population for more than 20 to 30 years. Its huge initial size amplifies even very small changes in fertility and mortality. For instance, between estimates made in 1994 and 1998, the United Nations Population Division had to revise its population projection for China for the year 2050 by 128 million. This uncertainty is unavoidable, and must be taken into account in predicting food demand and other population-related changes.

There is also considerable uncertainty concerning China's future economic development. Current trends indicate that China will experience a rapid transition from an agricultural to an industrial and service society. This would lead to expansion of infrastructure and areas of settlement, encroaching on valuable cropland, especially around urban-industrial agglomerations in coastal provinces.

An increase of productivity in both industry and agriculture could also lead to massive unemployment—particularly among the agricultural population. China already has an excess agricultural labor force of at least 120 million, which will increase the potential for rural-urban migration. Policy measures could probably slow down urbanization to some extent, but most projections assume a massive increase in urban population from the

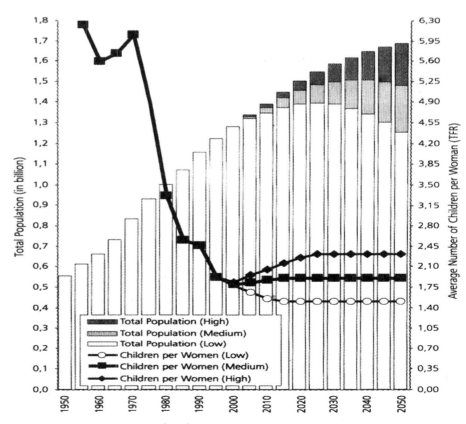

Figure 1: Population projection for China.

current 370 million to 800 million people. All these trends would—directly or indirectly—influence land use in China. With growing income and urban lifestyle, for instance, Chinese consumers will further change their diet (see Figure 2).

According to FAO estimates, there was already a massive increase in the domestic supply of meat from 6.6 to 47.9 million metric tons between the mid-1960s and the mid-1990s—with a significant effect on crop production patterns. The harvest of maize (the major feed crop) increased from 25 to 113 million tons between the mid-1960s and mid-1990s. Use of cereals for feeding animals grew from 12 million tons in 1964–66 to 107 million tons in 1994–96. Similar trends can be observed with other food commodities. Between the mid-1960s and mid-1990s, China's vegetable production increased from 37.9 to 182.7 million tons, and the production of fish increased from 3.3 to 24.3 million tons. These consumer-driven changes in agricultural production caused most of the land-use change in China during the last few decades.

Some general observations resulting from LUC's research

LUC's various models and analyses indicate that China is certainly facing a critical phase of its development during the next three decades. In that period, the demographic momentum will inevitably lead to further population growth, in the order of some 260 million people. Their demand for food, water, and shelter will greatly increase pressure on China's land and water resources. The country's urbanization and economic development will amplify this pressure. Without effective control of pollution and farmland loss, China faces serious problems of land degradation and productivity decline that might threaten the country's food security.

The LUC project has analyzed how global climate change might affect natural vegetation, agriculture, and water management systems in China. This is essential for a long-term water and land resource strategy. China will only be

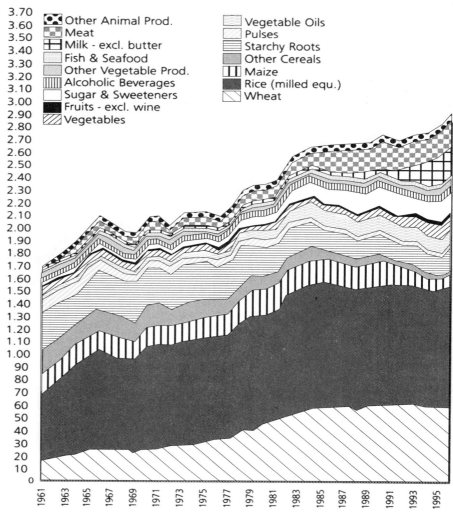

Figure 2: Per capita food consumption by commodity.

able to cope with its serious problems of flooding and drought if it takes into account the interactions between human and natural factors. For instance, flood damage has increased in recent years because farmers moved into areas that should have been reserved for flood control. The serious water deficit in downstream areas of the Yellow River is amplified by a massive increase in (rather inefficient) agricultural water consumption upstream. As the LUC analysis on *Human Impact on Yellow River Water Management* (IIASA IR 98-016) has shown. China will need intense water management activities to secure a reliable water supply.

Despite recent progress, great data deficits and uncertainties still exist concerning both biophysical and socioeconomic conditions and trends. While LUC was able—with the help of Chinese colleagues—to improve, cross-check, and update several (georeferenced) data sets on China (climate, soils, water, cultivated land, grassland, natural vegetation, population, etc.), project staff still see the need for additional sources of information. In particular, remote sensing data could improve land-use analyses and planning for certain regions. Due to the rapid change in China, available statistical or mapped data are not always up to date. For instance, the conversion of arable land due to infrastructure expansion and urban sprawl in recent years could be analyzed more accurately with satellite images.

Helping the world's poorest

Jeffrey Sachs, a top academic economist, argues that rich countries must mobilise global science and technology to address the specific problems which help to keep poor countries poor

IN OUR Gilded Age, the poorest of the poor are nearly invisible. Seven hundred million people live in the 42 so-called Highly Indebted Poor Countries (HIPCS), where a combination of extreme poverty and financial insolvency marks them for a special kind of despair and economic isolation. They escape our notice almost entirely, unless war or an exotic disease breaks out, or yet another programme with the International Monetary Fund (IMF) is signed. The Cologne Summit of the G8 in June was a welcome exception to this neglect. The summiteers acknowledged the plight of these countries, offered further debt relief and stressed the need for a greater emphasis by the international community on social programmes to help alleviate human suffering.

The G8 proposals should be seen as a beginning: inadequate to the problem, but at least a good-faith prod to something more useful. We urgently need new creativity and a new partnership between rich and poor if these 700m people (projected to rise to 1.5 billion by 2030), as well as the extremely poor in other parts of the world (especially South Asia), are to enjoy a chance for human betterment. Even outright debt forgiveness, far beyond the G8's stingy offer, is only a step in the right direction. Even the call to the IMF and World Bank to be more sensitive to social conditions is merely an indicative nod.

A much more important challenge, as yet mainly unrecognised, is that of mobilising global science and technology to address the crises of public health, agricultural productivity, environmental degradation and demographic stress confronting these countries. In part this will require that the wealthy governments enable the grossly underfinanced and underempowered United Nations institutions to become vibrant and active partners of human development. The failure of the United States to pay its UN dues is surely the world's most significant default on international obligations, far more egregious than any defaults by impoverished HIPCS. The broader American neglect of the UN agencies that assist impoverished countries in public health, science, agriculture and the environment must surely rank as another amazingly misguided aspect of current American development policies.

The conditions in many HIPCS are worsening dramatically, even as global science and technology create new surges of wealth and well-being in the richer countries. The problem is that, for myriad reasons, the technological gains in wealthy countries do not readily diffuse to the poorest ones. Some barriers are political and economic. New technologies will not take hold in poor societies if investors fear for their property rights, or even for their lives, in corrupt or conflict-ridden societies. *The Economist's* response to the Cologne Summit ("Helping the Third World", June 26th) is right to stress that aid without policy reform is easily wasted. But the barriers to development are often more subtle than the current emphasis on "good governance" in debtor countries suggests.

Research and development of new technologies are overwhelmingly directed at rich-country problems. To the extent that the poor face distinctive challenges, science and technology must be directed purposefully towards them. In today's global set-up, that rarely happens. Advances in science and technology not only lie at the core of long-term economic growth, but flourish on an intricate mix of social institutions—public and private, national and international.

Currently, the international system fails to meet the scientific and technological needs of the world's poorest. Even when the right institutions exist—say, the World Health Organisation to deal with pressing public health disasters facing the poorest countries—they are generally starved for funds, authority and even access to the key negotiations between poor-country governments and the Fund at which important development strategies get hammered out.

Jeffrey Sachs is director of the Centre for International Development and professor of international trade at Harvard University. A prolific writer, he has also advised the governments of many developing and East European countries.

The ecology of underdevelopment

If it were true that the poor were just like the rich but with less money, the global situation would be vastly easier than it is. As it happens, the poor live in different ecological zones, face different health conditions and must overcome agronomic limitations that are very different from those of rich countries. Those differences, indeed, are often a fundamental cause of persisting poverty.

Let us compare the 30 highest-income countries in the world with the 42 HIPCS (see table below). The rich countries overwhelmingly lie in the world's temperate zones. Not every country in those bands is rich, but a good rule of thumb is that temperate-zone economies are either rich, formerly socialist (and hence currently poor), or geographically isolated (such as Afghanistan and Mongolia). Around 93% of the combined population of the 30 highest-income countries lives in temperate and snow zones. The HIPCS by contrast, include 39 tropical or desert societies. There are only three in a substantially temperate climate, and those three are landlocked and therefore geographically isolated (Laos, Malawi and Zambia).

Not only life but also death differs between temperate and tropical zones. Individuals in temperate zones almost everywhere enjoy a life expectancy of 70 years or more. In the tropics, however, life expectancy is generally much shorter. One big reason is that populations are burdened by diseases such as malaria, hookworm, sleeping sickness and schistosomiasis, whose transmission generally depends on a warm climate. (Winter may be the greatest public-health intervention in the world.) Life expectancy in the HIPCS averages just 51 years, reflecting the interacting effects of tropical disease and poverty. The economic evidence strongly suggests that short life expectancy is not just a result of poverty, but is also a powerful cause of impoverishment.

All the rich-country research on rich-country ailments, such as cardiovascular diseases and cancer, will not solve the problems of malaria. Nor will the biotechnology advances for temperate-zone crops easily transfer to the conditions of tropical agriculture. To address the special conditions of the HIPCS, we must first understand their unique problems, and then use our ingenuity and co-operative spirit to create new methods of overcoming them.

Modern society and prosperity rest on the foundation of modern science. Global capitalism is, of course, a set of social institutions—of property rights, legal and political systems, international agreements, transnational corporations, educational establishments, and public and private research institutions—but the prosperity that results from these institutions has its roots in the development and applications of new science-based technologies. In the past 50 years, these have included technologies built on solid-state physics, which gave rise to the information-technology revolution, and on genetics, which have fostered breakthroughs in health and agricultural productivity.

Science at the ecological divide

In this context, it is worth noting that the inequalities of income across the globe are actually exceeded by the inequalities of scientific output and technological innovation. The chart below shows the remarkable dominance of rich countries in scientific publications and, even more notably, in patents filed in Europe and the United States.

The role of the developing world in one sense is much greater than the chart indicates. Many of the scientific and technological breakthroughs are made by poor-country scientists working in rich-country laboratories. Indian and Chinese engineers account for a significant proportion of Silicon Valley's workforce, for example. The basic point, then, holds even more strongly: global science is directed by the rich countries and for the rich-country markets, even to the extent of mobilising much of the scientific potential of the poorer countries.

The imbalance of global science reflects several forces. First, of course, science follows the market. This is especially true in an age when technological leaps require expensive scientific equipment and well-provisioned research laboratories. Second, scientific advance tends to have increasing returns to scale: adding more scientists to a community does not diminish individual marginal productivity but tends to increase it. Therein lies the origin of university science departments, regional agglomerations such as Silicon Valley and Route 128, and mega-laboratories at leading high-technology firms including Merck, Microsoft and Monsanto. And third, science requires a partnership between the public and private sectors. Free-market ideologues notwithstanding, there is scarcely one technology of significance that was not nurtured through public as well as private care.

If technologies easily crossed the ecological divide, the implications would be less dramatic than they are. Some technologies, certainly those involving the computer and other ways of managing information, do indeed cross over, and give great hopes of spurring technological capacity in the poorest countries. Others—especially in the life sciences but also in the use of energy, building techniques, new materials and the like—are prone to "ecological specificity". The result is a profound imbalance in the global production of knowledge: probably the most powerful engine of divergence in global wellbeing between the rich and the poor.

Consider malaria. The disease kills more than 1m people a year, and perhaps as many as 2.5m. The disease is so heavily concentrated in the poorest tropical countries, and overwhelmingly in sub-Saharan Africa, that nobody even bothers to keep an accurate count of clinical cases or deaths. Those who remember that richer places such as Spain, Italy, Greece and the southern United States once harboured the disease may be misled into thinking that the problem is one of social institutions to control its transmission. In fact, the sporadic transmission of malaria in the sub-tropical regions of the rich countries was vastly easier to control than is its chronic transmission in the heart of the tropics. Tropical countries are plagued by ecological conditions that produce hundreds of infective bites per year per person.

Different ecologies 1995	HIPCs* (42)	Rich countries (30)
GDP per person, PPP$†	1,187	18,818
Life expectancy at birth, years†	51.5	76.9
Population by ecozones, % in:		
tropical	55.6	0.7
dry	17.6	3.7
temperate and snow	12.5	92.6
highland	14.0	2.5

Source: J. Sachs *Highly indebted poor countries †Unweighted averages

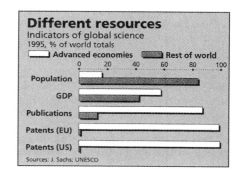

Different resources
Indicators of global science
1995, % of world totals
Advanced economies / Rest of world

Population, GDP, Publications, Patents (EU), Patents (US)

Sources: J. Sachs; UNESCO

46. Helping the World's Poorest

Mosquito control does not work well, if at all, in such circumstances. It is in any event expensive.

Recent advances in biotechnology, including mapping the genome of the malaria parasite, point to a possible malaria vaccine. One would think that this would be high on the agendas of both the international community and private pharmaceutical firms. It is not. A Wellcome Trust study a few years ago found that only around $80m a year was spent on malaria research, and only a small fraction of that on vaccines.

The big vaccine producers, such as Merck, Rhône-Poulenc's Pasteur-Mérieux-Connaught and SmithKline Beecham, have much of the in-house science but not the bottom-line motivation. They strongly believe that there is no market in malaria. Even if they spend the hundreds of millions, or perhaps billions, of dollars to do the R&D and come up with an effective vaccine, they believe, with reason, that their product would just be grabbed by international agencies or private-sector copycats. The hijackers will argue, plausibly, that the poor deserve to have the vaccine at low prices—enough to cover production costs but not the preceding R&D expenditures.

The malaria problem reflects, in microcosm, a vast range of problems facing the HIPCS in health, agriculture and environmental management. They are profound, accessible to science and utterly neglected. A hundred IMF missions or World Bank health-sector loans cannot produce a malaria vaccine. No individual country borrowing from the Fund or the World Bank will ever have the means or incentive to produce the global public good of a malaria vaccine. The root of the problem is a much more complex market failure: private investors and scientists doubt that malaria research will be rewarded financially. Creativity is needed to bridge the huge gulfs between human needs, scientific effort and market returns.

Promise a market

The following approach might work. Rich countries would make a firm pledge to purchase an effective malaria vaccine for Africa's 25m newborn children each year if such a vaccine is developed. They would even state, based on appropriate and clear scientific standards, that they would guarantee a minimum purchase price—say, $10 per dose—for a vaccine that meets minimum conditions of efficacy, and perhaps raise the price for a better one. The recipient countries might also be asked to pledge a part of the cost, depending on their incomes. But nothing need be spent by any government until the vaccine actually exists.

Even without a vast public-sector effort, such a pledge could galvanise the world of private-sector pharmaceutical and biotechnology firms. Malaria vaccine research would suddenly become hot. Within a few years, a breakthrough of profound benefit to the poorest countries would be likely. The costs in foreign aid would be small: a few hundred million dollars a year to tame a killer of millions of children. Such a vaccine would rank among the most effective public-health interventions conceivable. And, if science did not deliver, rich countries would end up paying nothing at all.

Malaria imposes a fearsome burden on poor countries, the AIDS epidemic an even weightier load. Two-thirds of the world's 33m individuals infected with the HIV virus are sub-Saharan Africans, according to a UN estimate in 1998, and the figure is rising. About 95% of worldwide HIV cases are in the developing world. Once again, science is stopping at the ecological divide.

Rich countries are controlling the epidemic through novel drug treatments that are too expensive, by orders of magnitude, for the poorest countries. Vaccine research, which could provide a cost-effective method of prevention, is dramatically underfunded. The vaccine research that is being done focuses on the specific viral strains prevalent in the United States and Europe, not on those which bedevil Africa and Asia. As in the case of malaria, the potential developers of vaccines consider the poor-country market to be no market at all. The same, one should note, is true for a third worldwide killer. Tuberculosis is still taking the lives of more than 2m poor people a year and, like malaria and AIDS, would probably be susceptible to a vaccine, if anyone cared to invest in the effort.

The poorer countries are not necessarily sitting still as their citizenry dies of AIDS. South Africa is on the verge of authorising the manufacture of AIDS medicines by South African pharmaceutical companies, despite patents held by American and European firms. The South African government says that, if rich-country firms will not supply the drugs to the South African market at affordable prices (ones that are high enough to meet marginal production costs but do not include the patent-generated monopoly profits that the drug companies claim as their return for R&D), then it will simply allow its own firms to manufacture the drugs, patent or no. In a world in which science is a rich-country prerogative while the poor continue to die, the niceties of intellectual property rights are likely to prove less compelling than social realities.

There is no shortage of complexities ahead. The world needs to reconsider the question of property rights before patent rights allow rich-country multinationals in effect to own the genetic codes of the very foodstuffs on which the world depends, and even the human genome itself. The world also needs to reconsider the role of institutions such as the World Health Organisation and the Food and Agriculture Organisation. These UN bodies should play a vital role in identifying global priorities in health and agriculture, and also in mobilising private-sector R&D towards globally desired goals. There is no escape from such public-private collaboration. It is notable, for example, that Monsanto, a life-sciences multinational based in St Louis, Missouri, has a research and development budget that is more than twice the R&D budget of the entire worldwide network of public-sector tropical research institutes. Monsanto's research, of course, is overwhelmingly directed towards temperate-zone agriculture.

People, food and the environment

Public health is one of the two distinctive crises of the tropics. The other is the production of food. Poor tropical countries are already incapable of securing an adequate level of nutrition, or paying for necessary food imports out of their own export earnings. The HIPC population is expected to more than double by 2030. Around one-third of all children under the age of five in these countries are malnourished and physically stunted, with profound consequences throughout their lives.

As with malaria, poor food productivity in the tropics is not merely a problem of poor social organisation (for example, exploiting farmers through controls on food prices). Using current technologies and seed types, the tropics are inherently less productive in annual food crops such as wheat (essentially a temperate-zone crop), rice and maize. Most agriculture in the equatorial tropics is of very low productivity, reflecting the fragility of most tropical soils at high temperatures combined with heavy rainfall.

High productivity in the rainforest ecozone is possible only in small parts of the tropics, generally on volcanic soils (on the island of Java, in Indonesia, for example). In the wet-dry tropics, such as the vast savannahs of Africa, agriculture is hindered by the terrible burdens of unpredictable and highly variable water supplies. Drought and resulting famine have killed millions of peasant families in the past generation alone.

Scientific advances again offer great hope. Biotechnology could mobilise genetic engineering to breed hardier plants that are more resistant to drought and less sensitive to pests. Such genetic engineering is stymied at every point, however. It is met with doubts in the rich countries (where people do not have to worry about their next meal); it requires a new scientific and policy framework in the poor countries; and it must somehow generate market incentives for the big life-sciences firms to turn their research towards tropical foodstuffs, in co-operation with tropical research centres. Calestous Juma, one of the world's authorities on biotechnology in Africa, stresses that there are dozens, or perhaps hundreds, of underused foodstuffs that are well adapted to the tropics and could be improved through directed biotechnology research. Such R&D is now all but lacking in the poorest countries.

The situation of much of the tropical world is, in fact, deteriorating, not only because of increased population but also because of long-term trends in climate. As the rich countries fill the atmosphere with increasing concentrations of carbon, it looks ever more likely that the poor tropical countries will bear much of the resulting burden.

Anthropogenic global warming, caused by the growth in atmospheric carbon, may actually benefit agriculture in high-latitude zones, such as Canada, Russia and the northern United States, by extending the growing season and improving photosynthesis through a process known as carbon fertilisation. It is likely to lower tropical food productivity, however, both because of increased heat stress on plants and because the carbon fertilisation effect appears to be smaller in tropical ecozones. Global warming is also contributing to the increased severity of tropical climatic disturbances, such as the "one-in-a-century" El Niño that hit the tropical world in 1997–98, and the "one-in-a-century" Hurricane Mitch that devastated Honduras and Nicaragua a year ago. Once-in-a-century weather events seem to be arriving with disturbing frequency.

The United States feels aggrieved that poor countries are not signing the convention on climatic change. The truth is that these poor tropical countries should be calling for outright compensation from America and other rich countries for the climatic damages that are being imposed on them. The global climate-change debate will be stalled until it is acknowledged in the United States and Europe that the temperate-zone economies are likely to impose heavy burdens on the already impoverished tropics.

New hope in a new millennium

The situation of the HIPCS has become intolerable, especially at a time when the rich countries are bursting with new wealth and scientific prowess. The time has arrived for a fundamental rethinking of the strategy for co-operation between rich and poor, with the avowed aim of helping the poorest of the poor back on to their own feet to join the race for human betterment. Four steps could change the shape of our global community.

First, rich and poor need to learn to talk together. As a start, the world's democracies, rich and poor, should join in a quest for common action. Once again the rich G8 met in 1999 without the presence of the developing world. This rich-country summit should be the last of its kind. A G16 for the new millennium should include old and new democracies such as Brazil, India, South Korea, Nigeria, Poland and South Africa.

Second, rich and poor countries should direct their urgent attention to the mobilisation of science and technology for poor-country problems. The rich countries should understand that the IMF and World Bank are by themselves not equipped for that challenge. The specialised UN agencies have a great role to play, especially if they also act as a bridge between the activities of advanced-country and developing-country scientific centres. They will be able to play that role, however, only after the United States pays its debts to the UN and ends its unthinking hostility to the UN system.

We will also need new and creative institutional alliances. A Millennium Vaccine Fund, which guaranteed future markets for malaria, tuberculosis and AIDS vaccines, would be the right place to start. The vaccine-fund approach is administratively straightforward, desperately needed and within our technological reach. Similar efforts to merge public and private science activities will be needed in agricultural biotechnology.

Third, just as knowledge is becoming the undisputed centrepiece of global prosperity (and lack of it, the core of human impoverishment), the global regime on intellectual property rights requires a new look. The United States prevailed upon the world to toughen patent codes and cut down on intellectual piracy. But now transnational corporations and rich-country institutions are patenting everything from the human genome to rainforest biodiversity. The poor will be ripped off unless some sense and equity are introduced into this runaway process.

Moreover, the system of intellectual property rights must balance the need to provide incentives for innovation against the need of poor countries to get the results of innovation. The current struggle over AIDS medicines in South Africa is but an early warning shot in a much larger struggle over access to the fruits of human knowledge. The issue of setting global rules for the uses and development of new technologies—especially the controversial biotechnologies—will again require global co-operation, not the strong-arming of the few rich countries.

Fourth, and perhaps toughest of all, we need a serious discussion about long-term finance for the international public goods necessary for HIPC countries to break through to prosperity. The rich countries are willing to talk about every aspect except money: money to develop new malaria, tuberculosis and AIDS vaccines; money to spur biotechnology research in food-scarce regions; money to help tropical countries adjust to climate changes imposed on them by the richer countries. The World Bank makes mostly loans, and loans to individual countries at that. It does not finance global public goods. America has systematically squeezed the budgets of UN agencies, including such vital ones as the World Health Organisation.

We will need, in the end, to put real resources in support of our hopes. A global tax on carbon-emitting fossil fuels might be the way to begin. Even a very small tax, less than that which is needed to correct humanity's climate-deforming overuse of fossil fuels, would finance a greatly enhanced supply of global public goods. No better time to start than as the new millennium begins.

Index

A

Abida Hussain, H. E. Syeda, 183
abortion services, U.S. aid and, 187
Acer Group, 47
adaptation, to weather, 69
aerial photographs, as USGS resource, 157
Africa, and AIDS, 189–192
African American children, at risk of lead poisoning, 40
agriculture: European Union policy on, 107; and poverty, 223–224; precision, 166; trade and disease, 177
agroterrorism, 176, 177–178
AIDS, 189–192, 223
air quality, urban, 215
algal blooms, 22–23
alluvial fan flooding. See flash flooding
Amundsen, Roald, 30
animals: climate and, 27; disease in, 175–179
Arctic Icebreaking Program, 30
Arctic Ocean, 30, 31
Arctic oscillation, 121
Arctic pack ice, 119–122
area studies (or regional) tradition, in geography, 12, 13–14
"armoring," 71
Asian clam, 25
Atlanta, Georgia, 76–80
automobile suburbs, 145–146, 148
automobiles, emissions of, 27

B

Babbitt, Bruce, 70–73, 75
Balkans, 116–118
Balling, Robert, on global warming, 52, 53
Bangladesh: flooding in, 162–163; population growth in, 182–188
Bashkortostan, 112–115
behavioral revolution, 199
bioinvasion, 175–179
biological diversity, loss of, 22
biotechnology, poverty and, 223, 224
BMW, 126–127
Bosnia, 116–118
bosque, 132–138
Brazil, 168, 169
Broecker, Wallace, 57, 59
Bush administration, 65, 66

C

Cairo conference, 1994, on population and development, 185, 186–187
California, dams in, 70–74
Canada, 29, 30
Canadian Wildlife Service, 31
carbon dioxide, 22, 23–24; global warming and, 27–29, 50, 63–69
Caribbean, and AIDS, 192
Carson, Rachel, 95, 97
Central America, 63–64
central business district (CBD), 144, 147, 148, 149
Chechen war, 111–112, 113
Cherokee County, Georgia, 76–80
Chicago, inner-ring suburbs of, 212–214
children: and AIDS, 190, 191; at-risk, 40; lead poisoning in, 32–45
China: climate change in, 219–220; food deficits in, 218; land use of, 218–220; population growth in, 218–220; and Taiwan, 46–47
cities, as habitat for humans, 215–217
Ciudad Juarez, 98–99, 134
climate, and poverty, 222, 223–224
climate change, 63–69; Arctic pack ice and, 119–122; carbon sinks and, 27–29, 66; opinions on, 50–53; El Niño and, 120; flip-flops in temperature and, 54–62; flood magnitude and, 208–210; polar ice cap and, 30–31; poverty and, 222, 223–224; satellite imaging of, 163; science and, 63–67
Clinton administration, 66, 81, 197
coastal changes: in Maine, 95–97; satellite imaging of, 163
Cologne Summit, of the G-8, 221
community, sense of, 17–18
conservation subdivisions, 79
conservationism, 84
contamination level, in fresh water, 204
contraceptives, population growth and, 182, 184–188
contract killing, of Galina Starovoitova, 108
corruption, in municipal elections, 109
cosmetics, lead acetate in, 34–35
cottonwood trees, 132–138
county, income by, 170–171
"creeping regional autonomy," 111
cropland, 85, 86, 87
cross-Strait relations, 46–47

D

dam-based ecosystem, 73–74
dams, 70–75
deforestation, 28, 29, 66
democracy, and human vulnerability to weather, 68–69
Des Groseilliers, 119
desalinization, 203
developing world, climate change and, 67–68
digital raster graphics (DRGs), 158–159
Digital Revolution, and the need for skyscrapers, 173–174
disease, 175–179; in poor countries, 222–224
Dongguan, China, 46
dropsy, 36
dry mantle events, 129

E

earth science tradition, in geography, 12, 14–15
Earth Summit, in Rio de Janeiro, 1992, 65
ecology, of underdeveloped countries, 222
education: family planning and, 186; formation of cities and, 216
Egypt, 183; satellite imaging of, 165
El Niño, 59–60, 224
El Paso, Texas, 98–99
Ellis Island, 150–153
eminent domain, 86–87
emissions, carbon dioxide, 27–29, 50, 64–66
energy use, 197
environment, 63–69; crude oil substitutes and, 197; lead in, 33, 35
Environmental Defense Fund, 64, 65
Environmental Protection Agency, (EPA), 40, 43, 64
environmentalism, 84
environmentalists: on carbon sinks, 28; strategy of, 65–66
Europe, carbon sinks and, 28; North Atlantic Current and, 55; polar ice cap and, 30–31
European Union, expansion of, 106–107
Evergreen Group, 47
externalities, 84
extinction, as natural process, 25

F

family planning services, population growth and, 182–186
Federal Cave Resources Protection Act, 158
Federal Emergency Management Agency (FEMA), 131
Federal Energy Regulatory Commission, 74, 75
Federal Residential Lead-Based Paint Hazard Reduction Act of 1992, 43
fertility rate, 199–200, 201
fisheries, 22, 25
flash flooding, 129, 130, 131
flooding: advantages, of, 134; hazards of, 129; in Bangladesh, 162–163; in the Midwest, 129–131; predicting, 207–211
Food and Drug Administration, 34
food production, and poverty, 223–224
forest fires, in Indonesia, 161
forests, 28, 29, 66, 85, 86–88
fossil-fuel burning, climate change and, 27–29, 50, 64, 66
Franklin, Benjamin, 36
Franklin, John, 30, 31
fresh water, 203–206; systems, 25

G

gas, natural, 197
gasoline, lead in, 35, 37
G-8, 224; Cologne summit on poverty, 221
general circulation models (GCMs), 66
geographical concentration, 139
geography: earth sciences and, 14; social studies and, 14
GIS technology, 150–153, 166
glaciers, 67
Glen Canyon, 73–74

global capitalism, 222
global economy, 200
global positioning system (GPS), 166
global trade/travel, and disease, 175–179
global warming. *See* climate change
Goodman, Oscar, 93–94
Gore, Al, 64, 66, 69, 81
Gorton, Slade, 72–73
grasslands, 85, 88
Gray Dawn, as global aging crisis, 198–202
Gray, William M., on global warming, 50, 52, 53
Great Salinity Anomaly, 59
greenhouse gases, 27–29, 50, 63–69; emissions, 121–122; and polar ice cap, 31
Greening Earth Society, 65–66
Greenpeace, 67
Greenville County, South Carolina, 123–128
Guangdong, province of China, 46, 103

H

habitat status, satellite imaging of, 163
Highly Indebted Poor Countries (HIPCs), 221–224
HIV (human immunodeficiency virus), 189–192, 223
Honduras, 63
Hong Kong, 46–47
Hubbert, M. King, 184, 195
Hughes, Howard, 93
human activity: fresh water and, 204; global warming and, 51–52, 65, 67
human processes, 10–11
human-dominated planet, 21–26
humans, diseases in, 175–179
Huntington, Samuel, 201
hurricane(s), 67; Mitch, 63–64, 68, 224
hydroelectric dams, 71, 74
hydrogen fuel cells, 28

I

ice dams, 58–59
ice, polar, 30–31, 119–122
ice-albedo feedback, 120
Illinois, 212–214
immigrant workers, 199
income, per capita, by county, 170–171
India: AIDS and, 192; population growth programs in, 185–186
Indonesia, 69, 161
industrialized countries, global warming and, 28, 64, 67
industry, emissions of, 27
intellectual property rights, 224
Intergovernmental Panel on Climate Change, 28, 65
invasions, of floras and faunas, 25
Iran, fertility rates in, 185
irrigation schemes, in China, 218
Isthmus of Panama, 57–58

J

Japan, 103–104, 105
Jetty Jacks, 134, 137
"Just Cause," 109

K

karst landscapes, 154, 160
KFOR, NATO-led peacekeeping force in Kosovo, 117–118
King, Angus, 97
knowledge, geographical, 16–20
Kosovo, 116–118
Kyoto Protocol of 1997, 27–29, 65, 68

L

Land and Water Conservation Fund, 80, 81
land cover, 83–91
land development, 85, 88; effects of, on arable land, 90; vs. grassland and forest, 89–90
land reclamation, 84
land transformation, 21–26, 83–91
land use, 83–91; and flood recurrence, 210
Las Vegas, Nevada, 92–94; Development Services Center, 92; gambling in, 93
law enforcement, environmental, by satellite imaging, 163
lead poisoning, in children, 32–45
lead-based paint, 32, 34, 37, 43
life expectancy, 199, 222
limestone, 154, 156, 157
Lindzen, Richard S., on global warming, 51, 53
liquid fuels, 197
LUC project, 218, 219, 220

M

Macau, 47
Macedonia, 116–118
Maine, 95–97
Makashov, Albert, 109
malaria, 222–223
mangrove ecosystems, 22
Manifest Destiny, 18
man-land tradition, in geography, 12, 14
mapping, art of, 13
maps: symbols of, 160; use of, for education, 154–160
maquiladoras, 98, 99
McPhee, Miles G., 119, 120, 122
Medicare reform, 199
metropolis, American, and growth of transportation, 142–149
Mexico, 98–99
Michaels, Patrick J., on global warming, 53
Middle East, 183; and oil, 193–194, 195, 197
Mielke, Howard W., 32–45
Mielke, Paul, 37
Milwaukee, Wisconsin, 214

Minneapolis/St. Paul, Minnesota, 213–214
models, climate, 66
Montenegro, 116–118
Multi-Response Permutation Procedures, 37
multiyear ice, 30, 31

N

National Academy of Sciences, 67
National Center for Atmospheric Research, 67
National Research Council, on global warming, 50–51
natural gas, 28, 197
negative externalities, 84
Nevada, flood hazards in, 129
New Jersey, 81; Department of Environmental Protection (NJDEP), 150–151
New Mexico, 132
Nicaragua, 63
nitrogen, 24, 29
North America Free Trade Agreement (NAFTA), 98–99, 103, 104
North Atlantic Current, 55, 57, 59
North Atlantic oscillation, 58, 59–60
North Atlantic, polar ice cap and, 30–31
North Pole, ice cap of, 30
Northeast Passage, 31
Northwest Passage, 30–31
no-till agriculture, 28, 29
Nriagu, Jerome, 36–37

O

oceans, climate change and, 27–29, 30–31, 52, 67, 119–122
Ogallala Aquifer, 87
oil, 193–197
OPEC (Organization of Petroleum Exporting Countries), 193–194, 195, 197
open space, 76–82
Oppenheimer, Michael, on global warming, 51, 53, 65

P

pack ice, Arctic, 119–122
Pakistan, population growth in, 182–183, 184–185
parametric statistical models, 37
Perovich, Donald K., 119, 120
Petropavlovsk-Kamchatsky, 108, 109
physical processes, 10–11
place, sense of, 17–18
planning: for floods, 129–131, 209–211; for urban growth, 212–214
plants, climate and, 27–29
polar bears, and ice cap, 30, 31
politics: in the Balkans, 116–118; and formation of cities, 216; of reducing carbon dioxide emissions, 27
pollution, 97
population, aging of, 198–202
Population and Community Development Association, 185

population growth, 67, 182–188; in cities, 215–217; cost of, in Las Vegas, 92–94; and loss of fresh water, 204; and immigration, 199; in the Middle East, 183–184
Portland, Oregon, 214
positive externalities, 84
Potemkin village, 109
poverty: failure of international system to reduce, 221–222; urban, 213–214
precision agriculture, 166
preservationism, 84
promotional campaigns, population growth and, 185–186
proximate determinants, 218
Putin, President Vladimir, 114

R

rangeland, 86
Ray, Charlie, 72, 73
Reagan, Ronald, 187
recovery factor, of oil, 196
region states, 102–105
religion, family planning and, 185
Río Grande, 132–138
riparian mapping program, 150
river valley bottom flooding, 129, 130
rivers, 77, 79, 95, 132–138; damming, 70–74; flooding, 129–131, 207–211
Rocky Mountains, 19
Russia, 29; crisis in, 108–113; federation, 110, 114–115

S

satellite imagery, 21, 161–169
science and technology, and poverty, 221–224
sea ice, 120
Senegal, and AIDS, 190–191, 192
Serbia, 116–118
sex, formation of cities and selling of, 216
sexually transmitted disease (STD), 191
SFOR, NATO-led peacekeeping force in Bosnia, 116–117
shaded relief map, 155
SHEBA project, 119–122
shipping, 31
Shumikhin, Andrei, 111
"silent epidemic," 32
Singer, S. Fred, on global warming, 51, 52, 53

sinks, carbon, 27–29, 66
skyscrapers, digital revolution and the need for, 172–174
Social Security, 199
soil, climate change and, 27–29
soil erosion. *See* physical processes
Southern Nevada Water Authority, 93
spatial, or geometric, tradition, in geography, 12, 13
sprawl, 76–80, 82, 92–94, 95–97, 213
Starovoitova, Galina, 108–113
state map indexes, topographic maps and, 155
Stommel, Henry, 57, 58
storms, and polar ice cap, 31
suburbs, inner-ring, 212–214
"sugar of lead," 34
Summit on Global Aging, 202
Surface Heat Budget of the Arctic (SHEBA), 119–122
systematic land reclamation, in China, 218

T

Taiwan, and China, 46–47
Tatarstan, 112, 115
tax, on fossil fuels, 224
technology, and poverty, 221–224
terrestrial biosphere, 27, 28
textile industry, in Greenville County, South Carolina, 124–125
Thailand, 185
thematic maps, 158
toxic waste, 98
trade, formation of cities and, 216
transportation, urban, 142–149
travel, international, and disease, 175–179

U

Uganda, and AIDS, 190–191
Ukraine, and AIDS, 192
United Nations, 224
United Nations Framework Convention on Climate Change, 63–64, 65, 69
United States, 10–11; controlling carbon dioxide emissions and, 27–29, 66; and global poverty, 224
U.S. Centers for Disease Control and Prevention (CDC), 32
U.S. Coast Survey, 152–153
U.S. Department of Agriculture, 37
U.S. Department of Energy, 27

U.S. Department of Housing and Urban Development (HUD), 43
U.S. Global Change Research Program, 65, 69
U.S. National Resource Inventory, 85–86
USGS: climatic maps and, 158; as an educational tool, 154, 155; geologic maps and, 157; publications of, 158; topographic maps and, 154–160

V

vaccine: for AIDS, 192, 223, 224; for malaria, 223, 224
vegetation, climate change and, 27–29
Viravidaiya, Mechai, 185
volcanoes, 67
vulnerability, human, to weather and climate, 63–69

W

Wallace, John Michael, on global warming, 51–52, 119, 121
water, 203–206; quality, 130–131; conflict over, in the Middle East, 183; regulation of, 205
watershed approach to conservation, 97
Watson, Robert T., 28, 29
wealth, personal, by county, 170–171
weather, 68–69; polar ice cap and, 31; satellite imaging of, 162–163
West Coast, of United States, 16
wetlands, 85, 88
Whitman, Gov. Christine Todd, 81
winter flooding. *See* river valley bottom flooding
Wirth, Timothy, 65
women, population growth and, 184, 186
wood, 28
world order, 200–201
World Wildlife Fund, 28

Y

Yellowstone, 86
Yeltsin, Boris, 108–113
Younger Dryas, 55–56
Yugoslavia, elections in former, 116–118

Z

Zone A flood zones, 129, 130, 131

Test Your Knowledge Form

We encourage you to photocopy and use this page as a tool to assess how the articles in **Annual Editions** expand on the information in your textbook. By reflecting on the articles you will gain enhanced text information. You can also access this useful form on a product's book support Web site at **http://www.dushkin.com/online/**.

NAME: DATE:

TITLE AND NUMBER OF ARTICLE:

BRIEFLY STATE THE MAIN IDEA OF THIS ARTICLE:

LIST THREE IMPORTANT FACTS THAT THE AUTHOR USES TO SUPPORT THE MAIN IDEA:

WHAT INFORMATION OR IDEAS DISCUSSED IN THIS ARTICLE ARE ALSO DISCUSSED IN YOUR TEXTBOOK OR OTHER READINGS THAT YOU HAVE DONE? LIST THE TEXTBOOK CHAPTERS AND PAGE NUMBERS:

LIST ANY EXAMPLES OF BIAS OR FAULTY REASONING THAT YOU FOUND IN THE ARTICLE:

LIST ANY NEW TERMS/CONCEPTS THAT WERE DISCUSSED IN THE ARTICLE, AND WRITE A SHORT DEFINITION:

ns depend on two major opinion sources: one is our Advisory Board, listed in the front of this volume, which works with us in scanning the thousands of articles published in the public press each year; the other is you—the person actually using the book. Please help us and the users of the next edition by completing the prepaid article rating form on this page and returning it to us. Thank you for your help!

ANNUAL EDITIONS: Geography 01/02

ARTICLE RATING FORM

Here is an opportunity for you to have direct input into the next revision of this volume. We would like you to rate each of the 46 articles listed below, using the following scale:

1. **Excellent: should definitely be retained**
2. **Above average: should probably be retained**
3. **Below average: should probably be deleted**
4. **Poor: should definitely be deleted**

Your ratings will play a vital part in the next revision. So please mail this prepaid form to us just as soon as you complete it. Thanks for your help!

We Want Your Advice

RATING | **ARTICLE**

1. Rediscovering the Importance of Geography
2. The Four Traditions of Geography
3. The American Geographies
4. Human Domination of Earth's Ecosystems
5. Seas and Soils Emerge as Keys to Climate
6. Plying a Fabled Waterway: Melting May Open the Northwest Passage
7. Lead in the Inner Cities
8. Uneasy Collaborators: To China, Taiwan Investors Are Both Welcome and Suspect
9. Global Warming: The Contrarian View
10. The Great Climate Flip-Flop
11. Breaking the Global-Warming Gridlock
12. Beyond the Valley of the Dammed
13. A River Runs Through It
14. There Goes the Neighborhood
15. The Race to Save Open Space
16. Past and Present Land Use and Land Cover in the USA
17. Operation Desert Sprawl
18. Turning the Tide: A Coastal Refuge Holds Its Ground as Sprawl Spills Into Southern Maine
19. A Greener, or Browner, Mexico?
20. The Rise of the Region State
21. Continental Divide
22. Russia's Fragile Union
23. Beyond the Kremlin's Walls
24. The Delicate Balkan Balance

RATING | **ARTICLE**

25. Sea Change in the Arctic
26. Greenville: From Back Country to Forefront
27. Flood Hazards and Planning in the Arid West
28. The Rio Grande: Beloved River Faces Rough Waters Ahead
29. Does It Matter Where You Are?
30. Transportation and Urban Growth: The Shaping of the American Metropolis
31. GIS Technology Reigns Supreme in Ellis Island Case
32. Teaching About Karst Using U.S. Geological Survey Resources
33. Gaining Perspective
34. Counties With Cash
35. Do We Still Need Skyscrapers?
36. Bio Invasion
37. Before the Next Doubling
38. A Turning-Point for AIDS?
39. The End of Cheap Oil
40. Gray Dawn: The Global Aging Crisis
41. A Rare and Precious Resource
42. Rethinking Flood Prediction: Does the Traditional Approach Need to Change?
43. Risking the Future of Our Cities
44. The End of Urban Man? Care to Bet?
45. China's Changing Land: Population, Food Demand and Land Use in China
46. Helping the World's Poorest

(Continued on next page)

ANNUAL EDITIONS: GEOGRAPHY 01/02

NO POSTAGE
NECESSARY
IF MAILED
IN THE
UNITED STATES

BUSINESS REPLY MAIL
FIRST-CLASS MAIL PERMIT NO. 84 GUILFORD CT

POSTAGE WILL BE PAID BY ADDRESSEE

McGraw-Hill/Dushkin
530 Old Whitfield Street
Guilford, CT 06437-9989

ABOUT YOU

Name _____ Date _____

Are you a teacher? ☐ A student? ☐
Your school's name _____

Department _____

Address _____ City _____ State ____ Zip ____

School telephone # _____

YOUR COMMENTS ARE IMPORTANT TO US!

Please fill in the following information:
For which course did you use this book?

Did you use a text with this ANNUAL EDITION? ☐ yes ☐ no
What was the title of the text?

What are your general reactions to the Annual Editions concept?

Have you read any particular articles recently that you think should be included in the next edition?

Are there any articles you feel should be replaced in the next edition? Why?

Are there any World Wide Web sites you feel should be included in the next edition? Please annotate.

May we contact you for editorial input? ☐ yes ☐ no
May we quote your comments? ☐ yes ☐ no